Advanced Technologies and Societa

T0237718

This series covers monographs, both authored and edited, conference proceedings and novel engineering literature related to technology enabled solutions in the area of Humanitarian and Philanthropic empowerment. The series includes sustainable humanitarian research outcomes, engineering innovations, material related to sustainable and lasting impact on health related challenges, technology enabled solutions to fight disasters, improve quality of life and underserved community solutions broadly. Impactful solutions fit to be scaled, research socially fit to be adopted and focused communities with rehabilitation related technological outcomes get a place in this series. The series also publishes proceedings from reputed engineering and technology conferences related to solar, water, electricity, green energy, social technological implications and agricultural solutions apart from humanitarian technology and human centric community based solutions.

Major areas of submission/contribution into this series include, but not limited to: Humanitarian solutions enabled by green technologies, medical technology, photonics technology, artificial intelligence and machine learning approaches, IOT based solutions, smart manufacturing solutions, smart industrial electronics, smart hospitals, robotics enabled engineering solutions, spectroscopy based solutions and sensor technology, smart villages, smart agriculture, any other technology fulfilling Humanitarian cause and low cost solutions to improve quality of life.

Velliangiri Sarveshwaran · Joy Iong-Zong Chen ·
Danilo Pelusi

Editors

Artificial Intelligence and Cyber Security in Industry 4.0

Springer

Editors
Velliangiri Sarveshwaran
Department of Computational Intelligence
SRM Institute of Science and Technology
Chennai, Tamil Nadu, India

Joy Iong-Zong Chen
Department of Electrical Engineering
Day-Eh University
Changhua, Taiwan

Danilo Pelusi
Department of Computer Science
University of Teramo
Teramo, Italy

ISSN 2191-6853   ISSN 2191-6861 (electronic)
Advanced Technologies and Societal Change
ISBN 978-981-99-2117-1   ISBN 978-981-99-2115-7 (eBook)
https://doi.org/10.1007/978-981-99-2115-7

This Springer imprint is published by the registered company Springer Nature Singapore Pte Ltd.
The registered company address is: 152 Beach Road, #21-01/04 Gateway East, Singapore 189721,
Singapore

Preface

This book can provide a wide-ranging overview of how AI and cybersecurity can be used and solve the problem in Industry 4.0 using advanced tools and applications available in the market. This book is a collection of 16 chapters authored by academics and practitioners worldwide. The contributions in this book aim to provide significant theoretical backgrounds, state-of-the-art research findings and case studies from around the world.

This book describes AI and cybersecurity frameworks to make the Industry 4.0 process efficient and straightforward. Recently, the power of computing machines made a big advance, particularly in the AI area. Using AI, we could significantly increase the degrees of accuracy and robustness in the detection of attacks and operate detection systems without requiring deep security expert knowledge as before.

All the books published earlier by different authors only address artificial intelligence and cybersecurity in general. Industry 4.0 requires strong cybersecurity schemes that balance the desired communication and mechanism effect with the computational result. However, many proposed research methods and techniques are not yet suitable for use in cybersecurity for Industry 4.0. Hence, it is decided to provide a book that not only discusses general and but also solves those problems with the help of artificial intelligence and cybersecurity. This book is starting point for the researchers on how they can apply AI and cybersecurity in Industry 4.0. So, this is very useful for students, academicians and research scholars to explore further their field of study. It is very much opted for readers who seek to learn from examples.

Chennai, India

Changhua, Taiwan

Teramo, Italy

Velliangiri Sarveshwaran

Joy Iong-Zong Chen

Danilo Pelusi

Contents

1   Introduction to Artificial Intelligence and Cybersecurity for Industry 1
Shipra Rohatgi, B. Abinash, Sunidhi Joshi, Shami Sushant, and Sachil Kumar

2   Role of AI and Its Impact on the Development of Cyber Security Applications 23
A. Anandita Iyer and K. S. Umadevi

3   AI and IoT in Manufacturing and Related Security Perspectives for Industry 4.0 47
Rohit Kumar and Shanmugam Sundaramurthy

4   IoT Security Vulnerabilities and Defensive Measures in Industry 4.0 71
Koppula Manasa and L. M. I. Leo Joseph

5   Adopting Artificial Intelligence in ITIL for Information Security Management—Way Forward in Industry 4.0 113
Manikandan Rajagopal and S. Ramkumar

6   Intelligent Autonomous Drones in Industry 4.0 133
Kriti Dwivedi, Priyanka Govindarajan, Deepika Srinivasan, A. Keerthi Sanjana, Ramani Selvanambi, and Marimuthu Karuppiah

7   A Review on Automatic Generation of Attack Trees and Its Application to Automotive Cybersecurity 165
Kacper Sowka, Vasile Palade, Hesamaldin Jadidbonab, Paul Wooderson, and Hoang Nguyen

8   Malware Analysis Using Machine Learning Tools and Techniques in IT Industry 195
N. G. Bhuvaneswari Amma and R. Akshay Madhavaraj

9 Use of Machine Learning in Forensics and Computer Security 211
   Nitish Ojha, Avinash Kumar, Neha Tyagi, Preetish Ranjan,
   and Abhishek Vaish

10 Control of Feed Drives in CNC Machine Tools Using Artificial
   Immune Adaptive Strategy 237
   A. Lavanya, S. Revathi, N. Sivakumaran, and K. Rajkumar

11 Efficient Anomaly Detection for Empowering Cyber Security
   by Using Adaptive Deep Learning Model 253
   Balasubramanian Prabhu Kavin, Jeeva Selvaraj,
   K. Shantha Kumari, Rashel Sarkar, S. Rudresha,
   and Hong-Seng Gan

12 Intrusion Detection in IoT-Based Healthcare Using ML
   and DL Approaches: A Case Study 271
   Priya Das and Sohail Saif

13 War Strategy Algorithm-Based GAN Model for Detecting
   the Malware Attacks in Modern Digital Age 295
   S. Rudresha, Alim Raza, Vivek Anand, Himanshu Payal,
   Kundan Yadav, and Balasubramanian Prabhu Kavin

14 ML Algorithms for Providing Financial Security in Banking
   Sectors with the Prediction of Loan Risks 315
   T. R. Mahesh, V. Vinoth Kumar, H. K. Shashikala,
   and S. Roopashree

15 Machine Learning-Based DDoS Attack Detection Using
   Support Vector Machine 329
   V. Kathiresan, Vamsidhar Yendapalli, J. Bhuvana, and Esther Daniel

16 Artificial Intelligence-Based Cyber Security Applications 343
   Sri Rupin Potula, Ramani Selvanambi, Marimuthu Karuppiah,
   and Danilo Pelusi

Chapter 1
Introduction to Artificial Intelligence and Cybersecurity for Industry

Shipra Rohatgi⬤, **B. Abinash**⬤, **Sunidhi Joshi**⬤, **Shami Sushant**⬤, and **Sachil Kumar**⬤

Introduction

According to Atiku et al. [1], cybersecurity is a phenomenon that deals with ways to safeguard computer systems, networks, and electronic data from disclosure, theft, service disruption, and unauthorized access. *Cybersecurity* is defined broadly as "the study of the security of anything in the cyber environment," including information security, network security, operational security, application security, Internet of Things (IoT) security, cloud security, and infrastructure security [2]. Cybersecurity's primary goal is to shield consumers from threats as much as feasible. Additionally, it serves the purpose of promptly and effectively completing the requirements for detection before, handling during, and recovery following the accident. According to predictions, the cybersecurity sector may increase by more than $1 trillion USD between 2016 and 2021. The modern generation now relies heavily on the Internet in their daily lives. The amount of data we trade every day is massive and vast. However, the rate at which cyberattacks are occurring is also dramatically rising. Every few months, fraudsters reduce the cost of their customized attacks while doubling their effectiveness. Furthermore, cybersecurity becomes ineffectual as cyberattacks get more complex and automated [3, 4]. Traditional cybersecurity techniques, such as network protection and computer security systems, are ineffectual against cyberattacks' always-changing, transformative, and inventive attempts [5].

S. Rohatgi · S. Joshi · S. Sushant
Amity Institute of Forensic Sciences, Amity University Noida Campus, Sector-125, Noida, Uttar Pradesh, India

B. Abinash · S. Kumar (✉)
Department of Life Sciences, CHRIST (Deemed to Be University), Bengaluru, Karnataka, India
e-mail: sachil.kumar@christuniversity.in

Classification of Cyberattacks (See Fig. 1.1)

1. **Depending on the goal**: It can target a person or an organization. Government
    and commercial sector cyberattacks are more harmful since they cause enormous
    losses and frequently start with altering important data for financial gain.
2. **Web-based attack:** A hacker attempts to obtain unauthorized access to a website
    by exploiting vulnerabilities in it. It may be done by:

    *   **SQL Injection Technique**—allows hackers to read, change, and delete tables
        from databases by manipulating a regular SQL query on a database-driven
        website.
    *   **Phishing**—Hacker's spoof emails, in particular, to get recipients to accept
        them and follow instructions that typically request personal information. Some
        phishing scams involve the download of malware.
    *   **Man in the Middle**—The hacker gains access to the information stream
        between the user's device and the website server. The hacker's computer
        assumes control of an IP address, so the communication channel between

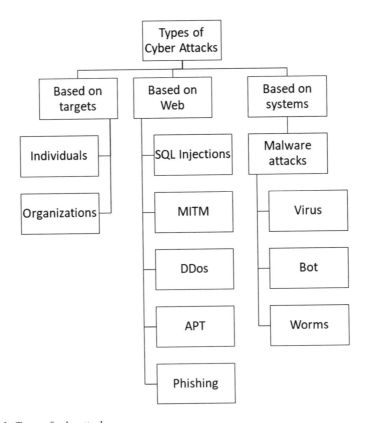

Fig. 1.1 Types of cyberattack

the user and the website is covertly interrupted. This frequently occurs on unprotected Wi-Fi networks.

- **Password Attack**—Typically, the system passwords are broken by utilizing either widely used passwords or by attempting all conceivable alphabetic combinations. It is among the most straightforward techniques to break into a system.
- **APT**—Hackers gain access to networks over an extended period to obtain sensitive information.

3. **A System-Based Attack**: It inserts itself into the target computer system, where it will replicate and compromise the system to harm the computer system or its network.

- **Malware Attack**—Typically, the hacker sends phony emails that seem to originate from a reliable source. This is done to implant malware or steal sensitive data, such as credit card numbers. Any form of malicious software, regardless of its operation, purpose, or delivery, is called malware. Some of them are as follows:
- **Virus**—In addition to their malicious behavior, viruses can spread to other systems and infect other programs. When a file is opened, the virus is launched along with it. The virus then moves, deletes, corrupts, or encrypts data and files. An enterprise-level antivirus solution gives you centralized management and visibility while safeguarding all your devices from one location regarding virus protection. Regularly do comprehensive scans and update your antivirus definitions.
- **Bots**—are computer programs that carry out automated activities without human interaction. Attacks can be executed by bots a lot quicker than by humans. A computer that has a bot infection can propagate the bot to other computers, forming a network of computers known as a botnet. This network of infected workstations may be managed and utilized to carry out extensive attacks like DDoS attacks and brute-force attacks. Many times, device owners are ignorant of their contribution to the attack. On specific hardware, bots are also employed for cryptocurrency mining. Using tools to distinguish traffic from actual users and bots is one technique to manage bots. For instance, you could add a CAPTCHA for her to stop bots from flooding your form.
- **Worms**—Like viruses, worms can replicate on different hardware and software. In contrast to viruses, worms do not require human intervention to spread over a network or system. Worms frequently target computer hard drives and memory. One should ensure that all their devices have the most recent fixes installed to protect themselves from worms. Technology can also assist in identifying files and links that can contain worms, such as firewalls and email filters.
- **Trojan Horse**—Trojans impersonate desired software or programs. The Trojan can hijack the victim's machine if it is downloaded by an unwary user and used for malicious purposes. Trojans can be inserted into phishing

email attachments, games, apps, and even software patches. Trojans are malicious programs that pose as trustworthy ones. Trojan horses must be spread by the victim, frequently via social engineering techniques like phishing, as they cannot spread themselves as viruses and worms can. Trojans rely on social engineering to proliferate, leaving the user responsible for defense. Unfortunately, 82% of breaches in 2022 had a human component. Security awareness training is crucial since employees are the first line of defense against these types of assaults and the target of Trojans.

Types of Cybersecurity

Hardware, software, and infrastructure are all part of cyberspace security. Consequently, it can also be classified as.

Hardware and Software Tools

For network security, shield infrastructure and networks from disruption, unlawful access, and abuse. Effective network security shields corporate assets from several risks inside and outside the business.

Information Security

Safeguards private data against review, modification, recording, and illegal activities like interruption or destruction. Its goal is to guarantee the security and privacy of sensitive data, including financial information, intellectual property, and customer account information.

Cloud Security

Utilizing cloud service providers like Amazon Web Services, Google, Azure, and Rack Space to build secure cloud architectures and apps for your business is referred to as cloud security.

Security for Applications

Application security entails putting in place a variety of defenses in a company's software and services against various attacks. This includes security testing by either white-box, black-box, or gray-box testing. This helps to improve the application's security aspects by making its code more secure, reducing vulnerability by reducing the likelihood of unwanted access or alteration, or implementing strong data input validation.

Tools Used in Cybersecurity

Firewall

A *firewall* is a safety measure that guards a network against unauthorized access to sensitive information. A firewall establishes a barrier between your secure internal and unreliable external networks and safeguards your computer against harmful malware. Depending on the amount of security required by the customer, firewalls provide various levels of protection. Firewalls often welcome incoming connections that are authorized access to the network. Depending on the security rules in place, the security system either permits or denies data packets. Web traffic is filtered at checkpoints set up by firewalls. These solutions allow us to examine and respond to unauthorized network activity before it negatively impacts the network being attacked.

Honeypots

Honeypots are security tools or decoy systems used by security experts to lure or attract attackers by making the target vulnerable or attractive. They are used to detect, deter, and investigate unauthorized access to the target system. They can act as high-value targets for attackers, like a server online. The primary purpose of honeypots is to gather attacker data by allowing for early detection of infiltration attempts while giving a few footprints of attackers. To minimize the threats, honeypots often employ a secure operating system with added security features.

Penetration Testing

Simulation of a cyberattack on a computer system to identify its vulnerabilities is known as penetration testing or Pentesting. This technique's most commonly implemented use is to improve the security of web applications and their firewalls (WAFs). Pentesting targets various systems, such as front-end and back-end systems/servers. APIs and other components to detect any security weaknesses. This testing process also checks for potential injection vulnerabilities. The results of penetration testing can be used to enhance the security settings of WAFs and help address any vulnerabilities.

Encryption Tools

Data are transformed from readable to encrypted forms using encryption in cyber-security. Only after it has been decrypted, an encrypted data can be read or processed.

A crucial element of data security is encryption. This is the most straightforward and crucial approach to stop someone from stealing or reading your computer system's data with harmful intent.

Data Protection Individual and large enterprises frequently employ encryption to safeguard user data between browsers and servers. Everything from payment information to personal information is included in this information. Data encryption software is used to create encryption schemes that, in theory, can only be broken by powerful computers. This software is sometimes referred to as encryption algorithms or ciphers. Symmetric and asymmetric encryptions are the two most popular types.

- Private key encryption also refers to using a symmetric encryption key. It is perfect for single-user and closed systems because the encryption key is also needed for decryption. If not, the recipient should receive the key. If a third party gains access, the risk of compromise rises. B. Cybercriminals are stopped. Compared to the asymmetric method, this one is quicker.
- Asymmetric cryptographic keys: These employ public and private keys that are mathematically tied to one another. Keys are essentially big numbers that are paired but not identical, hence the word "asymmetric." Public keys are distributed to approved recipients or made available to the public, while private keys are kept hidden by their owners.

Packet Sniffers

During a network's Transmission Control Protocol/Internet Protocol (TCP/IP) layer, a packet sniffer is a program or utility that reads data packets. These technologies are used by network administrators to "sniff" Internet traffic and keep an eye on data in real time. The data can be interpreted to evaluate and identify server, network, hub, and application performance issues. Network administrators can utilize one of several techniques to find sniffers on their networks when hackers use packet sniffing to monitor Internet activities illegitimately. Utilize this early warning and take action to safeguard your information from unauthorized listeners.

One strategy that can be employed successfully in cybersecurity is artificial intelligence (AI) [6]. AI is regarded as the intelligence being added in digital devices and computer-operated devices to perform n number of tasks like a human. It mainly focuses on studying the brain's cognitive process and activities to employ in developing devices and software with human intelligence. Fighting cyberattacks has been made much easier thanks to two recently developed fields of AI: machine learning (ML) and deep learning (DL). John McCarthy first proposed the concept of AI in 1956; it is the science and engineering of creating intelligent autonomous security systems. The primary goal of AI is to teach computers to think, learn, act, and behave intelligently and cognitively like humans.

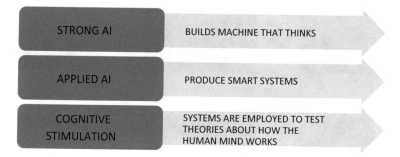

Fig. 1.2 Significance of AI

In the current technological period, AI offers services in many fields, including computer vision, pattern recognition, expert systems, language processing and translation, speech recognition, biometric systems, robotics, the Internet of Things (IoT), and other related domains [7]. A significant area of computer science called AI works with autonomous, intelligent systems resembling the human brain [8]. The importance of AI in cybersecurity is more significant than ever in the digital age (see Fig. 1.2). Cybersecurity is a popular topic. Several industries, including natural language processing, gaming, health care, manufacturing, and education, swiftly adopted artificial intelligence once it was initially presented. AI can analyze massive amounts of data quickly, effectively, and accurately thanks to its robust analytics capabilities. In contrast to previous systems, AI systems can anticipate upcoming cyberattacks based on current threats, even if those threats change. As a result, it is only possible to use AI to counter security risks.

Two Approaches in Artificial Intelligence: Symbolic Versus Connectionist

There are two basic approaches taken up in AI: the symbolic or top-down approach and the connectionist or bottom-up approach. The top-down approach is about analyzing the cognitive process and replicating it to analyze the symbols or symbolic labels. It is not dependent on the biological constitution of the brain. In contrast, the bottom-up or connectionist approach focuses more on brain structure. It involves the creation of an artificial neural network that imitates the brain.

It can be easily understood straightforwardly, considering the task of building a system that can recognize the letters of the alphabet. To demonstrate the distinction between these methods, a bottom-up approach is considered. A typical bottom-up strategy involves gradually "tuning" an artificial neural network to improve performance by presenting letters to it one at a time. The responsiveness of various neural pathways to various stimuli can be altered through tuning. A top-down approach, on the other hand, typically entails creating a computer program to compare each letter to geometric descriptions. The bottom-up approach relies on neural activities, whereas the top-down approach relies on symbolic descriptions.

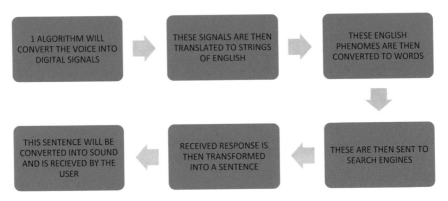

Fig. 1.3 Working mechanism

How Does Artificial Intelligence Work? (See Fig. 1.3)

AI works by implementing human intelligence and a set of algorithms in order to answer several questions. This combination allows AI to analyze the data and generate the results. These data are then further used to generate the results and build expertise.

For example, if someone asks an AI system—how to make chapatis?

AI Implementation Methods

Machine Learning

The ability to learn in AI is obtained through machine learning. It exploits the algorithms to obtain insight into patterns and generate a profile.

Deep Learning

Deep learning tries to mimic the event in layers of neurons to learn in a real sense to identify a pattern in the given text. These methods are only sometimes better than traditional supervised approaches as deep learning performance is subjected to the correct selection of algorithms and numbers of hidden layers and a feature representation technique. Deep learning models show a promising future in text mining as it depends entirely on the neural network with extra depth.

To emphasize the significance of AI systems and technologies as a defense against cybersecurity, this chapter is being offered. This chapter will provide a significant response to the essential subject of how AI-based solutions might help to safeguard cybersecurity. Additionally, we have demonstrated the limitations of AI in cybersecurity as well as some potential future research topics in this chapter.

Background Information on AI Methods and Cybersecurity Applications

The opposite is also true: Cybercrimes and attacks are rising rapidly due to the rapid advancement of Internet technology and systems. We require AI-based strategies in our cybersecurity systems to improve the security of cyberspace more effectively in order to combat and defeat these crooks and their cunning methods.

Table 1.1 provides short descriptions of AI methods and applications (see Table 1.1, Fig. 1.4).

Table 1.1 Techniques and applications of artificial intelligence (AI) in cybersecurity

Techniques of AI	Applications in cybersecurity
Neural nets	1. For intrusion detection and prevention system 2. Very high-speed of operation 3. For Denial-of-service (D0S) detection 4. For forensic investigation 5. Warm detection 6. Fuzzy logic
Intelligent agents	1. Proactive and reactive 2. Agent communication language 3. Defense against Distributed Denial-of service (DDoS)
Expert systems	1. For network intrusion detection 2. For decision support 3. Knowledge base 4. Inference engine
Application of learning	1. Machine learning and deep learning 2. Data mining 3. Supervised and unsupervised learning 4. Intrusion detection and malware detection 5. Self-organizing maps

Fig. 1.4 Applications of AI
in cybersecurity

The Significance of Cybersecurity

- Cybersecurity has grown significantly in importance in society due to the rise in cybercrime; antivirus software is no longer adequate to safeguard our system network and its data.
- The cybersecurity sector nowadays is primarily concerned with defending systems and devices against intruders. It is challenging to picture the bits and bytes that go into these efforts, but it is far simpler to consider their meaning. Due to frequent denial-of-service attempts, many websites would be almost entirely inoperable if not for the diligent efforts of cybersecurity experts.
- Only authorized individuals can access confidential data and operations thanks to cybersecurity, military secrets, for instance.

The integrity principle states that only authorized individuals and agents can add, modify, or delete sensitive information and functionality. Example: The database contains inaccurate data that a user entered.

Availability: According to the availability, concept, systems, functions, and data must always be accessible per predetermined guidelines based on service levels.

How AI Can Be Applied on Cybersecurity Issues

To address cybersecurity issues, AI offers many advantages, some of which are listed below:

1. As opposed to previous technology, which was primarily concerned with the past and solely relied on known cyberattacks, conventional systems fail to recognize

changes when a new cyberattack occurs, creating a blind space for unorthodox attacks. AI can recognize novel and intricate variations in attack flexibility. Future AI systems will better detect similar changes. AI machines are more capable of learning and adapting, and they can identify abnormal operations more quickly and accurately. This ability of AI systems is more critical when cyberattacks are becoming more refined, and cybercriminals are coming up with new and inventive methods [3, 4].

2. AI can handle many security data [3, 4], because AI has built-in security mechanisms that can recognize and respond to threats. Security employees must deal with an intolerable number of data breaches daily, but automatically identifying and responding to threats have lessened their workload. Furthermore, AI is the only approach that can effectively handle these intrusions. Network security analysts will find it increasingly challenging to detect and monitor attack elements accurately and promptly as more and more security data are generated and transferred over the network daily. AI can aid in this situation by boosting the frequency with which suspicious behavior is highlighted and recognized. This can help network security officials to respond to circumstances.

3. Over time, AI security systems examine application behavior, and routine network AI creates a baseline of typical patterns by detecting risks over time. The AI security system will identify the attacks if there is any modification or divergence from the usual routine.

AI Techniques Used for Cybersecurity

Some AI security models that can effectively counter threats and cyberattacks include neural networks, expert systems, machine learning, deep learning, and data mining. In the cybersphere, intelligent decisions can be made using AI-based techniques. All of these AI strategies have been emphasized in this essay and are briefly discussed in the parts that follow.

Artificial Neural Network

Neural networks represent deep learning and analysis of data by AI. Some scenarios and situations are too much and out of scope for machine learning algorithms to cope with. ANN can be connected to biological neurons within the body, which performs a specific function and carries stimulus resulting in action. It consists of layers of interconnected artificial neurons powered by different activation functions, which help in on/off mechanisms. Each neuron receives a unique version of input and random words, which are added with fundamental barriers and unique layers; this then passes to an activation function that depicts the final value of the neuron. The output is generated from the final neural layer, the loss function. It will input

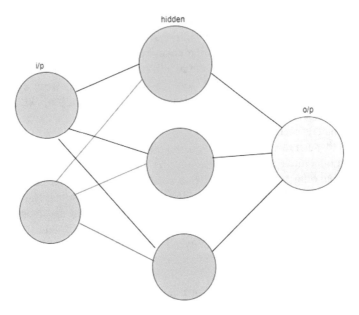

Fig. 1.5 Different layers of neurons in a neural network

versus output is calculated, and backward propagation of error is performed. The weights are designed to make the loss minimum.

1. **Weights**—values that are multiplied by the input. In the backpropagation of errors, they are modified to reduce the losses. Weeds are machine-learned values from the neural network (NN) that evaluate self-adjust depending on the difference between foreseen output and input. *Activation functions* are a mathematical formula that helps in the on–off of neurons.

The following figure shows three main layers of a neural network (see Fig. 1.5):

- **Input layers** represent the dimensions of input vectors.
- **Hidden layer** represents the media notes that divide the input into the area with boundaries 8, takes weighted inputs, and produces output through an activation function.
- **Output layer** provides the output of the neural network.

Detection Using Neural Networks in Tweets

1. **CNN**—A convolutional neural network consists of three layers naming

   - Convolution layer.
   - Pool layer.
   - FC layer.

2. **The convolution layer**—This layer is an essential element in CNN. It extracts features from data (input).
3. **Pool layer**—This layer performs a function to reduce the dimensions of input it gives you say the number of parameters.

There are several types of functions.

- Max pooling.
- Average pooling.

Max pooling is used widely as it only takes some maximum value in the window.

For example, the view layer takes a 4 * 4 feature map as input from the C layer and performs max pooling with 2 * 2 window size I am Stroud of 2 for each window max pooling takes a maximum value of 4 consequently exchange rates 2 * 2 feature outputs which are sent to fully connected layer.

Security Expert Systems

A computer program known as an expert system, or AI, aids a human expert in making decisions. The system consists of an inference engine and a knowledge base, which together produces security rules [9]. Security standards provide the basis for judgments made by cybersecurity expert systems. Applications for expert systems' modeling can be found in cyberspace, finance, and medical diagnosis. Expert systems can come in various sizes, from simple hybrid systems to massive, complex ones that handle complex concerns and issues. The knowledge base phase of the cybersecurity expert framework describes domain knowledge and operational understanding of the rules governing security decisions. The inference engine phase retrieves information from the knowledge base and concludes new facts through expert systems. In one strategy known as the "case-based reasoning (CBR) approach," a particular problem is solved by recalling prior, analogous cases. Then, a solution is determined by adapting the previous solution to a new problem case. This method examines novel solutions to raise the system's accuracy and capacity for learning.

Another method for resolving issues is known as rule-based systems (RBSs), which rules established by experts characterize. The condition portion and the action are the two subsystems of the rule-based system. The difficulties are analyzed using condition part evaluation, and the appropriate course of action is decided. Cybersecurity expert system uses basic standards and rules to combat cyberattacks. For instance, it compares the process to the knowledge base; if the process is excellent and known, the security system considers it secure; otherwise, the system declares the process a threat and ends it. The system looks for the sets of rules in the inference engine to determine the machine's state if the knowledge base does not have such a procedure. Based on the machine's condition, followed by the inference identified by the knowledge base, the system informs the manager or user of the machine's status.

Thus, a rule-based cybersecurity expert system model can make decisions like a security expert in an intelligent cybersecurity framework designed to address challenging cybersecurity problems. Because of this, cybersecurity expert system modeling, based on its computational powers and capacity for reasoning, might be helpful in AI-based cybersecurity.

Intelligent Agents (IAs)

Intelligent agents (IAs) are autonomous systems with a decision-making process internal to them and a personal goal. Through sensors, it assesses risks, and actuators monitor the domain. It directs the activity until a specific goal is attained [10]. These systems exhibit proactivity and responsiveness, and when interacting with other autonomous agents, they can comprehend and adapt to changes in their environment. These intelligent agents can learn about and interact with their surroundings, making them adaptive. IAs are successful in thwarting distributed denial-of-service (DDoS) assaults. How can these agents be used to defend against decentralized cyberattacks? The solution is to create artificial "Digital police," which must comprise mobile intelligent agents. It was necessary to install infrastructure to give solid support.

Search

Every day, we employ the search technique as a heuristic for solving issues. Before using a search algorithm, a prior understanding of the search strategy is essential. Nearly, all intelligent programs now use or include these search algorithms, which favors the entire intelligent system. Several search security systems are used in AI, including the search estimation used in many projects. For computer chess, the search estimation was developed. For computer chess, the search estimation was developed. It uses the "isolate and vanquish" critical thinking technique, which is helpful in ad hoc leadership situations where two opponents decide on their most advantageous course of action.

Bio-inspired Computing Method

Using sophisticated algorithms and techniques, bio-inspired computing in artificial intelligence (AI) uses bio-inspired behaviors and attributes to address various challenging academic and environmental problems. These methods are frequently used in cyberspace and include Evolution Strategies (ESs), Ant Colony Optimization (ACO), Artificial Immune System (AIS), Particle Swamp Optimization (PSO), and Genetic Algorithms (GAs). This method is also employed to categorize computer malware. These methods are generally employed to improve the characteristics and parameters used by the classifiers in classifying computer malware. As an illustration, PSO and GA approaches were used to increase the effectiveness of the malware detection system [11]. Another study employed fuzzy logic and GA to detect intrusions. Using glow analysis to forecast network traffic behavior for a given period, the GA was used to construct a digital signature of a network section. Additionally, the fuzzy logic method determined the network instance's anomalousness. A university's network traffic was used for the study, and the outcomes showed 96.53% accuracy and 0.56% false notice.

Machine Learning (ML) and Deep Learning (DL) Methods

Artificial intelligence, known as "machine learning," focuses on teaching computers how to learn new things and use algorithms to make data-based decisions. Machine learning is strongly tied to mathematical methods enabling data extraction, pattern detection, and conclusion drawing. Regression and classification are two of ML technology's most crucial techniques. Supervised, unsupervised, semi-supervised, and reinforcement learnings are the first four types of learning.

Machine learning, also called "deep learning," is a skill that uses data to teach computers how to perform tasks that humans previously could only perform. This is performed by modeling the mechanism of data interpretation in the human brain. Deep learning is predicated on the idea that more extensive neural networks perform better as we train them with more data and scale them up.

It has been demonstrated that ML and DL are crucial for solving cybersecurity problems. The security system can use ML approaches in a variety of ways. Spam filtering, network abnormalities analysis, botnet tracking, and user behavior anomalies' tracking are a few examples.

Deep Learning Detection for Misinformation

What is misinformation? (See Fig. 1.6)

Misinformation is false or inaccurate information that deceives people by obscuring the truth. It can also be referred to as untruth, ambiguity, or deception. The spread of misinformation can harm relationships and undermine trust by presenting false sensations, leading to a negative breach of expectations or trust in society and the people in it, which is harmful.

Misinformation has many terms:

- **Rumor** is a story of the circulation of information from one user to another whose authenticity status is doubtful.
- **Fake news** is an article that misleads its readers and is false.
- **Spam** can be an unsolicited text sent over the Internet to spread advertising malware and other unresourceful complete data.

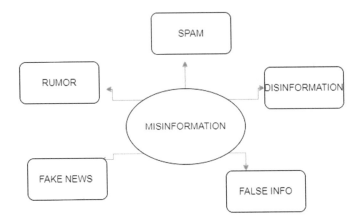

Fig. 1.6 Types of misinformation

- **Disinformation** is a piece of inaccurate information that people spread intentionally to mislead other readers
- **False information** is misinformation spread done intentionally.

Methodologies (see Fig. 1.7)

We have deep learning techniques in three main categories:

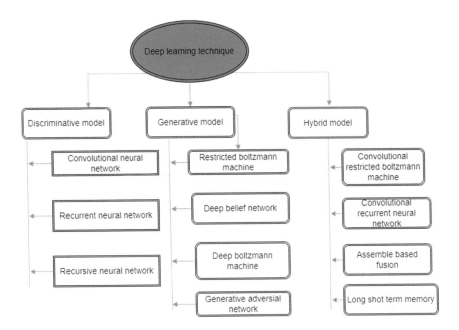

Fig. 1.7 Deep learning techniques for misinformation detection

1. **Descriptive model.**
2. **Generative model.**
3. **Hybrid model.**

Benefits of AI in Cybersecurity

Significant benefits have been received by those institutions that use AI techniques in their cybersecurity operations. Implementing AI in cybersecurity challenges, for instance, the ROI of some institutes has increased. Siemens AG developed the AI-based Siemens Cyber Defense Center (CDC), which is swift, autonomous, and adaptable. Amazon Web Services (AWSs) also utilize this system. The system was predicted to be under 60,000 attacks per unit of time due to AI application. With less than 12 individuals, the overall capability was quickly controlled, and system performance was well-maintained. Artificial intelligence in cybersecurity can detect emerging attacks by examining historical threat trends [12].

The time and effort spent on researching and identifying risks and assaults can be saved by using this AI approach. It has been discovered that AI is beneficial in identifying risks and responding to them at minimal cost (an average 12% cost decrease). AI may offer significant answers to cybersecurity concerns as the cybersecurity system shifts from conventional and manual procedures to automated algorithm mitigation. In contrast to traditional technology, AI can detect new, complex changes in the attack extensibility, which primarily relies on previously identified intruders and intrusions and thus leaves a blind spot during unusual intrusion activities. At present, AI technology has overcome the shortcomings of traditional security systems. For instance, it is now possible to monitor privileged Internet activity, and any change in how privileged access procedures are carried out is a potential threat. Security teams have an advantage thanks to AI predictive tactics, which are crucial to thwarting assaults before they cause any damage. For the detection of trends and dangers across a variety of sectors, including retail, manufacturing, energy, and transportation companies, Dark Trace (a UK company) uses ML technology. Through AI-based solutions, large amounts of data may be managed, as well as the development of network security systems. Security experts are overwhelmed by the vast number of open security issues. The workload of security organizations has lessened due to AI's autonomous detection and response to assaults. Security specialists need help for managing security data when it is produced and exchanged in large quantities daily. AI can therefore aid in accelerating the study of dubious procedures and activities.

Additionally, by eliminating the manual methods that take much time when responding to novel situations, security people can profit and react to new scenarios better. AI-based systems are equipped to learn over time and can react to threats and attacks more effectively. Using application characteristics and overall network activity, AI can help to identify attacks. As time passed, AI memorized the common and typical traffic situation and established a cap for the usual activities. As a result, the attack is noted when there is any abnormal deviation.

Detecting False Information Using Neural Networks

This section covers various methods and procedures for identifying false rumors and deceptive material. It also thoroughly examines the five most common phrases for misinformation and how they deceive social media users.

Multiple models are put out for detection based on various methods and annotations. This research on socially essential data is used in fake news identification to tell real news from phony by comparing the two. Convolutional neural networks with gated recurrent units extended short-term memory networks, and other deep learning networks were investigated in this. We also looked into the benefits of feature extraction and feature embedding in deep neural networks.

Using Neural Networks to Find Objectionable YouTube Content

The transfer learning method is used to extract frame features from videos. The video features are then processed through directional LSTM, where the model learns efficient video representation and forms video classifications. The deep learning-based framework is proposed for inappropriate video content detection and classification. All analyses are carried out by utilizing manually compiled YouTube video clip datasets. The following are some advantages of a system for detecting child-appropriate content based on deep learning:

- It operates by taking into account real-time circumstances and processing video at a speed of twenty-two frames per second while utilizing an effective network and bidirectional extended short-term memory-based framework.
- Any video-sharing site can benefit from using it to flag or remove videos with questionable content.

Detecting Hate Speech Using a Neural Network

Although significant legislative and law enforcement efforts and millions of dollars in investments from social media firms, the spread of hate speech on online platforms has been significantly expanding, it is commonly acknowledged that automated data mining techniques are essential for developing efficient defenses against various threats. The techniques for classifying hate speech utilizing deep neural models incorporating CNN and LSTM, as well as GRU, to increase classification accuracy were first introduced. Second, we compare and contrast the most extensive public dataset with the rest of the data to create new references for comparative research in the future. With all the information, distinct datasets categorized into different subclasses aid in classification.

Challenges Faced with Integration of AI in Cybersecurity

Classification Error

AI can possess severe classification errors; these errors can be due to the alteration of pixels. According to the published literature, classification errors of neural networks can occur with the change in pixels or modification of bytes. Hence, it can be concluded that if the data source is infected or altered, it can easily cheat the AI system.

Intensive Requirement for Resources

While AI can bring good automation for cybersecurity, one thing that can be a significant issue is the need and requirement for high-end equipment and servers for usage. Most small-scale businesses need help to afford this, which becomes a significant issue. Another thing is the need for personnel trained in using and maintaining AI.

Maliciously Modified Model

Implementing an AI model is a program that may have some vulnerabilities. These vulnerabilities may be due to the designer's unreasonable and careless design of the logical structure of the model. They may come from a specific high-level language, hardware-specific problems, or the back door embedded in the model. Gu et al. [13] implemented the backdoor in the neural network, which made the neural network's performance in the specific attacker sample very poor. These shortcomings also reflect that the given answers by the program are only sometimes accurate.

Cost

Integration of AI requires a large sum of investment to run and maintain. In the current market scenario, only a few companies in the IT field have been able to integrate AI into their systems.

Lack of Transparency

In the decision-making process of AI, all the participants, including programmers, do not know why the AI model gives the final decision results, i.e., the decision-making process of AI lacks transparency. The AI model is like a black box. In the process of its creation and self-improvement, it can realize the automatic configuration and adjustment of parameters without too much staff intervention, thus saving human resources. Nevertheless, at the same time, the problem is that its decision-making process needs to be explained clearly. Although the AI model can achieve high accuracy, the tests are implemented in the test set. Therefore, whether the AI model can achieve such a high accuracy remains to be verified when facing unknown events. When there are objections to the decision-making results given by the AI model, it is difficult to explain the decision-making process, so some people will be skeptical of the decision-making results, that will not be conducive to the rapid judgment of the network situation or even cause irreversible consequences. Some research teams have begun to conduct in-depth research on this issue.

Public Perception

Popular media and sci-fi movies have painted AI in a sense that puts fear into the minds of people unfamiliar with AI and machine learning and how it works. For them, AI can bring about the end of humanity if it has access to security systems.

AI Use Cases

A major issue with AI implementation is the use case. The primary reason for failure is the improper implementation of the AI use case, and often companies try to implement AI without taking baby steps.

Malware Signature

The method by which AI works on the malware is through its known signature. These signatures are like fingerprints, and changes in these signatures cause the fingerprint to change. With Scripts changing every moment, it becomes complicated for the AI to identify and work on this malware. The AI models need a lot of data to complete the training. Before using data, they may do operations that mainly include a series of steps such as data noise reduction, normalization, missing value filling. If a supervised method is used, it is necessary to label the data manually. However, due to the substantial heterogeneity of cyberspace, different cyber structures may produce different high-risk events, which have sudden characteristics. Therefore, each possible high-risk event must be estimated in advance before the design of the models, and these high-risk events should be analyzed and labeled in advance. Meanwhile, AI models have a high demand for data, which may need more time to make a timely judgment.

Discussion

Cybersecurity and artificial intelligence overlap in various transdisciplinary fields (AI). Deep learning and other AI technologies can be used in cybersecurity to create intelligent models for malware categorization, intrusion detection, and threat intelligence sensing. On the other hand, AI models will be exposed to various cyber threats, which will interfere with their decision-making, learning, and sampling. Therefore, cybersecurity defense and protection solutions are required for AI models to counter adversarial machine learning, safeguard machine learning privacy, secure federated learning, etc. We examine the interaction of AI and cybersecurity considering the two factors mentioned earlier:

1. We review the research on using AI to defend against cyberattacks, including adopting conventional machine learning techniques and current deep learning solutions.
2. We examine the counterattacks that AI itself might encounter, examine their traits, and categorize the related defense strategies.
3. We elaborate on the existing research on creating a secure AI system from the perspectives of developing encrypted neural networks and implementing secure federated deep learning.

Conclusion

New problems for cybersecurity have developed and emerged along with the ICT's quick improvements. Modern cyberattacks and threats are so complicated and advanced that traditional techniques and strategies can no longer help them. New procedures and strategies that are optimal, scalable, adaptable, and flexible are required to combat these sophisticated cyberattacks. We have provided an overview of AI applications in cybersecurity in this paper. Data learning, security expert systems, and bio-inspired methodologies are a few of the well-researched AI-based cybersecurity solutions that have been covered.

Additionally, areas, where AI is used in cybersecurity, are examined, including the prediction, detection, and prevention of intrusion and malware, defenses against distributed denial-of-service (DDoS), a method where digital police is used, and many other areas. AI applications in cybersecurity were also highlighted, along with some of the benefits and difficulties. These advantages include managing massive amounts of data quickly and accurately, lowering the cost of using AI approaches to address cybersecurity concerns, and boosting the return on investment for AI-powered cybersecurity technologies, among others. Adversarial machine learning and human self-approval are two significant difficulties with using AI-based applications for cybersecurity. Although there are more advantages and disadvantages, AI-based security solutions are still used in cybersecurity. Many industry professionals concur that AI and cybersecurity must be combined because humans depend on cybersecurity.

References

1. Atiku, S.B., Aaron, A.U., Job, G.K., Fatim, S., Yakubu, I.Z.: Survey on the applications of artificial intelligence in cyber security. Int. J. Sci. Technol. Res. **9**(10), 165–170 (2020)
2. Sarker, I.H., Furhad, M.H., Nowrozy, R.: Ai-driven cybersecurity: an overview, security intelligence modeling and research directions. SN Comput. Sci. **2**(3), 1–18 (2021)
3. Truong, T.C., Diep, Q.B., Zelinka, I.: Artificial intelligence in the cyber domain: Offense and defense. Symmetry **12**(3), 410 (2020)
4. Truong, T.C., Zelinka, I., Plucar, J., Čandík, M., Šulc, V.: Artificial intelligence and cybersecurity: Past, presence, and future. In: Artificial intelligence and evolutionary computations in engineering systems, pp. 351–363. Springer, Singapore (2020)
5. Kabbas, A., Alharthi, A., Munshi, A.: Artificial intelligence applications in cybersecurity. IJCSNS Int. J. Comput. Sci. Netw. Secur. **20**(2), 120–124 (2020)
6. Soni, V.D.: Challenges and Solution for Artificial Intelligence in Cybersecurity of the USA. Available at SSRN 3624487 (2020)
7. Shamiulla, A.M.: Role of artificial intelligence in cyber security. Int. J. Innov. Technol. Exploring Eng. **9**(1), 4628–4630 (2019)
8. Helm, J.M., Swiergosz, A.M., Haeberle, H.S., Karnuta, J.M., Schaffer, J.L., Krebs, V.E., Spitzer, A.I., Ramkumar, P.N.: Machine learning and artificial intelligence: Definitions, applications, and future directions. Curr. Rev. Musculoskelet. Med. **13**(1), 69–76 (2020)
9. Tyugu, E.: Artificial intelligence in cyber defense. In: 2011 3rd International Conference on Cyber Conflict, pp. 1–11. IEEE (2011)
10. Wirkuttis, N., Klein, H.: Artificial intelligence in cybersecurity. Cyber Intell. Secur. **1**(1), 103–119 (2017)
11. Fatima, A., Maurya, R., Dutta, M.K., Burget, R., Masek, J.: Android malware detection using genetic algorithm based optimized feature selection and machine learning. In: 2019 42nd International Conference on Telecommunications and Signal Processing (TSP), pp. 220–223 (2019)
12. LAZIĆ, L.: October. Benefit from Ai in cybersecurity. In: The 11th International Conference on Business Information Security (BISEC-2019), 18th October 2019, Belgrade, Serbia (2019)
13 Gu, F., Ma, B., Guo, J., Summers, P.A., Hall, P.: Internet of things and Big Data as potential solutions to the problems in waste electrical and electronic equipment management: An exploratory study. Waste. Manage. **68**, 434–48 (2017)

Chapter 2
Role of AI and Its Impact on the Development of Cyber Security Applications

A. Anandita Iyer and **K. S. Umadevi**

Introduction

In this era of technological advancements, every sector of the community is heavily and constantly relying on computer networks and different information solutions. With this exponential growth in networks, comes an equally drastic growth in cyber-attacks, thus making all the sectors vulnerable to these attacks. A cyberattack can be defined as an attempt to destroy, alter, expose, disable or steal data or gain unauthorized access to any system, organization, or any device. Since the first denial-of-service (DOS) attack, in 1988, there has been an astounding growth in the number of cyberattacks and its impact [1], thus paving a way to the urgent need of cyber security that will help in protecting and practicing safety measures for network devices and safeguard data from unauthorized access and attacks by a nefarious party.

In a traditional cyber security environment, the response to an attack is based on static control, which means that either the attack is responded after the attack has already taken place or the response will be based on certain rules predefined by the system admin based on the signatures of the previously occurred attacks. For example, in a network intrusion attack, the system will monitor incoming traffic based on a set of rules and will notify the admin after the attack has occurred or in certain cases will block those packets of data that are considered malicious. But since this method depends on the following certain rules for identifying an attack, there are many such possible scenarios where the attacks can be morphed into data packets that can pass the detection system unnoticed. Such an incident was reported by Equifax in 2017, where an attacker hacked the systems exposing data of more than 140 million customers [1]. The attackers discovered a flaw in the online portal of the company and uploaded a programming language to their server to gain remote

A. Anandita Iyer · K. S. Umadevi (✉)
Vellore Institute of Technology, Vellore, Tamil Nadu 632014, India
e-mail: umadeviks@vit.ac.in

© The Author(s), under exclusive license to Springer Nature Singapore Pte Ltd. 2023
V. Sarveshwaran et al. (eds.), *Artificial Intelligence and Cyber Security
in Industry 4.0*, Advanced Technologies and Societal Change,
https://doi.org/10.1007/978-981-99-2115-7_2

access and uncovered database credential of Equifax. Then used it to search for all the user sensitive document and stole the data. Apart from this, threats like advanced persistent threat (APT), zero-day attacks, etc., are attacks where attacker camouflages their activities, and before the admin can even discover the vulnerability in the system, the system is attacked and the sensitive data leveraged. In such an unpredictable environment, a different approach to prevent attacks from occurring is a priority rather than waiting for notifications of attacks that have already taken place.

We live in digital world, where data is vital and its security has become more important than ever. With every passing day, hackers are evolving, getting smarter and finding innovative ways to exploit vulnerable data leading to new attacks, data breaches, data crashes, data poisoning, etc. More and more complex yet sophisticated attacks are being launched every day making cyber security measures a very serious and crucial requirement to be implemented by organizations and individuals as well. Main security objectives have always been confidentiality, availability, non-repudiation, authentication and integrity [2]. These goals can be achieved by the use of artificial intelligence (AI), whose main motive is to mimic the cognitive behavior of humans and carry out tasks that would usually be conducted by a human being. Artificial intelligence is an independent entity that behaves like human, thinks like human and performs tasks like a human being. In today's time, we have been surrounded by AI, in the form of personal assistants, automated transportation, facial recognition, computer gaming experience, aviation, voice recognition, etc. In cyber security, an AI can identify risk, prioritize the security requirements, detect malware in a network, notice intrusion in a network, before it even starts and also prepares a proper incident response plan.

Overview of Artificial Intelligence

Artificial intelligence was officially born in 1956 as a summer research project by John McCarthy at Dartmouth, where his goals were to explore various ways where a machine can simulate few aspects of intelligence, which is still the basic idea behind the continuous development of artificial intelligence. Fast forward to eighteenth century where Thomas Bayes designed a framework for reasoning based on the probability of collective events and continuting to nineteenth century, logical reasoning was discussed by George Boole. Heuristic search was introduced by Newell and Simon, and later, Samuel's self-play, self-improved checkers' playing program laid out the first instance of machine learning (ML) in work. Later on, Rosenblatt's perceptron model based on biological neurons served as the basis for artificial neural network (ANN) [3]. In 2004, NASA created autonomous driving for Mars rover, a self-driving and navigation system based on machine learning. From here on, artificial intelligence has grown and developed to become one of the most indispensable technologies that we know of today.

A precise definition of artificial intelligence, that matches its sophisticated and popular image is quite difficult. In its simplest form, artificial intelligence can be defined as an activity that is committed in making machines intelligent, as intelligence is a feature that is required by an entity to work appropriately and on its own as well as make predictions about its environment. To this day, human intelligence is far superior when compared to artificial or biological world, and this is because human beings have the ability to understand, reason, perceive, achieve goals, generate language, create art, summarize information, etc. [3], thus making human intelligence a default choice to modify and train an AI. With the rapid growth of technology, matching a human ability can be considered as a sufficient condition and not a thumb rule, as there exist many systems which surpass certain levels of human intelligence, for example—speed.

Another way to define artificial intelligence, as described by John McCarthy, is "an approach which employs mathematical logic to formalize basic facts about events and their effects." An artificially intelligent machine can learn, understand and act on the basis of the information obtained from events and their effects [1].

The definitions presented in Table 2.1 concentrate on human behavior and knowledge representations to develop intelligent agents. That is comparing and understanding how the human mind functions, process knowledge, make decisions with respect to the computer program for e.g., using reason to conclude results, using logical approach to achieve goal etc. These agents (something that perceives and acts) grow and exchange knowledge with other agents and repeat the same process until they find an efficient solution to a problem.

This chapter will briefly discuss about artificial intelligence and its impact on cyber security as well as discuss various applications of AI in cyber security in detail. The chapter will be organized as follows: the chapter will start with an abstract, followed by first section: Introduction, which will establish the foundation of the chapter and its contents. Section "Literature Survey" will be a brief survey on the use of artificial intelligence by organization to secure their ecosystem. This section will talk about the existing research based on artificial intelligence and cyber security and how AI has evolved in providing better security to industries. Section "Artificial Intelligence Techniques for Cyber Security" elaborates AI techniques and AI-based use cases in cyber security. In addition, it explores the scenarios where artificial intelligence has proved to be an exceptional tool in terms of cyber security by using real-life examples. Section "Applications of AI in Cyber Security" enlists various applications of AI in security. This section will discuss in depth about various applications of AI

Table 2.1 Definition of AI [1]

	Humanly	Rationally
Thinking	A machine should think and solve problems like a human would	A machine must use proper logic and arguments and facts to find a correct solution
Acting	Machine must act like human. It should have the ability of natural language processing, reasoning and knowledge representation	Machine should proceed based on rational factors. It must act in order to produce a more efficient outcome based on the given scenario

in cyber security and how these applications are deployed. Section "Limitations of AI in Security" will discuss about limitations of using AI for cyber security. Using examples, the section will talk about how AI can be manipulated to nullify security measures. To conclude the chapter, section "Conclusion": Conclusion will be at the end, summarizing the whole chapter leading to the last section which will cite the references.

Literature Survey

AI in cyber security is targeting issues like malware detection, network intrusion, spam and phishing detection. Ongoing researches show different combinations of existing AI algorithms or different AI techniques combined together to solve various security problems. These combinations have generated great results, some better than the other. Reference [1] mentions that as AI has proved to be a boon for mitigating threats, AI will also prove to be a curse with the cases of AI-based attacks and threats increasing.

The combinations to develop a new security model are decided based on the data properties in a system. The learning algorithms must be first trained accurately using the security data and target information acquired from the system, before it can start intelligent decision-making. Reference [4] has reviewed various deep learning and some standard neural network approaches that can be used to build a security model, including supervised learning, semi-supervised learning, unsupervised and reinforcement learning. They included convolutional neural network (CNN), self-organizing maps (SOM), recurrent neural network (RNN) and many more. They concluded that deep neural network models and their hybrid combinations can intelligently resolve many existing security issues and also will help in predicting unknown attacks. Similarly, [5] discusses application-level security by integrating AI and cyber security models. The paper talks about how to identify and avoid network intrusion attacks and detect suspicious activities in network by using artificial intelligent models and malicious activities in server which is monitored by cyber security methodology, hence reducing the load on network. Cyber security is a vast sea, which holds many different security models for various threats that exist. Discussing one such model, Identity and Access Management (IAM) and its relationship with artificial intelligence. The study conducted in [6] explains that even though organizations are applying AI tools for security measures, they have still not matured enough in their approach to information and access management. If artificial intelligence is applied to IAM along with an appropriate monitoring and reporting tools, visualization of network connectivity and data access will become possible, thus reducing network breaches.

Various studies have been performed to compute the effectiveness of AI in cyber security ecosystem. One such work conducted in [7] presents an extensive review to compare different machine learning algorithms in cyber security applications: malware analysis, intrusion detection, spam and phishing detection. Commercial products were not included in the study as there is a high possibility that the vendor will not reveal their original algorithm and there may be some cases where they might overlook certain limitations in their systems. The study included analysis of botnet and Domain Generation Algorithm (DGA) for intrusion detection via ML. For malware analysis, the paper includes polymorphic and metamorphic features of the malware and how ML algorithms can help to mitigate them. Lastly, they suggest ML algorithms from the family of supervised and unsupervised learning algorithms that can be useful in addressing various attacks discussed. Another study focused in Iraq talks about effectiveness of AI models against cyber threats [8]. Research data was collected from 468 employees from IT industry, and basic analysis of model, discriminant validity, confirmatory factor analysis were carried out. First, an expert system analysis of data was carried out by the team without including any AI or cyber security techniques. Later for the same test cases, a smart PLS (variance-based structural equation modeling using the partial least squares (PLS) path method) was applied. It was seen that there was a huge impact on the results when AI was used.

Technology is slowly depending on artificial intelligence day by day as it can be seen in different sectors like education, health, transport and economic sectors. Most seen applications of AI are personal assistant, precise health consultation and treatment and cyber security. Reference [9] performs a rigorous survey on impact of AI in today's digital society and concludes that AI has shown tremendous success in the area of security by detecting and preventing serious threats and AI will emerge to be a very useful tool in future for cyber security. AI will come into aid for manufacturing and e-commerce by making the plants self-reliant. Research in neural networks proves that in future, AI can match human- like thinking. Machine learning methods are used in most sector nowadays, and social media is being one of it. An article on application of machine learning on social media discusses how ML algorithms have become a very convenient tool to measure sentiment of the user, and based on it sorting news and other digital data for the user to view [10]. Apart from this, the article speaks about how ML can be used to identify fake news and malicious content on social media. Moving further, it discussed the integration of cyber security with AI and the need for same in battling adversarial machine learning and privacy problems.

Federated machine learning has taken over the buzz now, as it provides a way for participants to learn a shared machine learning model, without exposing their local data. But, studies show that an intruder can still exploit shared parameters and compromise user's local data, for example medical smart watch, self-driving cars, etc., which can be very dangerous. Reference [11] proposes a privacy-enhanced federated learning model (PEFL) scheme for industrial artificial intelligence. The proposed method is meant to safeguard the local information of the user by considering them as local gradients and using deep learning algorithms along with their proposed model, thus providing postquantum security (next-generation security). On one hand, cyber security industry has successfully incorporated some AI techniques, that are used to mitigate threats, forensic analysis, intrusion detection, prevent sensitive data leak, predicting unknown attacks, critical infrastructure protection, malware detection, etc. On the other hand, rise of adversarial AI is evident, whose key idea is to break down artificially intelligent tools and systems for profit and even for fun [12]. With the use of AI to provide security, new AI vulnerabilities are coming into light, like system manipulation and data poisoning. Reference [13] says, adversaries are using these vulnerabilities to attack AI and alter the system behavior. Reference [14] concludes that every technology comes with its own drawbacks and issues, and after exploring the balance between AI-based security and AI-based threats, it is safe to say that AI security is better than no security as well as AI will provide better strategic plannings to mitigate new attacks.

Artificial Intelligence Techniques for Cyber Security

Information technology is a hotspot today and therefore an easy target to commit crimes and also being used as a medium for committing crimes as well. Readily available devices and high-end products have made it even easier to execute attacks from any location and be untraceable. Digital crime or Computer crimes or Network crimes come under Cybercrime, where the intent of the criminal is to steal sensitive data like hack bank servers for monetary gains, steal personal data and so on. Cyber-crimes encompass offenses like online extortion, misuse of intellectual property rights, international money laundering, economic espionage [15]. As these crimes have become common, they have also grown to become more threatening. Traditional security methods fail to prevent and at times even identify these attacks until it has already happened and then it is too late, as the sensitive data is either out in the open or they are misused.

Hence, the introduction of artificial intelligence in cyber security has proved to be a game changer. As presented above in the survey section, there are organizations and researchers working tediously to develop new and innovative ways to incorporate AI with security, and this has proven to be very useful in many organizations which will be mentioned further in this section. As already discussed, AI is a tool that finds ways to push machines to be more intelligent and replicate human-like behaviors like learning, planning, thinking, reasoning, etc. To counter this issue of trying to be more intelligent, a simplified approach was stated that will divide the main goal, i.e., making the machine intelligent into smaller sub-goals, thus breaking down the roles into different characteristics that should be mimicked by the system [15].

The different characteristics are as follows: Deduction, reasoning and problem solving—agents; neural networks, statistical approaches to AI; Knowledge representation—ontologies; Planning—multi-agent planning and cooperation; Learning—machine learning; Natural language processing—information retrieval, text mining; Motion and manipulation—mapping, navigation, localization; Perception—facial recognition, speech recognition; Social intelligence—empathy simulation; Creativity—artificial imagination; General intelligence—strong AI.

Now that it is discussed how AI strives to become more and more intelligent, and let us take a look at how AI works. AI works in three ways [1]:

1. Assisted intelligence: this helps people to improve what they have already been doing.
2. Augmented intelligence: provides help with things that cannot be easily done by people.
3. Autonomous intelligence: machine learning features that are a separate entity and that act on its own.

Keeping this outline in mind, it can be said that AI focuses on solving tasks which can simply be a support to the existing system to dealing with some most difficult issues on its own, like cyber security since cyberattacks have proved to potentially catastrophic.

Various Artificial Intelligence Tools and Techniques Are Mentioned Below

Expert system (ES):

An expert system is built to solve complex problems and provide human-like decision ability. Based on the user query, an expert system extracts knowledge from knowledge base and uses inference rules to make sound reasoning and returns the result [16]. Expert system aids in decision-making by combining both heuristics and facts just like a human expert. To solve any complex issue, this system extracts knowledge from its knowledge base and uses available facts to come to a certain conclusion. The name expert systems came to life because it stores expert knowledge about a specific domain and helps to solve complex issues in that particular domain. The accuracy and performance of expert system come from the knowledge it stores in this knowledge base. More knowledge saved in the knowledge base, the better the system improves its performance. Figure 2.1 represents how the expert system functions.

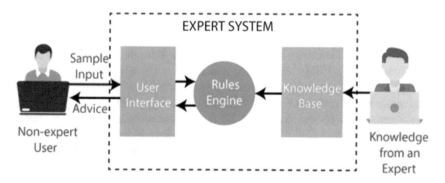

Fig. 2.1 Working of expert system in AI [16]

Expert systems have following characteristics: high performance—expert system solves complex issues with high performance, accuracy and efficiency; understandable—takes input from the user in human understandable language and returns output in same, and thus, it is easily comprehended by the user; reliable—can easily rely on this system for accurate results; highly responsive—complex queries can be solved within a very small-time frame.

ES can be bifurcated into two important components: knowledge and inference engine [1]. Knowledge is the heart of *knowledge-based* systems and stores all the information gathered by the machine as experiences which will be used to train the system. Knowledge base can be understood as a database where all the expert information (data received from the experts of a domain) regarding the domain is stored. Every data is stored with its corresponding attributes or feature, thus making it more efficient. Again, knowledge base contains two types of knowledge: factual knowledge—based on facts and accepted by the experts, and heuristic knowledge—based on experience, practice and its ability to learn.

Reasoning of these stored knowledge is done by *inference engine*. Inference engine can otherwise be referred to as the brain of the expert system. Inference rules are applied by the inference engine to the knowledge base in order to generate error-free solution to a posted query or to find new information based on the query, which is stored back to knowledge base for future reference. Rules are generated on the basis of two types of inference engine: deterministic inference engine—deductions are totally depended on facts and rules and are always assumed to be true, and probabilistic inference engine—contains uncertainty as the conclusions are based on probability.

These expert systems help in solving two scenarios, case-based reasoning and rule-based reasoning [1].

Case-based reasoning—These reasoning techniques pull out problem cases similar to the current problem scenario and assume that the solutions derived to solve the past cases can be applied to the current scenario to obtain a suitable solution. And, this solution is evaluated and revised until needed and then added to knowledge base.

Rule-based reasoning—Experts define rule to solve the problem. Rules are composed of two parts: condition and action. The problem is evaluated based on the condition, and then, necessary actions will be taken for the same. This technique cannot modify its existing rules or learn new rules.

Expert systems are considered as one of the best tools for making decisions as it has no memory limitations, has high performance, accuracy and efficiency, is an expert in a domain, takes into consideration all the facts, experiences and regular knowledge update which make it even more accurate and efficient. However, it also can give wrong results if the knowledge is wrong, cost of development and maintenance is too high, and these systems cannot self-learn.

Machine learning:

Machine learning is a part of artificial intelligence which enables a system (or machine) to learn from data, enhance its performance by inferring from its experience

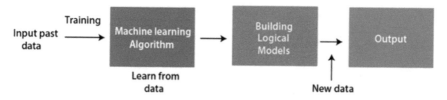

Fig. 2.2 Typical machine learning algorithm working [17]

and predict new outcomes. The goal of machine learning is to develop mathematical models and make predictions based on historical information. First, the machine is trained with the help of historical or pre-existing or in other words training data. Using this, the machine builds a decision-making or prediction model which as name suggests is used for taking a decision and/or predicting new outcomes. From here, whenever the machine receives a new input, it refers to the model it built and generates the result. Figure 2.2 depicts the work flow in a machine learning algorithm.

Machine learning helps system to understand the underlying connection between data and how to learn from data itself and past experiences without the need to be explicitly programmed. Machine learning uses statistical methods to discover new patterns, extract information from raw data and deduce connections even when dealing with huge amount of data. Machine learning algorithms are discussed further in the chapter. Commonly used algorithms in machine learning are: Support Vector Machine, Decision Tree, K-means clustering, Random Forest, K-nearest neighbors, etc.

Deep learning:

As it is pre-established that machine learning is a subset of AI, similarly deep learning is a subset of machine learning. Deep learning has the same goals as machine learning, and only difference is that deep learning is heavily influenced by human brain. Deep learning hopes to achieve the similar way of thinking and deductions, as is done by the human brain, and in order to achieve this, deep learning uses neural networks. A neural network is inspired by the biological neural system in human body. It can be taught to neural network to identify as well as classify patterns and information in the same way as the human brain would. Each layer of neural network can be understood as a filter that sorts the data to achieve an accurate result. A neural network contains different layers as depicted in Fig. 2.3, which contains input layer, three hidden layer and an output layer, where x is input data and y is the outcome of the neural network.

Artificial neural network (ANN) has the needed properties that equip deep learning models with capabilities to solve complex problems, that would be impossible by a machine learning algorithm to solve. The main factor that distinguishes deep learning algorithm from machine learning algorithm is feature extraction method. Machine learning performs feature extraction to draw the needed attribute from a dataset. Feature extraction is a complex process that requires in-depth knowledge about the domain, whereas deep learning does not require feature extraction as that is supported

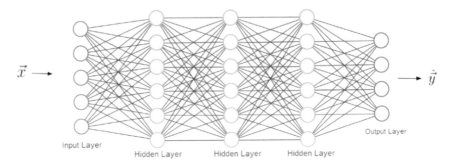

Fig. 2.3 Typical neural network [18]

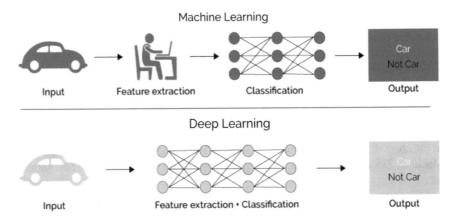

Fig. 2.4 Machine learning versus deep learning [18]

by ANN itself, thus making deep learning superior. This difference can be seen in Fig. 2.4.

Deep learning helps the computers to learn from their experiences in the same manner a human being learns new things from experience. Deep learning enables system to learn tasks that can be done by a human being, without any human intervention. It mimics human brains and nervous system by generating new patterns for the machine to use decision-making techniques. It keeps creating extensive neural networks and trains these networks with huge datasets, thus increasing the performance of neural networks.

Artificial immune systems (AISs) are models for computational analysis that are able to learn continuously and dynamically from its environment and easily adapt to ever-changing surrounding. This concept is based on the biological immune system, which detects and discards intruders (bacteria or virus) in a living organism; similarly, AISs imitate this biological system when applied for network security and act as a constantly evolving intrusion detection system [15].

Another AI technique inspired by biological system is *Genetic Algorithm*, which is a learning methodology that follows the process of natural selection. Genetic algorithm helps in solving complex mathematical problems that has a large set of input variables and possible outcomes. These algorithms are used for generating classification rules for attack patterns and can even generate pattern-specific rules as well. The basis of this algorithm is natural selection process that means poor solutions will be replaced by offspring of best possible solutions from the mix until a precise outcome is guaranteed [15].

Machine Learning Algorithm Used to Train a Machine

When AI or ML is used for security, one of the usual problems that still stand is to identify attack patterns efficiently, understand it, classify and take necessary actions. As attack signatures keep getting modified by the attacker to surpass the security of the system, it is an utmost important job to find every deviation of the attack pattern. A negative detection of the network traffic is as problematic as numerous false-positive detections. To train the machine, three learning algorithms are used, as mentioned in detail below:

Supervised learning

Supervised training is a form of learning algorithm, that trains a machine by utilizing *labeled data,* and using this data, possible outputs are predicted. A labeled data is nothing but predefined outputs tagged to their respective inputs. It can be understood in this way, that the labeled training data used to train the machine is a supervisor, guiding the machine to accurately predict an output. The machine is provided with input along with its correct output and the machine should be able to map the new input with an output correctly, hence being the goal of the supervised learning algorithm, i.e., to being able to map input variable (x) with output variable (y) $\{f(x) \to f(y)\}$.

Supervised learning functions as follows: it uses two types of datasets, training data and testing data. Training data is the labeled data, which helps the machine to learn patterns within the dataset, and based on this, the model is tested using the testing data and the machine predicts the output.

Figure 2.5 shows how the machine is trained with labeled dataset (shapes and their names), and after the training, it is tested with test dataset and the machine has to predict the names of the shapes. In this form of learning, the machine understands how to map any input data to its corresponding output based on given sample data. The machine is trained till it becomes so accurate that it can precisely calculate outputs from the new inputs.

This learning algorithm is classified into two main groups: classification problems and regression problems [19, 20].

Classification algorithm is used for categorical output data, which means that when the output of a training set is binary: Yes/No, True/False, Male/Female, etc.

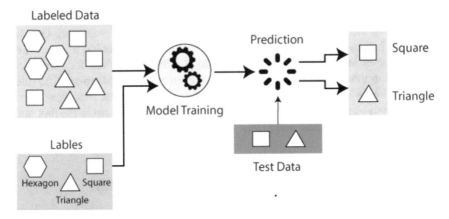

Fig. 2.5 Working of supervised learning [19]

Classification algorithm returns discreet outputs, e.g., classification of a book based on color black or green. Here are few of the classification algorithms—Random Forest, Decision Tree, Logistic Regression, Support Vector Machines.

Regression algorithms are sought when there exists a relationship or dependency in between the input and the output variables. This algorithm returns prediction values such as cost, weight, and it can be a continuous variables like market trends and weather forecasting, etc. Some of the regression algorithms are Linear regression, Nonlinear regression, Regression Trees, Bayesian Linear regression and Polynomial Regression.

There are enough datasets containing malware files, thus allowing easier training to these algorithms, which lead to improved detection of attacks and less numbers or false positive and false negative. These algorithms have contributed in smarter and more advanced approach in detecting cyberattacks than traditional string-matching techniques or address blacklisting techniques. Supervised learning can predict outcomes based on experience, and in real-world issues, it has proved quite useful in spam filtering, fraud detection, etc. However, this algorithm is not good for complex tasks and requires huge computational time.

Unsupervised learning

As opposed to the supervised learning where the machine is trained with labeled dataset, in unsupervised learning, the machine is trained with no sample labeled data. As the name suggests, the machine is not supervised using labeled training data, rather the machine tried to identify hidden patterns and knowledge from the provided dataset.

The aim of machine learning is to understand the structure of the given dataset and find similarities among the data, divide it into similar group and present the outcome. For example, if the machine is given a dataset that contains the images of fishes and birds but no label to it, the machine will have no clue about the features of the dataset.

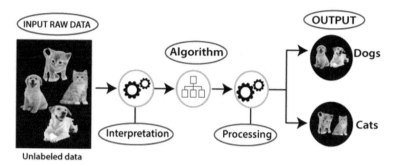

Fig. 2.6 Unsupervised learning algorithm [21]

Now, the goal of the machine is to learn on its own and differentiate the features and categorize them into groups (clusters) based on the similarities between the features. Unsupervised learning aids in identifying important insights from a given dataset, as it works very similar to how human being thinks and understands based on their own experience thus making this a real AI. The machine is expected to understand the data and learn from data alone regarding the underlying patterns and deduce the correct output for the new inputs, as can be seen in Fig. 2.6 with an example of dogs and cats. Unsupervised learning can be categorized into three types, namely: clustering, association and dimensionality reduction.

Clustering: A method in unsupervised learning model which divides data into different groups based on the similarities in the features and sorts the features with less to no similarity in another group.

Cluster analysis finds common items/features between data objects and categorizes them accordingly. When used for security, the algorithm identifies and groups similar type of network traffic data into relevant clusters and in the process isolating abnormal network traffic like shown in Fig. 2.7, for e.g., failed login events, data accessed by a user which is usually not accessed by him, connections from unusual locations, etc.

Association: These unsupervised methods categorized inputs based on association rules, which are based on finding relationship between the input variables in a large dataset. Association rule learning methods attempt to establish rules and relationship between large databases. It discovers set of items that usually occur simultaneously or are dependent on each other. For example, let us say a person who goes to buy candles, also can buy matches. These rules majorly benefit marketing strategies.

Dimensionality reduction: Dimensionality reduction methods serve very well in real-time intrusion analysis. Dimensionality reduction works by reducing the number of features (an attribute of various elements that exist in a dataset) of data which will help in solving the problem easily [20]. Suppose a network traffic has n number of features like source IP, destination IP, port number, protocol, routing information, MAC address, etc., analyzing each and every feature is cost ineffective and takes huge time to process, thus rendering real-time applications useless. However, if we

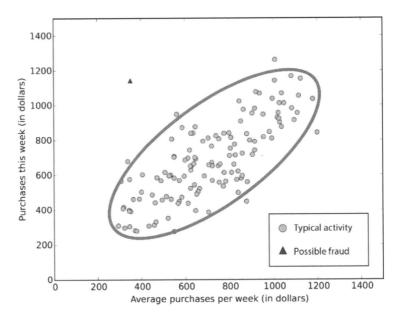

Fig. 2.7 Detection of online fraud using clustering (Microsoft 2017) [20]

are able to reduce the number of features to more relevant ones and divide them into manageable sets, the problem will be easier and quicker to solve.

Some of the algorithms of unsupervised learning are: K-means clustering, K-nearest neighbor (KNN), Apriori algorithm, hierarchal clustering, anomaly detection, etc. Unsupervised learning algorithm is used to perform more complex tasks and is more preferred as it uses unlabeled training data, which are much easier to acquire. However, the result of this learning algorithm can be less accurate and is intrinsically difficult.

Reinforcement learning:

A feedback-based learning algorithm where the agents (something that perceives and acts) learn how to behave in a given environment by the feedback they receive based on their actions. Positive feedback is given for a good or correct action and negative feedback is given for a wrong action as shown in Fig. 2.8. This algorithm also trains with no labeled data, and its actions and learning are purely based on experience. The agent associates with the environment and explores the same on its own. An agent under reinforcement learning aims to enhance the performance by receiving maximum positive rewards. Reinforcement learning can be described as a hit and trial method of learning, where every hit is rewarded (positive feedback) and every miss is punished (negative feedback). It is a combination of both supervised and unsupervised learnings which learns the next moves based on rewards or punishment. This learning algorithm is used when there is very little to no data provided.

Fig. 2.8 Machine learning
via reinforcement learning
algorithm [22]

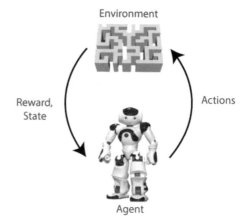

Why AI is Preferred Over Current Anomaly Detection and Prevention Systems?

The current system of anomaly prevention and detection system offers ways to detect unknown attacks, but there are certain limitations that entail them. One of the major issues faced by these traditional methods is difficulty in establishing a proper model that can encase list of acceptable behaviors and rules for attack patterns, as they generate a high number of false positives, which is in reality caused by a normal network traffic but might have some deviations with respect to the defined rules. It is very common for behaviors to change and modify depending on the environment, what is difficult here is that the detection and prevention system has to be updated regularly for the same so that it enables machine to recognize normal traffic from malicious ones [15]. Other limitations of these systems are:

- Any legitimate activity that gets marked as malicious one by the detection/prevention system (known as false positives) results in defective communication or exchange of data, as the prevention system will attempt to stop or change the activity it identified as malicious.
- Attackers can easily shutdown the intrusion detection systems, if they can learn how the system functions.
- Integrating data from heterogeneous environments proves to be an issue.
- Any intrusion detection and prevention system should adhere to certain legal regulations and/or service level agreements.

Artificial intelligence on the other hand is ever-learning and ever-growing technology that will keep updating its knowledge base, depending upon the new inputs and its environment [23]. The rules are updated and modified as and when required with minimum involvement of the system admin, thus paving a path toward robust and secure systems.

Use Cases of Artificial Intelligence in Cyber Security

Everyone is connected via internet today. This means, there are numerous personal and public data available on internet that can lead to many cyber security issues. Manual analysis of this data is impossible given its sheer volume, and for the same reason, cost to prevent attacks on such large volume of data is also too high. Designing and implementing new algorithms for every new threat that uncovers itself require lot of effort, money and time. AI comes to the rescue here, by examining huge data accurately and more efficiently within a short period of time. Even with ever-changing threat patterns, AI can predict similar type of attacks that can happen in future and avoid it by using the threat history. The main reason that AI is a perfect fit for cyber security is because: it can handle huge data volume, it can predict and discover new possible attacks and it learns continuously from its environment to respond better to future threats. Artificial intelligence can single handily reduce the response time and improve the network security when an attack or any malicious activity is detected.

Use cases [24]:

(i) Network Threat Identification:

Protecting the network infrastructure of any organization is very vital, and this requires understanding every aspect related to the network topology and building necessary cyber security processes that will be tailored according to the network's needs. Keeping track of all the information coming in and going out of the network is difficult for human professional to address, and it will also require considerable amount of time. To add on to this, figuring out which system is under attack or device is malicious is a challenging task for any professional. And when it comes to large-scale organizations, it requires huge amount of time to identify malicious apps or activities in the network and human professionals do not always provide accurate services. Thus, an AI security system for the organization's network traffic will help with monitoring all incoming and outgoing traffic and detect suspicious patterns in it. The data that is combed by an AI is usually a large volume of dataset that is difficult for a network professional to go through and find threat patterns [25].

Example: Versive is an AI vendor that provides cyber security based on artificial intelligence, which identifies threats based on dissonant detection. Dissonant detection means any sound that is so annoying or disruptive that it puts any listener on edge (alarm going off, scream of a person, etc.). This organization provides its services to banks and other financial institutions to identify any security threats. Cyber security offered by Versive takes DNS, Proxy and NetFlow as inputs for their security engine. Anomaly detection is used to monitor network to find discrepancies in the data.

(ii) User Behavior Modeling:

There exist certain security attacks that steal the private login credential of a client in a company without their knowledge and use these credentials to access company's network. Since it is a user's login credentials, it does not raise any red flags, and thus, these attacks are very difficult to detect and stop. AI-based risk management systems play an important role in this area. These systems analyze any changes in password pattern of a customer in the organization. Any inconsistency is being notified to security team to take necessary action.

Example: An AI vendor called "Darktrace" provides a security software which is used as a machine learning algorithm to inspect the information on network traffic and understand the user behavior in a company. This software alerts the company as soon as it detects any suspicious behavior that is contrary to normal user behavior.

(iii) Fighting AI threats:

As researchers and system admins are using AI to establish better cyber security, hackers are utilizing this technology to detect entry points within an organization's network. Thus, in future, one of the biggest security defenses against AI-adopted attacks will be AI-based software. NotPetya and WannaCry are the firms that suffered badly against cyberattacks and ransomware in past few years. These attacks are only going to grow here after, and these attackers can make a bigger impact on the technological sector by using AI to launch attacks.

Example: A security software called, Falcon Platform developed by CrowdStrike, a cyber security technology company, uses artificial intelligence to protect against ransomware attacks. The software uses anomaly detection for endpoint security.

Applications of AI in Cyber Security

As mentioned above, AI was introduced to learn human behavior and mimic them and to understand real-world issues from a human point of view. AI was required to make machines more intelligent and learn from its experiences. AI has made it possible as well as convenient to store and process large volume of data intelligently. AI has been using such processing capabilities to provide applications in different sectors like space exploration and defense [26].

Applications of AI can also be seen in healthcare sector where it is helping physicians to make a proper diagnosis, generate treatment plans based on patient's medical history and even help perform surgical treatments. Another most crucial application of AI can be seen in cyber security. Cyber security encompasses technologies, practices and process to protect devices, programs, network and data from unauthorized access or in case of attacks.

Various frameworks exist in cyber security such as ISO 27001/27002/27017, NIST, CCM, HIPAA, NERC [26]. These security standards control different security domains. Some of the crucial areas of study in cyber security are—Application security, Network security, Infrastructure security, Web security, Threat intelligence, IoT security, Identity and access management, Mobile security, Cloud security, Incident response plan, Human security.

Cyber security measures focus on managing risk and vulnerabilities to increase system's durability at the time of attack. Attacks are inevitable. An individual and especially an organization must always be prepared to bear and mitigate attacks. The key to effectively do this is to monitor network behaviors, anomalies and latest malwares. Cyber security is based on threat analysis, i.e., preparations to mitigate threats should be in place and enough protection should be provided for the system to endure the threat and come out of it with minimum to no loss. The major challenge for any cyber security methodology is to be able to function even when the system is under attack, rapidly end the attack and restore all the functionalities to the point (normal state) before the incident (attack) took place. An organization's security strategy is built on estimated risk and threat analysis.

Threat, risk and vulnerability are interconnected in cyber world. The system or a network is a valuable asset to the organization and has to be protected at all cost. Any organization will try to protect its system from malicious attacks/threat, mitigate risks and keep updating the system security to avoid vulnerability. Threat can be defined as any harmful activity, like an attack that puts a system at harm's way. Vulnerability points to weakness in the system, which might lead to an attack or increase the chances of an attack to occur, whereas risk is expected damage that can occur in the system [1].

Cyber security is interdisciplinary area including computer science, system and criminology. There are many factors that come into play while dealing with it, like people (user, admin, employee, etc.), network and application process and integration with other technologies, and therefore, problem can arise at any point in any factor associated with security, as shown in Fig. 2.9.

Lately, cyberattack numbers have increased exponentially along with the sophistication of the attacks. Cyber-criminals keep learning how to deploy latest tools and techniques to hack, attack and steal information from a user and/or a system, and traditional security measures have proven to be inadequate to prevent breaches resulting from such attacks. Therefore, artificial intelligence was introduced to build smart models to defend the networks from attacks. As AI can evolve rather quickly to understand complex situations, it has become an elementary tool in cyber security. AI techniques help in rapid identification of network intrusions, malware attacks, data breaches, phishing attacks, etc., as well as alert the necessary party when the attack occurs [1].

Use of AI can teach us how to enable expert security measures to monitor and analyze abnormalities in the system. Organizations apply AI in these four areas to enhance cyber security measures [2]:

Automated defense: Cyber security systems are divided into two types—Analyst-driven system and Automated systems. Analyst-driven systems are people dependent, and they are developed and operated by people. On the other hand automated systems make use of AI tools. These intelligent tools are self-learning systems. Humans alone cannot defend cyber space, and thus, a need of automation is very much necessary. With the unprecedented growth of data and network complexity, AI is a blessing for organizations to monitor secure their systems. These automation tools can easily be integrated with some of the existing security measures. Some of the functions are:

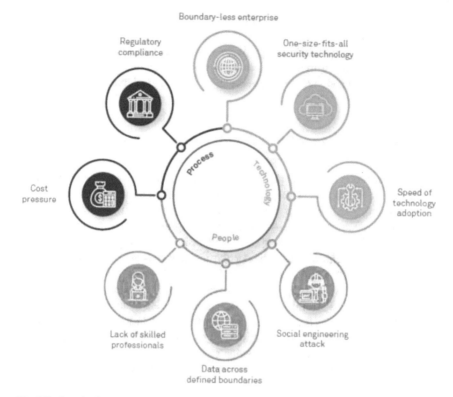

Fig. 2.9 Security issues associated with people, process and technology [2]

- Detecting threats and malicious activities using predictive analytics.
- Securing conditional authentication and access.
- Enhancing learning and analysis through natural language processing.
- Improving human analysis—from malicious attack detection to endpoint protection.
- Using automation in mundane security tasks.

Cognitive Security: This approach combines human intelligence as well as AI. It is an advanced form of AI that uses different forms of AI. Cognitive security is related to AI in a way that both push the boundaries of machine intelligence. The only difference between the two is in the way they react with humans. AI can be described as a technology that strives to return a most accurate result or action based on algorithms while requiring human intervention, whereas cognitive security aims at overcoming the boundaries of programming and unite with humans to help them make better decisions.

Adversarial training: An adversarial learning is associated with using artificial intelligence for malicious reasons. Using this learning, AI can be taught to misbehave and spill sensitive data. But, training AI with adversarial attack models can help detection of vulnerabilities that can be fixed in early stages and help to make the model more robust.

Parallel and dynamic monitoring: When a system is targeted with new learning abilities, it needs to be monitored constantly. This is done so that any deviation between the actual and expected outputs can be noted and addressed.

AI Solutions for Cyber Security [26]

- AI2: An artificial intelligent platform—AI2 was developed by MIT and PatternEx to predict cyberattacks. This platform resulted in 86% accuracy while detecting and predicting cyberattacks which was regarded as three times better than any previous prediction model. AI2 uses clustering algorithm from unsupervised learning technique and isolates suspicious activities, which are then passed to system analyst who decides whether the incident was an attack or not. Every outcome from the platform is added to a dataset by the analyst to enable future learning based on existing data (supervised learning). New models are also generated by this platform for better detection of future attacks.
- Darktrace: Darktrace is a security solution that helps in identifying and recognizing new cyber threats, that a traditional security model would normally miss. It detects anomalies in an organization's network by using Enterprise Immune System (EIS) technology and machine learning algorithms. EIS works with mathematical principles, which means that it does not employ rules or signatures and therefore can detect and respond to new and/or unknown security attacks which the system has not experienced before. Using machine learning and mathematical principles, Darktrace adapts and learns user behaviors, device and network behavior so that it can distinguish between genuine behaviors and behaviors that may indicate attack. Self-learning technology of Darktrace allows organization to view a detailed network analysis and proactively respond to threats to mitigate risks.
- CylanceProtect: Launched in 2018, CylanceProtect is an integration of artificial intelligence with information security tools that helps in threat prevention. Script-based attacks, memory-targeted attacks and attacks on external devices are protected by the use of information security and artificial intelligence which helps in identifying known and unknown malicious software. CylanceProtect helps in preventing unknown as well as known zero-day attacks. It also protects the device without causing any inconvenience to the end user.
- Deep Instinct: This software was developed to protect firm's mobile devices and services in real time against malicious attacks. Using artificial intelligence, Deep Instinct detects hostile activities on devices, office workstations and services. It then employs deep learning techniques to predict unrecognized cyberattacks. The aim behind this is to help the software learn different types of combinations of requests that occur when dealing with malicious systems. The aim of using deep

learning here is to slice the software code into smaller code snippets for future survey.

- Human security: Organizations prepare its network, system and devices to evade or tolerate external attacks, but they should also protect their resources from threats that may originate within the firm. Such threats can be originated from an employee that works in the company or an intruder who has assumed the role of an employee and cause damage from within, making the assets of the company weak. They can escalate their privileges to access more sensitive data from the company or they can steal customer information. Applying artificial intelligence to monitor the user behavior is must here. Any slight modification that might indicate potential attacks must be red flagged and notified to the admin immediately.

Limitations of AI in Security

As seen above, artificial intelligence methods are an indispensable tool to fight cyber-crimes, but it comes with its own fair share of consequences. There are certain limitations associated with AI such as intensive training, high resources and cost. If adequate resources are used for training and testing any machine learning model, it is a quite convenient process, but with limited resources, AI systems may take huge processing time since the volume of data required to train a system is equally huge. One of the big drawbacks is that hacker can also pursue AI methodologies to develop malware and train it to become unnoticeable from AI detection tools like inserting adversarial data into the mix, model stealing, data poisoning. Though AI provides extensive support in the field of cyber security, if the same is used against security, it can make the future of attacks more dangerous and unpredictable [2].

Ethical Issues Related to AI [27]

Data ethics plays an important role in determining the boundaries to ethical access of data. It is clearly established that AI requires data to train its models and therefore requires data collection and data processing. This means that the AI and data are completely intertwined. In cases like these, setting an ethical boundary for data and its use or access has become a very complex problem. Some concerns that come up are:

- Is the data safe?
- How AI manages sensitive data?
- What happens if a piece of data is modified?

According to researchers, in about 50 years there is a huge possibility that AI may become a serious threat to humanity. By year 2040 there is said to be a 50% similarity in the thinking and understanding process of a human and a machine which could jump to 95% by the year 2075, where one would not be able to distinguish between a human and a machine based on their thinking processes.

Ethical issues related to artificial intelligence can be divided into two types:

1. Data collected, its processing and data analysis.
2. AI making decisions that are based on generalized data.

The primary ethical problem arises with the collection and analysis of user data, which include social data, digital data, personal data. Companies need this data to train machines for better data mapping and generating self-resilient security models. Ethical issue that rises here is that all these data are stored in one dataset to train the machine, and any malicious activity with it and billions of user sensitive data is compromised, but imposing over the top restrictions on data access will generate poor security models and slow down development of AI in cyber security [27, 28].

Another issue is the making decisions based on generalized data. Let us take an example, AI used in the development of military needs like smart missile, etc., and these use geographical location as data input and make decision based on it. These intelligent weapons can help to save thousands of lives when deployed correctly; if there is even a slight tweak in the data, it may even target the creators. Seeing this network security point of view, an AI security model designed to prevent attack for an organization is modified by a hacker and all the hacker did was add some adversarial inputs in the dataset mix, then there is a huge possibility that the model developed to provide security may be a security threat to the system.

AI-Based Threat to Cyber Security [27]

After analyzing the trends in creation and use of artificial intelligence, two criminological risks come to the surface:

1. Direct risk: When the risk of using AI will have a direct effect on the user. These risks are; certain wrong data in a self-learning AI can lead to decisions taken by AI that may constitute crime; intentional actions taken by AI that can harm its user for example, AI personal assistant misusing the customer data; and AI which was created by hackers with the main aim of committing crimes.
2. Indirect risk: Unintended AI hazards that affect the user. These risks can be explained as AI system or software error, error made by AI during its operations, etc. These risks make system vulnerable and create a backdoor for the hackers to attack or gain access to the system.

IT experts classify threats that can be created by AI in three ways:

- Malware attacks.
- Attacks using social engineering techniques.
- Physical attack: defines attack on physical AI objects, like attack on drones.

Using AI for cyber security is a blessing but also a curse, although better has been done compared to misuse of AI. As easily as AI can provide strong security models, cyber criminals can use the same features of AI to launch threating attacks as well. Therefore, the above-mentioned challenges are the reason why AI is not completely used as a cyber security solution.

Conclusion

AI is an ever learning and ever developing as well as fast emerging technology. It has become a must have standard in the industry to defend cyberattacks. With the ever-growing data volume, it has become a difficult task for humans alone to factor and secure enterprise-level attack surface. Artificial intelligence comes to a much-needed rescue here by providing in-depth data analysis and threat identification, which will help security professionals to reduce security breaches in the organization and enhance the security of an organization. In cyber security ecosystem, AI has achieved successful discovery of attacks, prioritizing risks in a system, direct incident response plan, malware detection, prevention and analysis and identification and prevention of unknown attacks before they even occur.

Artificial intelligence is growing tremendously fast in every sector, especially in cyber security, as can be seen by various organizations launching a new intelligent tool to handle different types of cyber threats, both known and unknown. Despite of all the outstanding achievements made by AI in the field of security, one must still stay very cautious regarding the extensive dependency on AI when it comes to network security.

References

1. Morovat, K., Panda, B.: A Survey of Artificial Intelligence in Cybersecurity. Paper presented at the International Conference on Computational Science and Computational Intelligence 109–115 IEEE December (2020)
2. Das, R., Sandhane, R.: Artificial intelligence in cyber security. Int. J. Phys. **1964**(4), 042072 (2021)
3. Stone, P., Brooks, R., Brynjolfsson, E., et al.: Artificial intelligence and life in 2030: the one hundred year study on artificial intelligence (2022). arXiv:2211.06318
4. Ghillani, D.: Deep Learning and Artificial Intelligence Framework to Improve the Cyber Security. Authorea Preprints (2022)
5. Anitha, A., Paul, G., Kumari, S.: A cyber defence using artificial intelligence. Int. J. Pharm. Technol. **8**(4), 25325–25357 (2016)
6. Azhar, I.: The interaction between artificial intelligence and identity & access management: An empirical study. Int. J. Creat. Res. Thoughts 2320–2882 (2015)
7. Lubin, A.: Cyber law and espionage law as communicating vessels. Paper presented at the In 2018 10th International Conference on Cyber Conflict 203–226 (2018)
8. Alhayani, B., Mohammed, H.J., Chaloob, I.Z. et al.: Effectiveness of artificial intelligence techniques against cyber security risks apply of IT industry. Proc. Mater. Today (2021)

9. Raimundo, R., Rosário, A.: The impact of artificial intelligence on Data System Security: A literature review. Proc. Sens. **21**(21), 7029 (2021)
10. Thuraisingham, B.: The role of artificial intelligence and cyber security for social media. Paper presented at the IEEE International Parallel and Distributed Processing Symposium Workshops 1–3 May (2020)
11. Hao, M., Li, H., Luo, X., et al.: Efficient and privacy-enhanced federated learning for industrial artificial intelligence. IEEE Trans. Ind. Inform. **16**(10), 6532–6542 (2019)
12. Bertino, E., Kantarcioglu, M., Akcora, et al.: AI for security and security for AI. In: Proceedings of the Eleventh ACM Conference on Data and Application Security and Privacy 333–334 (2021)
13. Feng, X., Feng, Y., Dawam, E.S.: Artificial Intelligence Cyber Security Strategy. Paper presented in the In 2020 IEEE International Conference on Dependable, Autonomic and Secure Computing, International Conference on Pervasive Intelligence and Computing, International Conference on Cloud and Big Data Computing, Intl Conf on Cyber Science and Technology Congress 328–333 IEEE (2020)
14. Sahu, A., Harshvardhan, G.M., Gourisaria, M.K.: A dual approach for credit card fraud detection using neural network and data mining techniques. Paper presented in the In 2020 IEEE 17th India Council International Conference 1–7 IEEE (2020)
15. Dilek, S., Çakır, H., Aydın, M.: Applications of Artificial Intelligence Techniques to Combating Cyber Crimes: A Review (2015). arXiv:1502.03552
16. Javapoint What is expert system? Javapoint Available via https://www.javatpoint.com/expert-systems-in-artificial-intelligence
17. Javapoint Machine Learning Available via https://www.javatpoint.com/machine-learning
18. Artem Oppermann What is Deep Learning and How does it work? Towards Data Science. Available via https://towardsdatascience.com/what-is-deep-learning-and-how-does-it-work-2ce44b b692ac. Accessed 13 November 2019
19. Javapoint Supervised Machine Learning Available via https://www.javatpoint.com/supervised-machine-learning
20. Veiga, A.P.: Applications of artificial intelligence to network security (2018). arXiv preprint arXiv:1803.09992
21. Javapoint Unsupervised Machine Learning Available via https://www.javatpoint.com/unsupe rvised-machine-learning
22. Javapoint Reinforcement Learning Available via https://www.javatpoint.com/reinforcement-learning
23. Chan, L., Morgan, I., Simon, H., Alshabanat, et al.: Survey of AI in cybersecurity for information technology management. Paper presented in the In 2019 IEEE technology & engineering management conference 1–8 (2019)
24. USM (2020) AI & ML in Cybersecurity: Top 5 Use Cases & Examples. USM system. https://usmsystems.com/ai-ml-in-cybersecurity-use-cases-examples/. Accessed 05 June 2020
25. Gaurav Belani The Use of Artificial Intelligence in Cybersecurity: A Review. IEEE Computer Socitey. https://www.computer.org/publications/tech-news/trends/the-use-of-artifi cial-intelligence-in-cybersecurity
26. Vähäkainu, P., Lehto, M.: Artificial intelligence in the cyber security environment. In: Proceedings of the ICCWS 2019 14th International Conference on Cyber WarfarSe and Security: ICCWS, Stellenbosch, South Africa 431 (2019)
27. Khisamova, Z.I., Begishev, I.R., Sidorenko, E.L.: Artificial intelligence and problems of ensuring cyber security. Int. J. Cyber Criminol. **13**(2), 564–577 (2019)
28. Atiku, S.B., Aaron, A.U., Job, G.K., et al.: Survey On the applications of artificial intelligence in cyber security. Int. J. Sci. Technol. Res. **9**(10), 165–170 (2020)

Chapter 3
AI and IoT in Manufacturing and Related Security Perspectives for Industry 4.0

Rohit Kumar and Shanmugam Sundaramurthy

Introduction

The world recognizes artificial intelligence (AI) as a cutting-edge technology. Today, several businesses and people are attempting to use AI in practically every industry, including health care, education, manufacturing, smart cities, agriculture, etc. Many multinational corporations are using automation and intelligent robots to improve production, improve the quality of the completed product, and increase total efficiency as a result of the concepts of "Smart Factories" and "Industry 4.0." As a matter of fact, artificial intelligence is a crucial instrument to support manufacturing by enabling R&D, improving quality, lowering mistakes, and preserving the supply chain by projecting demand forecasts and simulating outcomes to nurture larger margins amid fierce competition. Manufacturing is about to enter a phase of significant innovation and alterations brought by the deeper integration of sensors and the IoT (Internet of Things), improved data accessibility, and developments in automata and robotics. This results in widespread digitization and helps to reassess the current activities and future strategic directions in the smart manufacturing. These new developments point to the future. As AI models get more intricate and bigger datasets, centralized training techniques cannot change to meet these new demands. Distributed learning strategies like federated learning have been developed by Google, enabling numerous clever terminals for collaborative learning of a shared model. Nevertheless, all training data are kept in terminal devices, which pose several security difficulties. How to check that the model is not being taken

R. Kumar (✉) · S. Sundaramurthy
Department of Computing Technologies, SRM Institute of Science and Technology, Kattankulathur, Chennai, TN, India
e-mail: rohitk@srmist.edu.in

S. Sundaramurthy
e-mail: shanmugs9@srmist.edu.in

with malice, with which it may build a distributed ML system. A significant area of research is privacy protection.

Similarly, IoT may be envisioned as a continuous connection between the physical and digital worlds that is all-pervasive. IoT technologies enable the remote monitoring and control of physical items across a network. This capacity makes way for a new class of applications when combined with the widespread connectivity of heterogeneous devices. The physical objects that can be connected range in size from small embedded electronics in items to enormous automated structures, machines, and everything in between. In order for networked items to work together to accomplish shared goals, an IoT application depends on a foundation built via the integration of sensors, actuators, and tracking devices like RFID tags with information and communication technologies. At the same time, IoT devices frequently have security flaws that are exceedingly challenging to fix. According to HP's research, there are an average of 25 security flaws per device, and 70% of IoT goods have them. Utilizing controls and weaknesses for nefarious purposes, the attacker committed several crimes. In conclusion, as IoT devices become more widely used, the development of security flaws will pose serious threats to user security and privacy, as well as the protection of people's lives and property.

The Paper focuses on the vision of smart manufacturing, specifically to investigate the critical role AI and IoT play to develop the manufacturing industry differently. The rest of the work is structured as follows: Role of AI in manufacturing is discussed in Sect. "Role of AI in Manufacturing". It provides related literature, technologies, need of AI security, applications, etc. In Sect. "IoT in Manufacturing", role of IoT in manufacturing is reviewed with a focus on related literature, technologies, need of IoT security, applications, etc. The problems and possibilities based on the evaluation of vulnerability analysis technologies are compiled in Sect. "Vulnerabilities and Challenges". Finally, the work is concluded in Sect. "Conclusion and Future Work" with a hotspot focus on further possibilities.

Role of AI in Manufacturing

AI researchers will study how machines can reason and solve problems independently of human input. It is easy to assume that AI is designed to mimic human thought processes, but this is not necessarily the case. Although humans excel at certain tasks, they are not without flaws. The most superior AI is one that can reason and act rationally. One of the finest examples of this is the inability of humans to make sense of enormous datasets and the complex patterns within them. But, an AI can quickly go through a manufacturing machine's sensor data and find outliers that mean the machine may need to be fixed in the next few weeks. Although it would take a human quite some time to carefully consider the evidence and draw a decision, artificial intelligence can complete the task in a matter of seconds. Some of these concepts appear to have been lifted directly from science fiction, and others are merely the next logical step in the development of existing technologies. First and foremost, there will be a renewed focus on information collection. The manufacturing sector is not taking full advantage of artificial intelligence technology and approaches. More data could be collected and used by AI platforms to improve various manufacturing operations as the use and efficiency of IIoT devices increase. However, as the development of AI applications continues, new developments like totally automated factories and automatically designed products with minimal human oversight may become a reality. But, if we do not keep up our rate of innovation, we will never get there. Just have a concept, and you are good to go. It could include combining several technologies or adapting an existing one for a different purpose. These changes change the way things are made and give businesses an edge over their competitors.

The progress in the field of artificial intelligence application development may one day allow for totally automated production and the creation of products with little to no human intervention. We will never succeed if we give up on trying new things. Consolidating technology or putting it to use in novel contexts could be part of the solution. Technologies in manufacturing provide a competitive advantage for businesses.

Related Works

The research discussed in [1] critiques current AI-based manufacturing and management solutions for long-life batteries. AI-based battery production and smart battery health are shown first. Next, the most popular AI solutions for battery life diagnosis, including state-of-health assessment and aging prediction, are discussed. Also shown are AI-designed battery longevity solutions. Finally, this field's biggest difficulties and tactics are discussed. This research will reveal practical, advanced AI for health-conscious battery manufacture, control, and optimization at diverse technology readiness levels. Similarly, the study presented in [2] discusses using information models to construct explainable artificial intelligence decision support systems. The article will review and outline manufacturing system design requirements to suit client goals and use cases. An information model is proposed to convey system design needs and transparently portray decisions and alternatives to improve the description of artificial intelligence-based decision support systems during the (re)design of manufacturing systems. The information model explores the requirements and technical solutions needed to improve manufacturing systems without losing track of alternatives and adapts them dynamically to market and production conditions.

The work explored in [3] highlights the use of AI systems in industrial manufacturing, purchasing, and supply management operations, resulting in smart factory and smart manufacturing concepts and the reorganization and digitalization of the production floor, formerly dominated by humans. In addition, the study presented in [4] explores the integration progress of AI and fragmented manufacturing industries, designs a supply-and-demand labor market with varied talents, and theoretically explains how AI technology affects manufacturing employment. Then, it designs the propensity score matching difference-in-difference model, categorizes intelligent manufacturing businesses, and examines the effects of AI technology on their employment structures before and after integration. Finally, this work provides effective manufacturing enterprise transformation and upgrading methodologies and employment structure solutions.

The authors examine an AI-enabled AN item enhancement work in [5]. AI, digital reality, IoT, blockchain, driverless cars, and other advancements depend on 5G's lightning-fast connectivity and low latency. 5G opens new IT opportunities, not just a generational shift. This article aims to combine AI with additive manufacturing systems based on the original goals and expectations of both areas. Furthermore, manufacturers state various obstacles in [6] when they innovate their business and operational processes to boost efficiency and adaptability. Manufacturers must adapt to new technologies while maintaining operational efficiency and sustainability to reach Industry 4.0. This report critically evaluates important issues affecting the next generation of manufacturers, obstacles, and implementation impediments. A new Industry 4.0 framework and UN Sustainability Goals alignment analyze these factors. Moreover, the work discussed in [7] shows how to automate delivering production orders to a manufacturing facility to optimize waste and plant capacity using industrial process automation. Intelligent process automation has quantitative benefits in production, and the research can be applied to worldwide organizations.

A comparative study of the above-discussed works has been provided in the form of Table 3.1.

Technologies

This section discusses the new technologies related to AI in the manufacturing sector. Some of the major technologies have been shown in Fig. 3.1 and are discussed below.

Machine Learning

Machine learning is an AI method for automatically analyzing large amounts of data in order to detect trends and generate predictions. Artificial neurons are used in the input layer of a neural network. Once the input is received, it is sent to a hidden layer, where it is given a weight before being sent on to the output layer. Deep Learning is a method of machine learning that uses a network of nodes called "layers" to model the way the human brain processes data, with information being passed from one layer to the next.

Natural Language Processing (NLP)

In manufacturing, AI support like chatbots using NLP helps to increase problem reporting and service requests. This AI field simulates natural human conversation.

Table 3.1 Comparative analysis of AI in manufacturing-related works

S. no	Authors and year	Theme	Inference
1	Kailong Liu et al. (2022)	AI-based manufacturing towards long lifetime battery	Feasible, advanced AI for the manufacturing, control, and optimization of batteries with a focus on health at various levels of technological preparedness
2	David S. Cochran et al. (2022)	Advance explainable AI-based decision support systems in manufacturing system	Examines the requirements and technical solutions required to improve manufacturing systems without losing track of alternatives and to dynamically adjust them to changing market or production conditions
3	Kehayov M et al. (2022)	Application of AI in supply chain, manufacturing, and purchasing process	Smart factory and smart manufacturing, as well as the reorganization and digitalization of the production floor, which has been dominated by human labor until now
4	Shao S et al. (2022)	Impact of AI in manufacturing	Effective strategies for the transformation and upgrading of manufacturing firms, as well as solutions for resolving employment structure issues
5	Shirwaikar R. D et al. (2022)	AI-enabled manufacturing using 5G and IIoT	Developed a method that integrates AI with additive manufacturing systems
6	Chatterjee S et al. (2022)	Predicting welding responses using GA and ANFIS	Multi-gene genetic programming accurately predicts welding responses in laser welding processes and can be employed to accurately anticipate performance metrics
7	Lievano-Martínez et al. (2022)	Intelligent process automation in manufacturing industries	This research provides the potential of intelligent process automation and its quantitative benefits in the production process, and the contribution may be applied in a worldwide context to a variety of businesses

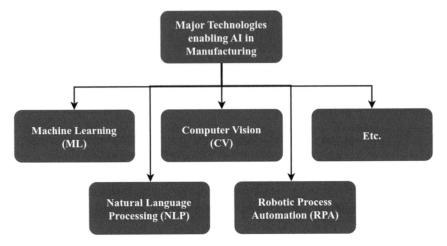

Fig. 3.1 Major enabling technologies for AI in manufacturing

If workers can utilize mobile devices to report issues and queries to chatbots, AI can help them to submit high-quality, understandable reports faster. This enhances worker accountability and minimizes supervisor and worker workload. The use of NLP can help manufacturers to improve their comprehension of the data obtained through the process known as web scraping. Artificial intelligence can scour the web for data on transportation, fuel, and labor expenses, as well as industry benchmarks. This has the potential to contribute to the overall efficiency of the business's operations.

Computer Vision

The field of computer vision in manufacturing is concerned with the development of artificial systems that can capture, process, and interpret visual inputs from the real world (primarily factories and other industrial sites) in order to prompt appropriate responses and aid humans in a wide range of production-related tasks. Using a rule-based method, such as by recognizing particular elements in the acquired visuals and evaluating if they fit with a set of supplied parameters, even the earliest iterations of computer vision may recognize specific items and trigger an answer in the manufacturing and other areas. This method, however, generates a large number of false positives and is not as effective when dealing with the complexities inherent in unstructured data types like images and videos.

Robotic Process Automation

The term "robotic processing automation" appears frequently in discussions of robots and artificial intelligence. Keep in mind, though, that hardware is completely irrelevant to this discussion. Instead, this is a software-related issue. Robotic processing automation focuses on software, not hardware, automation. Data extraction, form filling, file transfer/processing, and more are just some of the tasks that can be automated by this take on the concept of assembly line robots. The importance of these roles in manufacturing cannot be overstated because of the need to keep track of stock and other administrative duties. This is of paramount significance if your products require individual software installation.

Impact of AI in Manufacturing

Workers might imagine a world where machines operate autonomously and robots gradually solder and bolt together parts without human supervision because of the term "artificial intelligence." Humans are still playing a significant role in smart manufacturing, even as the number of such enterprises increases, and this trend is expected to continue for decades.

These erroneous beliefs may pave the way for novel solutions and cutting-edge technologies. Workers may fear that the introduction of AI may lead to fewer employment opportunities for humans. A common misconception among engineers, data scientists, and other technicians is that manufacturing is a dying industry. If these kinds of worries keep smart, young people from going into the industrial sector, it could lose talent for a whole generation.

About 426,000 manufacturing jobs in the world are unfilled due to a lack of trained workers. Eighty % of organizations lack AI project staff. Skilled people may not apply for open opportunities since AI technology may eliminate their jobs. Implementing that advanced technology is difficult due to a scarcity of competent workers. Hiring AI experts is difficult, but demand is tremendous. Thus, several manufacturers train them to create internal AI competence. Software and third-party resources can help you to develop experience.

Uses Cases of AI in Manufacturing

The important applications considered in this chapter related to AI-based solutions in manufacturing industries using industry 4.0 are discussed as follows:

Defect Detection

The entire AI quality inspection process may be automated with the help of machine learning algorithms, and AI flaw detection is founded on computer vision. Models used for defect detection are taught to visually inspect products as they come off the assembly line in order to identify surface irregularities and detect differences in size, shape, and color. This applies to a wide variety of produced goods, any of which may have surface flaws. The food business, the textile industry, the electronic industry, the heavy manufacturing industry, and many more are all good examples. However, standard machine learning presents its own challenges when applied to the development of flaw detection systems.

Quality Assurance

Electronics manufacturing demands extreme precision. Quality assurance has traditionally required a professional engineer to manually check electronics and microprocessors for correct manufacturing and circuit configuration. Today, image processing algorithms can automatically check for optimal production. By putting cameras in key spots on the production floor, this sorting can be done automatically and in real time.

Assembly Line Integration

Machines and sensors are ubiquitous in today's modern manufacturing. Although assembly lines have been pioneers in the use of many forms of automation technology, much development is still possible in this area. With full or partial automation, assembly lines can function to their fullest capacity with the help of artificial intelligence. Artificial intelligence is still a developing field of study, which could make some firms wary of implementing it. There is usually a considerable outlay of capital for AI systems, but the payoff might be substantial. AI can exploit the massive amounts of data generated by an assembly line to optimize the working conditions there.

Generative Design

The design and production of products can both benefit from AI. Designers and engineers input design objectives into generative design algorithms. After that, these algorithms produce design possibilities based on all of the possible solution variations. Each iteration of machine learning consists of testing and upgrading.

Predictive Maintenance

Manufacturers have implemented several preventative maintenance measures in an effort to keep up with the soaring demand. Reactive maintenance is one of them; it represents the "don't fix it until it's broken" mentality but ultimately leads to more unscheduled repairs and downtime. Preventive maintenance, in contrast to reactive maintenance, involves carrying out routine maintenance procedures in advance of a problem arising. Although it significantly lengthens the life of the machinery, the investment is very high. The possibility of a piece of machinery or its part failing unexpectedly remains, nevertheless. Predictive maintenance appears to be a superior choice for manufacturing organizations as it can reduce downtime and maintenance costs. Also, the manufacturing industry of today has tight deadlines and profit margins, which makes unplanned downtimes.

IoT in Manufacturing

The adoption of digital transformations from numerous contexts, such as customer attention, efficient productivity, automation, competitive advantages, and quick returns, has been made possible by the use of IoT by manufacturers. The quick growth of industrial processes has been characterized by high-speed and inexpensive electronic circuits, quick signal processing techniques, and inventive advancements in manufacturing technologies. The variety of applications for sensor systems as well as their usage are both continually expanding. These technologically advanced sensors and tools are capable of remote monitoring and control as well as Internet-based communication and collaboration.

Smart technology's forerunner, Industry 4.0 factories, permits the application of countless radical technologies, such as cloud computers, large data analysis, AI, 3D printing, and sophisticated robotics. Big data analytics turn collected data into usable information that may be used to improve procedures and are essential to many production processes, including predictive maintenance, asset management, etc., as well.

The manufacturing, transportation, oil and gas, health care, agrotech, energy, and utilities industries have benefited greatly from IoT technology. These sectors have intricate infrastructures that include numerous interconnected sensors, smart meters, industrial robots, and software for data transmission and communication. Industrial IoT systems have sensitive information of importance, making them attractive targets for hackers.

Related Works

A variety of manufacturing resources now have a new approach to manage resources more effectively and implement dynamic scheduling thanks to the Industrial Internet of Things, which unites the essential industrial communication, computing, and control technologies. In order to accomplish effective dynamic resource interaction, device-to-device communication technology, software-defined industrial networks, and OLE for process control technology are suggested in this article. In order to accomplish dynamic resource management, ontology modeling and multiagent technology integration are also included. Authors suggest a load-balancing system in [8] that concentrates on intelligent machinery in the smart industry and is based on Jena reasoning and Contract-Net Protocol technology. Complex resource allocation issues in contemporary manufacturing situations can be resolved by dynamic resource management for IoT-based manufacturing.

The Fourth Industrial Revolution idea has been realized by the Internet of Things (IoT), although its applications in the manufacturing sector are comparatively few and mostly researched without contextual characteristics. In order to fill this gap, the research in [9] conducts an intensive critical analysis to look into important IoT applications from the manufacturing Industry 4.0 perspective. By investigating the major contribution categories and identifying six significant contrasts between conventional and manufacturing Industry 4.0, authors outline the essential knowledge gaps in the body of literature and empirical investigations that exist.

Edge computing makes it possible to apply business logic to upstream data from the Internet of Things and downstream data from cloud services. Edge computing offers the further advantages of agility, real-time processing, and autonomy in the area of Industrial IoT to provide value for intelligent manufacturing. The work discussed in [10] suggests an architecture for IoT-based manufacturing with a particular emphasis on the idea of edge computing. Additionally, it examines the function of edge computing from four angles: edge hardware, network communication, information fusion, and joint cloud computing. The case study provided by the authors uses a prototype platform to implement active maintenance. The authors also offer a technical resource for edge computing implementation in the smart factory. Furthermore, IoT adoption in manufacturing allows for the conversion of conventional industrial systems into contemporary digitized ones, creating enormous economic prospects through the reshaping of industries. Modern businesses are now better equipped thanks to industrial IoT to adopt new data-driven strategies and more readily bear the strain of international competition. However, the IoT adoption raises the overall amount of created data, turning industrial data into industrial big data. The study presented in [11] shows how industrial big data will be produced as a result of IoT adoption in manufacturing, taking sensing systems and mobile devices into consideration. Additionally, a created IoT application is demonstrated to demonstrate how actual industrial data may be produced, leading to industrial big data.

In the work presented in [12], additive Manufacturing (AM) is a cutting-edge manufacturing technique and technology that uses the latest in sophisticated equipment and control systems. AM has been recognized as having a unique value to the business and finding several uses in a variety of sectors, including aerospace, health care, energy, and automotive. High-performance processing and computation will thus be crucial in AM. This study examines the cloud-based models and ideas of cloud computing, cloud manufacturing (CM), and IoT, as well as how these concepts relate to and affect the AM sector in the 4.0 age. This paper offers a thorough theoretical foundation and paradigm for AM integration. This paper also introduces CM applications and their integration with AM as well as suggests an integrated AM cloud platform. Additionally, Product Lifecycle Management (PLM) has primarily influenced product development and production engineering and contributed to a significant acceleration of procedures and operations. IoT, on the other hand, is now experiencing a very strong emergence, and public interest in it is steadily rising. Surprisingly, there is not much connection between PLM and IoT systems right now. When and when there was a high and definite return on investment, industrialists have facilitated a few linkages. Today, however, the cultural gap between PLM's engineering heritage and IoT's computer science background persists, making systems' integration a challenge. Through a thorough literature study that details the IoT's expanding perimeter over the past ten years, the research endeavor presented in [13] seeks to close the gap between PLM and IoT research. It also discusses how PLM and IoT information systems and people are approached in literature. Finally, a framework supporting the integration of PLM and IoT in the manufacturing industry is suggested and discussed.

Production processes have been hampered by the incorporation of smart devices in the industrial sector. There is a significant void in the literature on IIoT and Industry 4.0's implementation issues, particularly when it comes to India. The research has mostly concentrated on the technical elements of network architecture and development. The IIoT deployment issues faced by the manufacturing industry are exhaustively analyzed and categorized. With the aid of professionals, difficulties were divided into technical and organizational difficulties. The study offered in [14] tries to create a hierarchical framework that will assist policymakers to determine the most pressing issue, so they can decide with knowledge. The findings of this study are anticipated to identify the main issue on which the scientific community and business community should concentrate their strategic efforts. When applying IIoT techniques in the industrial sector, this will make it easier to solve any hidden problems.

A comparative study of the above-discussed works has been provided in the form of Table 3.2.

Table 3.2 Comparative analysis of IoT in manufacturing-related works

S. no	Authors and year	Theme	Inference
1	Wan et al. (2018)	Dynamic resources management for IoT-based manufacturing	Provides a load-balancing system based on Jena reasoning and Contract-Net Protocol technology in the context of smart factory
2	Kalsoom et al. (2021)	Systematic analysis on the impact of IoT on manufacturing Industry 4.0	Examines the major contribution categories and identifies six crucial distinctions between conventional manufacturing and Industry 4.0
3	Chen et al. (2018)	Edge computing analysis in IoT-based manufacturing	Explores the function of edge computing in relation to four different areas, including edge computing hardware, network communication, information fusion, and cloud computing cooperation
4	Mourtzis et al. (2016)	Industrial big data analysis in IoT-adopted manufacturing	Explains how the use of IoT in production would produce industrial big data by taking into account sensing systems and mobile devices
5	Haghnegahdar et al. (2022)	IoT-based cloud manufacturing in intelligent additive manufacturing:	Highlights the cloud-based models and ideas of cloud computing, cloud manufacturing, and IoT, as well as their connections to and impacts on the modern additive manufacturing sector
6	Barrios et al. (2022)	Integration of IoT and PLM in manufacturing industry	Intends to make clear all prior research on PLM and IoT through a thorough literature analysis that highlights the expanding boundaries of IoT over the past ten years
7	Malhotra et al. (2022)	Analysis of IIoT in manufacturing industries using the ISM approach	Seeks to create a hierarchical framework to assist policymakers to identify the most pressing issue to help them make an informed choice

Technologies

This section discusses state-of-the-art technologies related to IoT in the manufacturing sector. Some of the major technologies have been shown in Fig. 3.2 and are discussed below.

Radio-Frequency Identification (RFID)

RFID tags connected to items are automatically identified and tracked using electromagnetic fields to send data [9]. RFID readers and tags make up the RFID systems. RFID tags that are connected to the items contain information about the objects, and RFID readers that are not required to have a line of sight may read this information (including the unique IDs) and report it to the business information system. Because of this, readers might inadvertently monitor the movement of the objects to which the tags are connected as well as the actual physical movement of the tags in real time. RFID may be used in manufacturing for supply chain management, production scheduling, and other purposes.

Wireless Sensor Networks (WSNs)

The WSNs are networks of autonomous, geographically dispersed nodes that can detect their surroundings, do calculations, and interact with other nodes [10]. The sensor nodes function autonomously and decentralized, preserving the best connectivity for as long as feasible and sending data to the base station through multi-hop spreading. Since a single node cannot always sense the entire environment, they

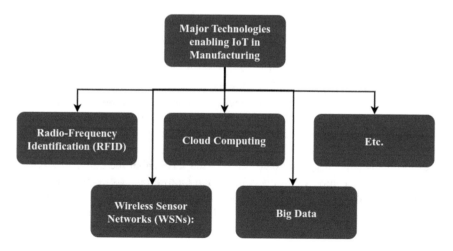

Fig. 3.2 Major enabling technologies for IoT in manufacturing

must work together and employ collaborative signal and information processing approaches to complete their jobs. Individual nodes, on the other hand, are tiny, energy-constrained devices with mediocre CPUs and little memory, which has a big impact on how WSNs are designed and implemented. Actuators that modify the world's physical properties might be found in the nodes. WSNs offer a broad range of potential applications in a variety of sensing-based industrial decision-making scenarios. WSNs and RFID are complementary technologies. RFID cannot be used to monitor an object's state, but it may be used to detect and identify things that are difficult to detect or differentiate using conventional sensor technologies. In contrast, WSNs offer multi-hop wireless communication in addition to monitoring the state of the objects and surroundings.

Cloud Computing

Cloud computing is based on virtualization technology and the Service-Oriented Architecture (SOA), enables effective management of a very sizable shared pool of reconfigurable computing resources (such as networks, servers, and storage), which can be swiftly provisioned and released with little management effort or service provider interaction. It features crucial traits including resource pooling (multi-tenant), on-demand access, quick flexibility, and measurable service (pay-as-you-go business model). The manufacturing industry may be transformed with help from cloud computing. The idea of cloud manufacturing (CMfg) (a new paradigm for service-oriented manufacturing) is one prominent initiative that has gained global attention.

Big Data

The cloud computing paradigm offers unheard-before capabilities for the practical management of big data generated by manufacturing IoT thanks to its vast computational resources. IoT success or failure depends on big data. Big data is a general term for datasets that are too big or complicated for conventional data processing techniques. Comparing it to typical datasets, it differs from them in five key ways: volume, variety, velocity, value, and veracity. Phases of data capture, extraction, integration, analysis, and interpretation are included in the lifespan of big data. Cloud computing's strong storage and computation capabilities play a key part in those phases of activity. The requirements from big data also hasten cloud computing's growth. In manufacturing, big data can be used throughout the entire lifecycle of products, having a significant impact on design innovation, cost reduction, quality, efficiency, and customer satisfaction. For instance, big data analysis can be used to design more precisely targeted products and create efficient promotion strategies.

Need of IIoT Security

IIoT operations are supported by one or more types of cloud-related technology. However, each sort of cloud has its own set of vulnerabilities. Public clouds pose the greatest danger since they rely heavily on Internet services for communication. When the data coming from the Industries are saved for data analytics and processing, this layer also includes mobile devices that may retrieve data from the cloud or the Internet. As the devices in OT are embedded devices made up of hardware, firmware/OS, and applications, the lower layers employ Intranet to prevent attacks from the public networks. Attacks often take use of flaws in the protocols the devices use to interact with one another, as discussed below.

Denial-Of-Service (DoS) Attack

A DoS attack uses network bandwidth or services to prevent a server from providing services to a client. Cloud DoS attacks are more harmful since they entail sending requests from zombie-like, innocent sites on the network, thus the name "Cloud Zombie Attack." Firewalls may be used to grant or deny access to requests, and stronger authentication and authorization systems, IDS for detecting DoS assaults, and firewalls themselves can help to prevent such attacks [15].

Side Channel Attacks

To carry out a side channel attack on a cloud, a malicious virtual machine is installed and set to target the cryptographic methods being implemented by the system. In the IIoT, cloud service providers often oversee security. The dangers posed by these assaults should be understood by the Industries.

Cloud Malware Injection

The attacker infects the cloud system with a worm or malicious service implementation on a virtual machine that is capable of infecting the cloud targets. This attack may be stopped by doing a service instance integrity check before utilizing a service instance for incoming requests.

Attacks on Authentication

The majority of cloud services still rely only on login and password authentication today. This attack can be stopped using account lockout, delayed response, and multi-factor authentication techniques.

Man-in-the-Middle Cryptographic Attack

In this assault, the attacker (MitM) places himself in the middle of a communication line between two users, where he may intercept or change that message. Using mutual authentication or an One-Time Password (OTP), such assaults can be stopped.

Attacks on Mobile Devices

Malware, data exfiltration, data manipulation, and data loss are all examples of mobile device attack vectors. Make sure that the program only receives the permissions it needs, and check the trustworthy signatures to make sure that it does not have any backdoors in order to prevent these mobile device assaults.

Phishing Attacks

Phishing attacks take place when attackers fool a user into interacting with genuine-looking bogus websites or emails in order to get the victim's private information. The strongest protection against phishing is to educate people about such attacks.

SQL Injection

The term "SQLI" refers to an injection attack in which the attacker inserts malicious input to get private information from databases, destroy databases, and evade authentication. By employing parameterized queries and stored procedures, SQLi may be avoided.

Malware

If the IT systems are not patched and no security policy is put in place, malware at layer IV can be utilized to propagate to the OT. Malicious code can transmit malicious payload to the lower layer devices in the OT network. Malware may be found using security firewalls and antivirus software that receive frequent updates.

DNS Poisoning

To carry out this attack, the attacker forges a Domain Name Server (DNS) response and transmits it to the DNS server, causing the corrupt data to be stored there. Network managers should restrict recursive inquiries and ensure that query replies only return data for the specified domain in order to prevent poisoning.

Remote Code Execution

In order to introduce malware that may remotely control the target machine, an attacker must exploit a system vulnerability known as remote code execution. Regular patching, checking for data buffer boundaries, and virus protection methods can all aid in preventing remote code execution.

Brute Force Attack

Attackers that use brute force test a variety of inputs until a legitimate one is discovered, to acquire access to sites that require a user's identification in order to access. Verifying the lockout function can aid in thwarting this assault.

Assaults Against Web Apps

These types of attacks target web applications that are hosted on web servers located in the DMZ. Using a reverse web proxy, online applications or services may be secured against threats. The traffic heading to a web application is intercepted by a reverse proxy, which then applies filters to find offensive material, improper grammar, and harmful instructions.

IP Spoofing

By adding a falsified IP address to the packet header, an attacker pretends to be another device on purpose. The Ethernet/IP Industrial Protocol, which employs IP addresses to identify devices, might be problematic when utilized for communication in industrial operations.

Data Sniffing

Since all of the industrial network's devices are connected to the same subnet, sniffing data that are being sent across it is simple. Encrypting all data traveling over the communication connection is the best technique to protect against sniffing attempts. The computational complexity necessary for encryption makes it impractical to encrypt all data, but we can do so for sensitive information like credentials that are often transmitted in plain text when industrial protocols are employed.

Data Manipulation

Data manipulation is simpler on industrial networks since earlier industrial protocols do not perform integrity checks, and even when they do, the data still travel in plain text, making it possible for attackers to alter the checksums in the packets.

Replay Attack

Attacks known as replays take place when a packet is recorded and delivered at a later time. The control systems may receive the incorrect values thanks to this assault. Replaying packets with normal values might mislead operators by providing inaccurate values. Timestamps should be supplied together with the packets, it is advised.

Wireless Network-Based Attacks

Attacks on wireless networks are feasible in a variety of ways. Wired Equivalent Privacy (WEP) assault, evil twin attack, man-in-the-middle attack, denial-of-service attack, etc. are a few of them. Using the most recent patches, best practices, and encryption technologies like Wi-Fi Protected Access with Advanced Encryption Standard (WPA2/AES) and correctly implementing the 802.1X authentication scheme, many of these attacks may be avoided.

Reverse Engineering

An attacker can uncover defects in the code and gain sensitive information, such as hard-coded credentials, by reversing the source code of the software or firmware used on the embedded device. To insert malicious code into the device, firmware may be reverse engineered and modified. RE may be prevented by avoiding hard-coded credentials and programming obfuscation.

Applications

There are numerous related applications with respect to IoT in manufacturing. Some of the common ones are listed below.

Efficiency and Automation

IoT gathers current status information from production floors (e.g., machinery, vehicles, materials, and environments). Without requiring human participation, these data may be utilized to automate workflows and procedures that maintain and improve design and manufacturing systems. The control software may automatically make judgments and command actuators to reduce deviations from the plan using real-time data collected, clever algorithms, and networked actuators. Intelligent algorithms (like machine learning algorithms) with a lot of multi-source data may automatically provide the best selections. The amount of autonomy, the ability to manage the production processes, and the ability to respond to various disturbances have significantly increased thanks to advances in machine learning technology. Large-scale IoT device management (control) necessitates multi-stakeholder participation and data access that goes beyond the traditional monolithic one-domain and task-specific development methodologies. The degree of centralization, the participating systems' optional independence from one another, and their autonomous evolution are additional difficulties.

Energy Management

With a third of the world's energy consumption coming from manufacturing and energy prices rising, energy management is a serious problem. Due to a lack of infrastructure for holistic mapping to business and fine-grained, continuous monitoring of energy consumption, traditional techniques are based on discrete plant states without a complete picture of the entire plant. By placing sensors at any area of interest, IoT may not only continually measure and correlate energy usage and business operations in real time, but also enforce online dynamic energy-aware control in the IoT-enabled "closed" loops. A more comprehensive approach to energy efficiency should go beyond straightforward stand-alone methods, such as single process/machine optimization. To develop effective strategies, cross-domain cooperation (physical world: machinery, materials, and vehicles; business world: enterprise information systems, manufacturing processes, and logistics) is required, as well as data collection and correlation. Real-time energy-related indices and statistical analyses should be merged altogether.

Preventive Upkeep

This promotes early diagnoses, and part replacement based on the forecast and monitoring of machine deterioration is generally acknowledged by manufacturers as a means of reducing expensive, unplanned downtime, and unexpected failures. Low-cost sensors, wireless connection, and BD tools can provide insightful information about the operation and state of the machine. Machine deterioration may be predicted and seen through modeling, correlating, analyzing, and visualizing historical and real-time data. Additionally, the product designer can use this information to inform closed-loop lifecycle redesign.

Connected Supply Chain Management

Through real-time information sharing on shop floors, inventory, purchasing and sales, maintenance, logistics, etc., IoT-enabled systems can link all parties in the supply chain so that everyone can understand interdependencies, the flow of materials/parts, and production cycle times, spot potential issues before they arise, and take the appropriate action (forming a closed control loop). This might have a significant effect on how effectively lean manufacturing is implemented. Information asymmetry will no longer exist since all stakeholders will have real-time access to information on demand, supply, and feedback. Common standards for data interchange, data privacy and security, and economic models for shared information and IoT infrastructure are the primary obstacles.

Vulnerabilities and Challenges

There are many options to communicate and trade data remotely thanks to smart gadgets. These technologies are vulnerable to hackers because they may be rapidly penetrated. Because there were not many technical security challenges in the past, most industries and manufacturers lacked specialist set regulations and procedures on how to address them. Organizations have used ad hoc solutions, outdated, flawed technology, internal client security alone, and third-party security service providers to manage security issues. These procedures have been shown to be insufficient for proper safety because of the shortcomings in a typical industrial IoT system [16].

Quality Control

To make sure that they work effectively and do not pose any hazards to endpoint security, smart devices should undergo thorough testing methods on a regular basis. If we want to attain a high degree of security against cyberattackers, prompt support and maintenance services are crucial.

Threat Recognition

A lot of devices are connected over a shared network at once. The precise origin of a danger is so difficult to pinpoint. Making specialized threat identification modules that assist in identifying weak spots in the ecosystem's software and hardware is a smart idea.

Configuration of the Hardware

Sometimes, the value of hardware integration is disregarded. However, this is a crucial component to take into account in order to guarantee the ecosystem's proper operation. The infrastructure may be made more secure and protected from unauthorized access by attackers aiming to alter the default configuration, steal data, and cause system disruption with the use of proper hardware setup.

Encryption of Data

Software developers utilize encrypted data to guarantee user privacy and enable safe storage of their personal information. Cryptography protocols are used to conform to defined data encryption rules. These techniques make interactions with the infrastructure secure and help safeguard sensitive data.

Confidentiality of Data

Applications for the Industrial Internet of Things (IIoT) that link machines, sensors, and actuators in sectors with high stakes will generate a significant amount of irregular data that must be handled in real time. A data breach in IIoT systems might have disastrous effects for business, the factory, and even vital infrastructures like the water supply and power grid. Therefore, one of the most serious security concerns in both IoT and IIoT is protecting data confidentiality.

Integration of Cyber-Physical Systems

Since the majority of IoT and IIoT devices are CPS systems, IoT and IIoT systems should facilitate the integrity verification of CPS systems. Attestation, which enables the identification of unauthorized and malicious software alterations, is a crucial tool for ensuring the integrity of a system's software configuration. The fundamental principle of using attestation for integrity verification is that the device that needs to be attested, called prover, sends an attestation to another device, called verifier, as a status report of its current software configuration to show that it is in a known and, therefore, reliable state. CPS integrity verification has been the subject of several studies in recent years, and it continues to be an open challenge.

Devices Pairing Key Establishment

Newly deployed IoT/IIoT devices must be able to securely pair with existing devices through the creation of cryptographic keys for communication security. Asymmetric key setup techniques are not appropriate for IoT/IIoT devices with limited resources since they necessitate expensive computing activities, including the modular exponentiation processes. Devices must be pre-loaded with shared secrets in order to use symmetric key setup protocols, and the produced key can be based on the shared secrets. However, many firms make IoT/IIoT devices; therefore, it is impractical to assume that they come pre-programmed with certain information when they leave the company.

Device Management

Managing massive deployments of devices effectively and securely is one of the growing issues in the IoT/IIoT environment. Device management must provide governance and tracking at every stage of an IoT/IIoT device's lifetime, including when instructions are sent, status is monitored, tasks are completed, the device's software is updated, and issues are dealt with in real time [6]. Along with managing qualities like key pairs, capabilities, and features, device management also handles connections like those between devices and people, devices and other devices, and devices and services. During the lifespan of an IIoT device, all of the aforementioned items might change. An efficient and effective administration of IoT devices is essential for IoT/IIoT systems and is an open challenge.

Conclusion and Future Work

Systems used in production today are becoming more sophisticated, dynamic, and connected. Highly nonlinear and unpredictable activity presents difficulties for production operations, because there are so many ambiguities and dependencies. Newly developed events particularly in artificial intelligence (AI), machine learning (ML) have demonstrated considerable promise. The use of sophisticated analytics techniques to revolutionize the industrial sector manufacturing generates enormous volumes of big data. This work offers some recent applications of AI to representative manufacturing, identifies obstacles, and talks about potential solutions to the numerous levels. Similarly, the Internet of Things (IoT), a promising technology that has seen increased adoption in recent years, realizes ubiquitous physical device interconnection through the Internet, opening the door for developing potent industrial applications by utilizing the advancements in sensor technology and wireless networks. IoT technologies may be seen as catalysts for Industry 4.0 and smart manufacturing. This work also discusses Industrial Internet of Things (IIoT) focusing on IoT applications in manufacturing. To that end, technologies including wireless sensor networks (WSNs), smart sensors, big data analytics, and cloud computing are examined as they relate to the application of IoT in manufacturing.

References

1. Liu, K., Wei, Z., Zhang, C., Shang, Y., Teodorescu, R., Han, Q. L.: Towards long lifetime battery: AI-based manufacturing and management. IEEE/CAA J. Autom. Sinica. (2022)
2. Cochran, D. S., Smith, J., Mark, B. G., Rauch, E.: Information model to advance explainable AI-based decision support systems in manufacturing system design. In: International Symposium on Industrial Engineering and Automation, pp. 49–60, Springer, Cham (2022)
3. Kehayov, M., Holder, L., Koch, V.: Application of artificial intelligence technology in the manufacturing process and purchasing and supply management. Proc. Comput. Sci. **200**, 1209–1217 (2022)
4. Shao, S., Shi, Z., Shi, Y.: Impact of AI on employment in manufacturing industry. Int. J. Financ. Eng. 2141013 (2022)
5. Shirwaikar, R.D., Tandon, A., Kumar, K.S., Nag, M.A., Jos, B.C., Jos, B.M.: Artificial intelligence enabled additive manufacturing system using 5G and industrial IoT. Int. J. Eng. Syst. Model. Simul. **13**(4), 235–240 (2022)
6. Chatterjee, S., Mahapatra, S.S., Lamberti, L., Pruncu, C.I.: Prediction of welding responses using AI approach: Adaptive neuro-fuzzy inference system and genetic programming. J. Braz. Soc. Mech. Sci. Eng. **44**(2), 1–15 (2022)
7. Lievano-Martínez, F.A., Fernández-Ledesma, J.D., Burgos, D., Branch-Bedoya, J.W., Jimenez-Builes, J.A.: Intelligent process automation: An application in manufacturing industry. Sustainability **14**(14), 8804 (2022)
8. Wan, J., Chen, B., Imran, M., Tao, F., Li, D., Liu, C., Ahmad, S.: Toward dynamic resources management for IoT-based manufacturing. IEEE Commun. Mag. **56**(2), 52–59 (2018)
9. Kalsoom, T., Ahmed, S., Rafi-ul-Shan, P.M., Azmat, M., Akhtar, P., Pervez, Z., Imran, M.A., Ur-Rehman, M.: Impact of IoT on manufacturing industry 4.0: A new triangular systematic review. Sustainability **13**(22), 12506 (2021)
10. Chen, B., Jiafu, W., Antonio, C., Di, L., Haider, A., Qin, Z.: Edge computing in IoT-based manufacturing. IEEE Commun. Mag. **56**(9), 103–109 (2018)
11. Mourtzis, D., Vlachou, E., Milas, N.J.P.C.: Industrial big data as a result of IoT adoption in manufacturing. Procedia cirp **55**, 290–295 (2016)
12. Haghnegahdar, L., Joshi, S.S., Dahotre, N.B.: From IoT-based cloud manufacturing approach to intelligent additive manufacturing: Industrial Internet of Things—An overview. Int. J. Adv. Manuf. Technol. 1–18 (2022)
13. Barrios, P., Danjou, C., Eynard, B.: Literature review and methodological framework for integration of IoT and PLM in manufacturing industry. Comput. Ind. **140**, 103688 (2022)
14. Malhotra, S., Agarwal, V., Kapur, P.K.: Hierarchical framework for analyzing the challenges of implementing industrial Internet of Things in manufacturing industries using ISM approach. Int. J. Syst. Assur. Eng. Manag. 1–15 (2022)
15. Rajasekar, V., Sarika, S., Velliangiri, S., Joseph, I.T.S., Kalaivani, K.S.: An efficient intrusion detection model based on recurrent neural network. In: 2022 IEEE International Conference

on Distributed Computing and Electrical Circuits and Electronics (ICDCECE), Ballari, India, pp. 1–6 (2022)

16. Balaji, R., Deepajothi, S., Prabaharan, G., Daniya, T., Karthikeyan, P., Velliangiri, S.: Survey on intrusions detection system using deep learning in IoT environment. In: International Conference on Sustainable Computing and Data Communication Systems (ICSCDS), Erode, India, pp. 195–199 (2022)

Chapter 4
IoT Security Vulnerabilities and Defensive Measures in Industry 4.0

Koppula Manasa and L. M. I. Leo Joseph

Introduction

The IoT is an evolving technology that is defined as connecting diverse types of things (objects) to the Internet, gathering data by means of sensors, controlling machinery or appliances by means of remote control, monitoring agriculture, environments, health, construction, traffic, and so on [1]. According to [2, 3], the IoT can be defined as the objects with wireless identifying devices can connect with one another to create a self-configuring network. In the year 1999, Kevin Ashton invented the phrase "Internet of things". The author described the IoT as the network connecting objects to the Internet in the physical world [4]. IoT allows us to interconnect billions of devices in order to perform communication and computing. Digital individuals such as radio frequency identifier (RFID), sensors, localization techniques, and the Internet make it viable to renovate normal objects into smart objects [3]. Monitoring, sensing, and data collecting can be done by sensors that are embedded into smart objects [5].

The likelihood of IoT can be implicit by an increasing number of objects linked with the Internet at a record rate. For example, consider smart or intelligent homes that can be endowed with sensors, heating ventilation air conditioning (HVAC) perceiving and control frameworks, regulators, thermostats, etc. There are so many circumstances and areas in which IoT can play a crucial job and make our lives easy. The revelation of IoT is on the way to connecting devices and persons in any place, at any time, to anything through the Internet [6]. So, by considering this revelation, IoT application areas will rise continuously and intensely in our life.

The quantity and diversity of IoT devices have hastily developed in the past ages, by means of a forecast of 75 billion devices associated with the Internet by the year 2025 [7]. This drastic upsurge in the number of IoT devices itself affirms it to be the

K. Manasa (✉) · L. M. I. Leo Joseph
Department of Electronics and Communication Engineering, SR University, Warangal, India
e-mail: Manasa436@gmail.com

© The Author(s), under exclusive license to Springer Nature Singapore Pte Ltd. 2023   71
V. Sarveshwaran et al. (eds.), *Artificial Intelligence and Cyber Security in Industry 4.0*, Advanced Technologies and Societal Change,
https://doi.org/10.1007/978-981-99-2115-7_4

main fourth coming technology to expanding the digital economy. The global IoT business was valued at coarsely 389 billion dollars in the year 2020 and is anticipated to nurture to over 1 trillion dollars in the year 2030, more than doubling up its returns in 10 years. Along with this, the number of Internet-associated devices globally is expected to increase in this period [8]. According to the authors of [9], operational safety issues arise from mistakes and failures combining people, infrastructure, tools, and machines within the cyber-physical production system (CPPS). The authors identified the "extensive attack vector of CPPS" as the greatest potential vulnerability to Industry 4.0. IoT devices assure a wide variety of applications such as smart homes, smart grids, smart cities, precision agriculture, and connected vehicles. So, this broad range of applications brings the issue of privacy and security. Evolving IoT applications cannot gain supreme demand without a trusted and protected IoT architecture.

Figure 4.1 depicts the IoT security attack model. The sensors or/and actuators are hooked up to the server/cloud via the gateways and the Internet. The attacker can attack IoT devices in distinct ways. The motive of the attacker may be anything like gathering sensitive information about the user or interrupting the service or controlling the devices or etc. The attacker can directly attack IoT devices, and they can get access to the gateways/server/cloud by attacking them. If the assailant procures access to the gateway/server/cloud, he can control all the IoT devices that are associated with the infected server.

The arrangement of this article is as follows: Sect. "IoT/IIoT Security Challenges and Requirements" provides the IoT/IIoT security challenges and requirements. Section "IoT Architecture and Security Attacks" describes four-layer IoT architectures along with IoT security attacks from an architectural perspective. In

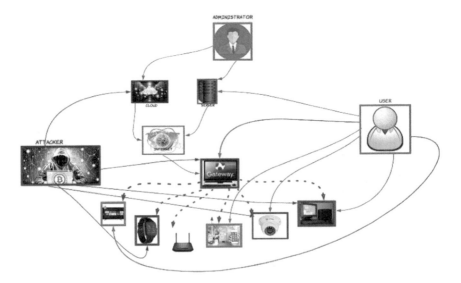

Fig. 4.1 IoT security attack model

Sect. "Preventive Techniques and Mechanisms of IoT/IIoT Security", we briefed the promising solutions to the aforementioned IoT security attacks along with useful mechanisms for providing the solutions. Section "Conclusion" concludes the overall study.

IoT/IIoT Security Challenges and Requirements

IoT/IIoT Security Challenges

Different types of heterogeneous components like sensors and actuators are used to develop IoT. Some of these IoT devices are equipped with more dominant 32-bit processors and some other devices use simple 8-bit microcontrollers. We can use diverse software for different types of processors. So, the software which is selected must be suitable for all ranges of devices. The selected processor and software will not only be able to compatible with the functionality of the devices but should also support the number of nodes participating in the network, storage capacity, power consumption, etc. IoT can be able to develop many applications in different fields. Utilizing constrained IoT devices in Industry 4.0 leads to various security challenges as mentioned in [10–14].

Heterogeneity: This means that IoT architecture should be able to support various devices from different families including sensors, actuators, gateways, networks, appliances, etc. Diverse categories of algorithms, protocols, and circuitry will be employed to build the various IoT devices. The selected software for a particular application must be operated with limited memory and low difficulty [12].

Data Privacy: In some specific applications, the privacy of user information is most important. User data privacy will be interrupted if the personal data provided to the network administrator is leaked to unauthorized persons [14].

Energy Consumption: It is one of the basic and important demanding restrictions in designing IoT applications. Any set of rules designed for IoT devices must be able to satisfy this constraint [7].

Communications: Technologies like ZigBee and low-power wide area network (LPWAN) may be used in IoT applications. Using these devices, authentication is also one of the challenges [7].

Interoperability: This means that a pre-determined and consistent data interchange format must be used to communicate heterogeneous IoT devices and share information with each other. Developing a consistent data interchange format is difficult as the IoT devices are distinct in nature [7].

Scalability: A vast amount of dynamic and new IoT devices are coming into the picture, and billions of devices are associated with the network globally. So, conventional security schemes are not suitable and scalable for the escalating number of IoT devices [11, 14].

Self-awareness: To accomplish pre-defined tasks in real-world applications without human interference, IoT devices should be able to self-organize themselves [7].

Programmability: The software must provide a typical application programming interface (API) and also be able to sustain the generally used programming languages like C and C++ [12].

Autonomy: The software should permit sleep cycles for preserving energy if the device is in an idle state, the selected software must be reliable, and the network protocol to be substituted at every layer [12].

Resource Limitations: Most of the IoT devices like sensors have inadequate resources like battery, memory, and computation power. As most encryption mechanisms are computationally lavish, utilizing them to guarantee high security while diminishing energy consumption is difficult [13].

Mobility: Actuators and sensors used in some applications like smart wearables will always move randomly. One of the core components of IoT is mobility. In this regard, the devices possibly will join or leave a network at any instance. So, the traditional security algorithms possibly will not serve this type of dynamic changes in topology [12, 13].

Data Management: Data produced by IoT devices will be very vast. Managing the vast amount of generated data like storing, controlling access to this information, and protecting its truthfulness and privacy is a very hectic challenge [13].

Security Patch: IoT devices may not obtain software updates (security patches) deprived of disturbing functional security if the devices are installed in a remote area. Hence, it is not possible to solve potential security problems [14].

Multi-protocol Networking: Non-Internet protocol, Internet protocol (IP), or both network protocols may be used at a time for communication of IoT devices. A typical security algorithm that can be designed to work with various communication protocols is difficult to develop [14].

Because of these challenges, diverse types of cyber threats can be created in IoT applications. A variety of privacy and security outbreaks have been produced on already mounted IoT applications. Mirai botnet in 2016 infected around 2.5 million IoT devices and initialized distributed denial-of-service (DDoS) assault [15]. Hajime and Reaper are the other large botnet next to the Mirai botnet in counter to a huge quantity of IoT devices [15]. As IoT devices are little power-driven and not as protected, they offer a gateway for attackers for arriving at its architecture and thereafter give simple access to the user's information.

Requirements for IoT/IIoT Security

In this section, IoT/IIoT security requirements have been discussed. The most important security requirements are confidentiality, integrity, and availability (CIA). Along with the CIA triad, some other important requirements have been briefed in the present section.

Confidentiality is to guarantee the concealment of the data communicated among sensor nodes by means of restraining the information accessible only to authorized persons. In simple words, we can say that third parties cannot understand the information. A trustworthy connection should be installed among IoT devices to interchange information. To maintain confidentiality, an appropriate encryption technique is needed. By means of proper encryption techniques, the information transferred can be made not understandable to unauthorized persons. It mostly depends on the utilization of encryption practices at the physical layer, where information is scrambled at the communicating node to circumvent data exposure to illegal operators [16]. Confidentiality is one of the main advantages of securing IoT. Protection of data from unauthorized persons can be done by using a key that is shared between sender and receiver to encrypt and decrypt information. In order to provide confidentiality, updating the secret key is necessary. But due to the constricted behavior of IoT, updating the security key is a challenging task. More efforts in research should be carried out to cross the challenges to provide automatic key management schemes. IoT can be able to withstand symmetric key schemes [17].

Integrity means that the data remains unaffected throughout the communication process. The integrity can be achieved by firm auditing access control, difficult encryption approaches, input authentications, interface limitations, etc. Message integrity check is an approach employed to cross-check the truthfulness of the information which is received. A symmetric cryptographic algorithm is also utilized to create signatures for the transmitted data. For resource-constrained IoT, an automatic security solution is projected [17]. According to this approach, if any data modification is noticed, the system must be able to disclose the way by producing activity logs. These activity logs are able to be preserved for a large period in local or central memory. A malevolent node in the network may be a cause for the breakage of data integrity. An automatic code update and recovery process were proposed to solve this issue [18].

Availability is crucial for achieving a completely working Internet-connected system. It guarantees that the devices are free to collect data and avoid service obstacles. It ensures the judicious right of entry to resources like networks, applications, and data. It is frequently required to monitor the handling capacities of such resources, uphold backup systems, smearing efficient security policies, and employ redundancy mechanisms. In 2017, the Department of Homeland Security (DHS), the U.S. gave an alert notification [19] specifying IoT users about the increase of everlasting denial-of-service (DoS) attacks, those attack devices with unique authorizations and open Telnet ports. In this intellect, an assailant can interrupt device functions by corrupting its memory capacity. Mitigation techniques also known as moving target defense (MTD) like modifying the default authorizations, retaining server clusters, and deactivating Telnet access have the ability to handle huge network traffic noted by DHS. Fault tolerance and scalability are the two security aims of availability to be counted for stable system and data obtainability. In error forbearance, in case of failure, the system must utilize the self-protection method in combination with the self-healing technique. IoT devices must be arranged hierarchically in order to facilitate scalability. The packet flow can be streamlined to allow for scalability [20].

Accountability is a device's capacity to make users answerable for their activities. A trustworthy scheme for securing an IoT network is accounting for the usage of the resource, auditing, and reporting. Its motive is to ensure the probability of mapping out movements to the corresponding system aiming to launch responsibility for activities. In the article [21], the authors examined possession guidelines and reliability monitoring capacities of different types of smart home devices. The main focus of the authors here is the lack of ability to trace activities which are happened and their sources.

Authenticity ensures confirmation and validates the users who have interacted in the network. In [22], the authors presented a widespread review of authentication mechanisms. Near field communication (NFC) and RFID are the authentication devices that can be used in IoT. An authentication scheme based on NFC has been proposed in [23]. Context authentication, user and device authentication, and trust management are also very much important factors in measuring authenticity [14, 24]. In context authentication, authentication must be done in terms of gathered data from sensors, and control information from the processors, states, and functional characteristics of devices. If any user requested certain activity, IoT devices and the main processing unit should authenticate the user by utilizing any sign-in mechanism for the user or device authentication. Trust management decreases the threat possibility and allows consumer acceptance.

Privacy means the condition with which the service will be accessed by the user. A tough privacy policy is compulsory to make the nodes scalable and for considering a variety of IoT applications. Some of the IoT devices are embedded using RFID tags, and these tags can be tracked effortlessly. Hence, the privacy of these devices is a very important factor. The goals of privacy are classified [17] as data privacy, device privacy, location privacy, and non-link ability. In smart healthcare applications, wearables will be connected to the human body and the Internet. In such a case, data should be kept secure. RFID tags will be used for user authentication purposes in some of the applications. But the devices with RFID tags are very easily traceable devices. Any communication in the IoT system must hide the device's identity. A decentralized identifier scheme is proposed in [25] to maintain IoT devices' privacy. This method can be used in resource-constrained small IoT devices. In the case of location privacy, the location of the IoT device should not be revealed to an unauthorized person. In [26], the authors proposed an effective privacy prevention framework to provide location privacy. Non-link ability ensures that private data should not be able to link to any authorized client in the IoT network. Dummy customers should not be able to make a contour from the personal information which is gathered from original users. To protect IoT devices from non-link ability issues, a group signature-based approach is proposed in [27].

Anonymity ensures that unauthorized persons should not aware of the identity of the data source. Protection of information can be achieved by hiding the source of data [13].

Resilience is defined as the network's ability to guarantee device security despite attacks and failures. The network can be able to work and assist properly if some nodes got compromised or the number of nodes increases [28].

Self-organization is the capability of an IoT system to operate normally even after the breakdown of some fragments due to sporadic malfunctioning or mischievous attacks. If any new sensor node joins the network also, the system should be self-organized by itself [28].

Non-repudiation guarantees that the message sent by the sender should not be denied. So that, it can reach on time. Otherwise, it may send in future causing a disturbance in the overall system [13].

Time synchronization can be crucial for a variety of reasons, including power maintenance and tracking the systems that calculate the end-to-end group synchronization of packets [28].

IoT Architecture and Security Attacks

Architecture of IoT

The IoT has the strength to link up millions of disparate devices through the Internet. As a result, adaptive layered engineering is required. So, what exactly do we want in an "end-to-end" or "full" IoT architecture? The following are some critical needs [29]:

Concurrent Data Collection: A huge number of sensors and actuators can be connected, analyzed, and controlled at the same time.

Open Standards and Interoperability: To achieve interoperability, communication between layers must be based on open standards.

Scalable: The system's component elements can be scaled up or down using a similar architecture.

Effective Data Handling: Reduce raw data and increase actionable data.

Flexible, Modular, and Platform-independent: Every layer must allow for the use of diverse vendors for capabilities, hardware, and cloud platform.

Defined API: Every layer should really have specified APIs that make it simple to integrate with legacy applications as well as other IoT solutions.

Quality of Service (QoS) and Availability: To obtain good QoS and availability, every layer in the architecture should have low latency and fault tolerance.

Connectivity and Communications: Offer network connectivity as well as assistance for versatile and resilient protocols among sensors or actuators and the cloud.

Device Management: If the layers of architecture have an automated/remote device management facility, then upgrades are possible.

Security: Encryption and monitoring of the layers from beginning to end can provide security.

The ever-increasing number of proposed concepts has yet to be united with a reference model [30]. However, there are a few approaches, such as IoT-A [31] which attempt to arrange a legacy practice using experts and market requirements research. From numerous proposed models, the fundamental approach is a three-layer design [32–34] made of perception, network, and application layers. The perception layer is also stated as the sensing layer, and it is incorporated as the bottom layer. The network layer is also known as the transmission layer, and it is incorporated as the middle layer. The application layer is also referred to as the business layer, and it is incorporated as the top layer. Several specialized models have been presented in recent studies to provide prodigious thought to IoT design [35–37], and [38]. The three-layer design, which has been developed and implemented in a number of systems, is crucial to IoT. For a large-scale IoT network, a self-configuring and scalable architecture was given in [39]. Internet-based, sensor and actuator-based, and knowledge-based are the three divisions of the IoT. IoT technology implementation generally reflects modern culture, where devices and society are actually coupled to data systems by wireless sensors [40].

The design of four-layer IoT architecture has also been presented as a result of advancements in IoT and issues in IoT-related security and privacy [10]. In fact, IoT architecture is a complex structure that includes sensors, cloud services, protocols, actuators, and layers. In addition, IoT architecture layers are distinguished in order to assess the uniformity of the system. Figure 4.2 demonstrates the basic four-layer IoT architecture of IoT [10, 41].

Sensing Layer: The IoT's internal sensors that require collected knowledge to be acquired and evaluated are referred to as the sensing or perception layer. This layer combines sensors and actuators to conduct a variety of tasks, including enquiring about the field, temperature, pressure, moisture, motion, velocity, and noise. To build heterogeneous devices, this layer should rely on institutionalized plug-and-play mechanisms. The massive data produced by the IoT begins at this layer.

Network Layer: This layer sends information created by the sensing/perception layer to the middleware layer via secure channels. Data can be sent using various technologies such as Zigbee, RFID, universal mobile telecommunications system (UMTS), Bluetooth low energy (BLE), global system for mobile communication (GSM), wireless fidelity (Wi-Fi), and so on. In addition,

Fig. 4.2 Four-layer architecture of IoT

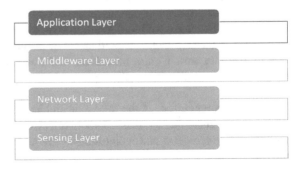

around this layer, several techniques such as cloud computing and data analytics are addressed.

Middleware Layer: Clients can rely on the middleware layer for names and addresses. This layer allows developers to work with a variety of devices without having to be concerned about a dedicated hardware ecosystem. This layer also analyzes the received data and performs critical computations. Following that, it sends the required services across communication protocols.

Application Layer: The responsibility of offering the user application-specific services falls under the purview of the application layer. It outlines several applications for the IoT, including smart buildings, smart logistics, smart healthcare, etc.

IoT Security Attacks in Architectural Perspective

IoT applications are vulnerable to a variety of attacks due to their wireless nature, inadequate physical shielding, and resource constraints. Attacks can be categorized as passive and active in general. A passive attack attempts to study or use system information deprived of causing any destruction to the system's resources. Traffic analysis or eavesdropping is a characteristic of passive attacks. Here, the opponent's intention is not to interrupt the information being delivered. An active attack attempts to alter the system's resources or disturb its functions. Active attacks entail altering the input information or generating false declarations. It is not always easy to distinguish between passive and active attacks because they can behave similarly. As a result, the majority of the study in this article focuses on vulnerabilities and security solutions specific to each layer of the design. We also provide a taxonomy of attacks in terms of IoT architecture. We also include assaults that can be performed against many layers for clarity. IoT sources of attacks are given below:

The *sensing layer/physical layer* is the most basic layer in an IoT architecture, and it specifies the physical parameters of signal transmission. Due to distributed behavior of wireless communication, IoT is vulnerable to jamming, node tampering, eavesdropping, device hacking, etc.

The *network layer's* principal responsibility is to communicate data from the sensing node to a processing device. It enables network connectivity by offering information routing channels. Attacks such as spoofing, selective forwarding, sinkhole, black hole, wormhole, node replication, and hello flood can be introduced by a rogue node inside the network.

The purpose of the *middleware layer* is to offer abstraction among the application and network layers. The scientific and cognitive capabilities of middleware are equally noteworthy. It verifies that APIs meet the application interface's requirements. Persistent storage systems, brokers, scheduling systems, artificial intelligence or compute clusters, and other components make up the middleware architecture [41]. Even though the middleware layer contributes to a secure and efficient IoT operation, it is also vulnerable to specific cyber assaults. Such attackers will seize control of the whole IoT application by poisoning the middleware. Storage and cloud protection are the security concerns that are more important in this layer. Distinct possible attacks investigated in the middleware layer are man-in-the-middle (MITM) attacks, signature wrapping attacks, replay attacks, etc. [42].

Client services are handled and provided by the *application layer*. Smart cities, smart grids, smart factories, smart homes, and other IoT capabilities are only a few examples of applications. This layer contains security vulnerabilities that are distinct from those of other layers, like identity, thievery, and privacy issues. Security issues are frequently particular to individual apps within the defined layer.

A *medium access control (MAC)* protocol is crucial in the IoT. Since it coordinates data transport among various IoT devices. Collisions and jamming are the two most frequent MAC protocol threats.

Across networks such as Wi-Fi, 3G, Bluetooth, local area networks (LAN), and NFC, the *transport layer* directs sensor data from the sensing layer to the middleware layer and conversely. The transport layer is a sublayer of the network layer. Synchronize (SYN) flooding and session hijacking are some examples of transport layer attacks [43].

Multilayer attacks can be launched at two or more layers of the IoT architecture. Examples are denial-of-service (DoS) and MITM attacks.

Depending on the attacks that occurred in different types of layers, a complete overview is summarized in Table 4.1 It discusses the attack types whether it is passive or active, which layer is affected by a particular attack and the main purpose of the attack.

To the best of the author's knowledge, 46 distinct attacks in different IoT architectural layers have been recognized. Further, there may be an increase in this number in the upcoming years due to the increase in connected IoT devices in the world. Concerning Table 4.1, we can recognize that the maximum (i.e., 26% of total attacks) type of attacks can be found in the network layer, and the minimum (i.e., 6%) can be found in both the MAC and transport layer. A pictorial representation of the whole scenario has given in Fig. 4.3, which represents the percentage of vulnerabilities that each layer can withstand. In Fig. 4.3, we can find the attacks developed in more than one layer (20%) which is the second-highest.

DDoS attacks are multilayer attacks as described in Table 4.1. According to the Kaspersky report [76], DDoS attacks, mining cryptocurrency, stealing confidential information, and Sabotage are the most employed attacks by cybercriminals. As per the Kaspersky 2022 report of DDoS attacks [77], in the 1st quarter of 2022, DDoS attacks triggered an all-time high record. At the same time, the interval of the DDoS attacks is also more. Some of the attacks continued for days and for weeks also. Many organizations could not have the ability to fight these attacks. By the Kaspersky DDoS attack report, one can notice that DDoS attacks are more dangerous and widespread.

Table 4.1 Summary of IoT security attacks

Attack name	Type	Layers affected	Description
Eavesdropping	Passive	Physical	To intentionally listen to personal IoT communication and saves data for future attacks [44, 45]
Physical damage	Active	Physical	This can destroy or disturb the infrastructure of IoT applications [46]
Hardware trojans	Active	Physical	The attacker inserts a triggering device during the fabrication process to actuate Trojan's malicious deeds [47]
Tag cloning	Active	Physical	The attacker designs a new unauthorized tag from the information of the original card. The cloned tag can be used by the attacker in future [48, 49]
Basic/RF jammers	Active	Physical	The main strategy of basic jammers is breaking the signal, producing congestion, and draining the energy of the node by generating radio frequency signals [50, 51]
Node tampering	Passive/active	Physical	In node tampering, the attacker physically replaces the complete node or a portion of the node to access and change sensitive data [28]
Node injection attack	Active	Physical	Injecting of the malicious node connecting more than one IoT node in the network topology. The inserted node participates in interaction and assumes control of network traffic [14, 52]
Node destruction	Active	Physical	Physical damage of a node by any method (e.g., with an electrical surge or physical force) so that the damaged node is no longer functioning [7]

(continued)

Table 4.1 (continued)

Attack name	Type	Layers affected	Description
Counterfeiting	Active	Physical	In counterfeiting, the adversary alters an item's identification, usually by tag tampering. A tag is partially modified in these assaults [47]
Access attack	Active	Network	This is a type of attack in which an unauthorized user gains access to the information system. For an extended period, the cyber attacker will persist undetected on the network. The goal of an access attack is to obtain information [41]
Phishing site attack	Active	Network	Phishing site attacks allow a single assailant to easily aim at a significant number of IoT devices. In the IoT, the network layer is enormously sensitive to phishing site attacks [10]
Selective forwarding or gray hole	Active	Network	The attacker drops selected packets with intent and selectively sends others toward the next node. Because IoT networks are inherently lossy, it is hard to pinpoint the cause of packet loss [53, 54]
Replay attack	Active	Network	A corrupted node caches transmitted data and then retransmits it at a later time or repeats transmission [55]
Routing attack	Active	Network	In this attack, mischievous nodes in an IoT system may effort to adjust routing pathways during data transmission [47]
Sinkhole	Active	Network	An attacker encourages a false best routing path and inspires nodes to utilize it to route traffic [14]

(continued)

Table 4.1 (continued)

Attack name	Type	Layers affected	Description
Black hole	Active	Network	A black hole attack is a sort of DoS attack where a router, rather than forwarding packets, dismisses them [14, 47]
Wormhole	Active	Network	A wormhole is a link that permits rapid packet exchange flanked by two nodes. An adversary can attempt to defeat the elementary security rules in an IoT system by generating a wormhole between a malicious node and a device on the network [56]
Hello flood	Active	Network	In this assault, an attacker can inject "HELLO PACKETS" with each other node in the network, pretending to be their neighboring node, using a malicious node with significant transmission power [14]
Sybil	Active	Network	The adversary uses Sybil nodes with false identities in a Sybil attack. Sybil nodes in the system can go against honest nodes [47]
Traffic analysis	Passive	Network	The traffic pattern of a network will be beneficial as the element of data packets to the adversary. Understanding traffic patterns can carry valuable data regarding the networking topology [7]
Sniffing attack	Passive/active	Network	Sniffers can be used by computer hackers to monitor the network traffic of the IoT system and to obtain confidential customer information [57]

(continued)

Table 4.1 (continued)

Attack name	Type	Layers affected	Description
Collision attack	Active	MAC	As soon as a genuine node of the network initiates sending data, an attacker transmits useless data on the same channel. As a consequence, the two transmissions strike, rendering the acquired data making no sense to the recipient. Subsequently, the receiver requests the retransmission of the identical packet [58]
Denial of sleep	Active	MAC	This assault can be carried out through collision assaults or recurring handshaking, which includes repeatedly altering clear to send (CTS) and request to send (RTS) flow control signals in order to avoid the node from entering the sleep mode [59]
6LowPAN exploit	Active	MAC	The fragment duplication attack, whereby an assailant inserts his individual fragments into the fragmentation sequence, is a unique 6LoWPAN attack. The fragment duplication attack exploits the truth that even a recipient cannot tell if a fragment came out of a similar origin as beforehand received pieces from the identical IPv6 packet there at the 6LoWPAN layer [14]
SYN flooding	Active	Transport	An attacker uses a flooding attack to deplete the memory and/or energy of a node by flooding it with fake messages [60, 61]

(continued)

Table 4.1 (continued)

Attack name	Type	Layers affected	Description
Session Hijacking	Active	Transport	This attack is defined as "exploiting" and "tampering" with an authorized data transmission (also known as a session key) to get illegal entry to a system's services and statistics [7]
Data integrity attack	Active	Transport	Data integrity refers to the process of altering data during transmission by modifying the information or injecting false information. The principle of operation of a network can be disrupted by misrepresenting routing data [28]
Signature wrapping	Active	Middleware	Extensible markup language (XML) signatures are utilized in the Web applications used by the middleware layer [62]. By finding loopholes in the simple object access protocol (SOAP), the adversary can compromise the signature scheme and conduct operations or alter the eavesdropped document in a signature wrapping attack [63]
SQL injection attack	Active	Middleware	An adversary can use mischievous structured query language (SQL) commands involved in a program to carry out such attacks [64]. Once in the database, the attackers have access to the personal data of any user and can even make changes to the records [65]

(continued)

Table 4.1 (continued)

Attack name	Type	Layers affected	Description
Impersonation attack	Passive/active	Middleware	An impersonation attack is a type of attack where the attacker assumes the identities of authorized parties in a technique or protocol [41]
Privileged insider attack	Passive/active	Middleware	Malicious users with permitted (insider) system access led to privileged insider attacks [41]
Path-based DoS	Active	Application	In this technique, an attacker floods an edge communication link with either generated packets or repeated packets [66]. As a consequence, every node on the path between the attacker and the recipient is infected
Constrained application protocol (CoAP) exploit	Active	Application	The deployment of CoAP poses a number of security challenges. It does not translate all of HTTP's capability, posing a security risk for multicast communications [67, 68]
Virus/malware	Passive/active	Application	The purpose of these assaults is to compromise the system's confidentiality. They are most commonly found in the form of worms, spam, Trojans, as well as other viruses [69]
Message forging	Active	Application	When a malicious node edits or produces a message, it delivers content that differs from the original [70]
Service interruption attack	Active	Application	By deliberately rendering the servers or network too saturated to react, such attacks prevent genuine consumers from using the services of IoT applications [71]

(continued)

Table 4.1 (continued)

Attack name	Type	Layers affected	Description
Service logging failure	Active	Application	Monitoring the status of installed services and detecting security activities can both benefit from logging activities. In cloud/edge systems, inefficient log tracking can restrict the ability to deploy security measures [72]
DoS	Active	Multi	IoT devices, programs, or networks could be switched off by a DoS attack, causing the service unavailable to its customers [47–49]
DDoS	Active	Multi	DDoS is much more destructive than the DoS outbreak, which uses many attacking platforms to compromise one or many systems [14]
Side-channel attack	Passive/active	Multi	Even though the information is encrypted, such assaults use cutting-edge tools to capture and manipulate information in order to obtain information from diverse patterns [47]
MITM attack	Passive/active	Multi	Without user awareness, the attacker sniffs the connection to capture the communication between two IoT devices during the interchange keys step to get access to personal and/or sensitive data [28]
Node capture	Active	Physical/network	The attacker will capture the node or may replace the node with the new one. The new node will be controlled by the attacker [73]
False data injection attack	Active	Physical/application	The attacker will inject erroneous data into the IoT network after capturing nodes. This may cause wrong output and failure of the total system [10]

(continued)

Table 4.1 (continued)

Attack name	Type	Layers affected	Description
Malicious code injection attack	Active	Physical/application	The attacker will inject malicious code into the processor's memory to gain control of the device [10]
Spoofing	Active	MAC/application	A hostile node impersonates a target node's MAC address, then builds many real identities from the target node, and uses them somewhere else in the network in a spoofing attack [74]
Desynchronization	Active	MAC/transport	Attacks toward the time synchronized channel hopping (TSCH) synchronization might occur whenever an attacker sends messages during the time frames allotted to certain other users. The packets collide and are destroyed as a result. By carefully examining the back-off periods, an attacker might set off a series of these incidents, eventually leading to the desynchronization of the neighboring motes. This attack can therefore be viewed as a more sophisticated version of the collusion attack [14, 75]

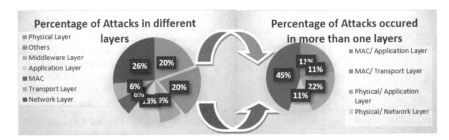

Fig. 4.3 Representation of the percentage of security vulnerabilities of different layers

Preventive Techniques and Mechanisms of IoT/IIoT Security

Preventive Techniques of IoT Security Attacks

One of the main study topics in IoT is security, which has garnered a large number of experts from different organizations. To date, there are numerous proposals to tackle security issues in IoT. In this section, we present various existing solutions to the aforementioned security attacks.

Solutions for Physical Layer Attacks:

Eavesdropping: In [78], a more complicated session key generation algorithm based on a hash function was used to protect the information from eavesdropping attacks. A MAKA technique was developed in [79] to address security challenges such as replay attacks, Sybil attacks, DoS attacks, impersonation attacks, and eavesdropping attacks. The authentication process is achieved through this technique, which prohibits an adversary from accessing the system moving in the network throughout transmission. Furthermore, identical session keys are used to encrypt all detected data. This technique ensures data privacy and prohibits the attacker from disclosing the data. The authentication credentials are encrypted by means of the agreed session key, and the hash function of the access privileges is computed and transferred jointly over the channel in the suggested technique [80]. As a result, extracting the user credentials from both the encrypted text and the hash function is almost impossible. So, the suggested technique is immune to eavesdropping. The research in [81] focused on achieving effective privacy in an IIoT-multiple output multiple antenna eavesdropping communications situations. For the IIoT network system, a closed-form equation for asymptotic regularized quick privacy rate is first developed. The research then goes on to look at how to build the best jamming settings by developing a model called the optimal counter eavesdropping channel approximation (OPCECA) approach for dealing with eavesdropping attacks in IIoT. The study [82] looked into eavesdropping problems in IoT nodes that leverage the present 5G non-standalone (NSA) network infrastructure, as well as mitigation and verification methodologies. The authors in [83] registered a patent in which MTD was used to protect the system from eavesdropping attacks. By monitoring traffic between the server and the client and examining the header of an IP packet using packet eavesdropping, the attacker can notice the IP address of the server. The server's address can be presumed to modify to any of the 10 addresses (A1 to A10) in the MTD technique. As a result, even if an attacker eavesdrops on packets and finds the server's address, the attacker cannot interrupt the system. Because the server IP address will change continuously with the MTD technique. But, the discussed MTD technique is for android devices, and it did not include practical IoT devices. Developing MTD in IoT applications is challenging due to the limited energy of IoT devices. Hash function-based encryption algorithms will serve better for eavesdropping attacks. Because of its lightweight nature, it is suitable for energy-constrained IoT devices.

Physical Damage: Mutual authentication, privacy protection, data provenance, and security against DoS and physical attacks were all objectives of the developed protocol [84]. To accomplish data provenance, anonymity, and mutual authentication, the suggested technique employs physical unclonable functions (PUFs) and fingerprints retrieved from the wireless link. The channel quality indicator data is used to create wireless fingerprints. The suggested technique can identify antagonistic channels with great accuracy, according to experimental verification using MICA Z motes. The software solution may deactivate the distant location and kill instruction, but it will not prevent physical damage to the device [17]. The only method to keep the connected devices safe is to cover them with a protective case. Since these threats are more physical, so the sensor nodes should be constantly monitored.

Hardware Trojans: To identify hardware Trojans in circuitry, path delay measurements for scheduling channels have been presented [85]. The timing channel's smallest delay pathway is used as a resource for Trojans to be inserted. A unique physical side-channel [86], also known as the backscattering side-channel, presented hardware Trojan and counterfeit IC identification approaches. On many circuit standards, these approaches can identify different forms of dormant hardware Trojans and counterfeit ICs despite accepting manufacturing fluctuation. In [87], the authors presented a new technique for detecting hardware Trojan insertion employing bitstream manipulation in FPGA-based IP.

Tag cloning: In [88], the authors suggested a probabilistic detection method and three protocols to identify tag cloning exploits in most RFID systems, namely BASE, DeClone, and DeCloneC. The authors of [89] employed PUF, and this approach can ensure that the tag will not be cloned. The PUF is a different property of a circuit that transfers a challenge to a response during the chip production process; the underlying structure offers a one-way activity, making PUF extremely difficult to clone.

Basic Jammers: Multi-agent reinforcement learning-based sub-band selection framework for anti-jamming utilizing Q-learning for WACR was considered by the authors [90]. However, the authors in [91] employed a deep Q-learning-based method to boost the effectiveness of [90] and saved nearly 66.7% time in comparison with Q-learning. Using an machine learning (ML) algorithm and received signal strength indicators (RSSI) data from simulation as well as real-world networks, the authors devised a low-overhead, non-node-centric, and passive network jamming detection scheme [92]. The results showed that employing RSSI from five anchor nodes resulted in 98.6% for simulated data and 89.7% accuracy for real network data, respectively. The improved detection approach based on multi-path profile data is then described, and simulation data is used to validate it. For 5 anchor nodes, the results indicated performance improvements of 99% accuracy over signal-strength-based detection of 98.6%. By taking into account exploration, evasion, causation, and priority violation attacks, a new method called autoencoder deep neural network was designed [93]. The developed technique classifies communication results in order to anticipate transmission conditions whether it is jamming data transmission or sensor data transmission. The research of the key primitive of decentralized message dissemination over multi-hop wireless connections under a powerful adversarial jamming paradigm commenced with the publication [94]. The authors introduced a decentralized randomized algorithm that can achieve message dissemination in specified time slots with a significant chance of performance guarantee using a feasible strong adversarial jamming model and the signal interference noise ratio (SINR) model, which eliminates the financial problem.

Node Tampering: To protect IoT sensors from node tampering threats, nodes might be well provided using tamper-resistant circuitry. This circuitry would erase the storage in the occurrence of such a tampering attack and prevent sensitive information from being exposed. Disabling the sensors' joint test action group (JTAG) interface and using a strong password for the protection of the sensor boards' bootstrap loader are two techniques to prevent node manipulation discussed in [61]. A simple solution to defend the nodes from such threats would be to camouflage them [95]. In [96], a two-way and two-stage authentication mechanism leveraging hardware security blocks termed PUF is described to tackle node replacement and node tampering exploitations. A distinct response is created under a challenging input is the principle of PUF. An IoT system with a PUF can therefore be uniquely determined and protected in opposition to node tampering threats.

Node Injection Attack: The study [97] suggested random forest, a special case of support vector machine (SVM) learning technique for detecting rogue nodes in an IoT context. The study used 17 diverse IoT devices to collect information, with a 99.49% overall accuracy for 110 successive sessions in different locations. The authors in [98] concentrated on detecting bogus IoT nodes for network security rather than illicit activities. The study proposed using an artificial neural networks (ANNs) to train an altered real-time traffic dataset. The ANN classifiers detected malicious IoT nodes with a success rate of 77.51%. The node injection attacks can be significantly detected by AI and ML techniques.

Node Destruction: To overcome node destruction, the nodes or physical devices must be protected from damage either by physical force or high voltage. The nodes can maintain a protective case to keep away from the physical force [17].

Counterfeiting: Hu-Fu, the very first physical layer RFID authentication technique that is resistant to significant vulnerabilities such as tag counterfeiting and replay attacks, was presented by the researchers in [99]. This mechanism is used for IoT devices that do not require batteries, such as passive RFID tags. The coupling and authentication aspects are successful in identifying counterfeit tags in this design. The suggested PUFs [100] can be employed as anti-counterfeiting identifiers, authentication credentials, and cryptographic key producers due to their special surface micropattern and temporally coded laser scattering. Since the private keys are inherently camouflaged in the sophisticated microscopic stochastic physical attributes of the PUFs. The security protection provided by the PUFs generated remains effective even though the attacker gains access to the information.

Solutions for Network Layer Attacks:

Access Attack: The authors of [101] developed an integrated user authentication mechanism for cloud-based IoT systems. The suggested technique uses a lightweight hash function in which the distant users and cloud servers are mutually validated and exchange a session key to ensure future connections are secure. This defends against various attacks like access attacks, replay attacks, insider attacks, and spoofing attacks. The authors suggested [102], a diverse sign-cryption technique for IoT systems that has the effectiveness and security robustness of a hyperelliptic curve. This scheme provides security features like replay attack protection, unforgeability, integrity, and non-repudiation.

Phishing site Attack: Deep belief networks are a detection model that was proposed [103] for detecting phishing Websites and achieved a true positivity rate of over 90%. To identify the similarity of page layouts and detect phishing pages, a learning-based aggregate analysis mechanism [104] was presented. This method automatically trains classifiers to detect similarity across Web pages based on CSS layout data. The authors in [105] investigated a strategy to randomize attribute ranking to create a large phishing detection mechanism, with a good understanding of ML algorithms and the availability of commonly utilized training datasets, which could indeed learn key aspects of phishing detection. The suggested technique allows for the creation of a meta-classification system using a variety of different models created from the same training dataset. Prevention of phishing site attacks is not possible. But detection is possible with ML algorithms.

Selective forwarding: The authors developed a real-time intrusion detection system (IDS) for identifying malevolent activities in contradiction of routing algorithms in an IoT ecosystem [106]. In [107], the authors suggested an efficient repeating game defense model combined with the TDMA protocol for detecting selective forwarding attacks and hardware failure in clustered WSN-based IoT systems.

Replay Attack: To mitigate replay attacks, the authors [101] designed a lightweight and secure algorithm that will use server and remote user mutual authentication and interchange of session key schemes. In [108], a three-layer IDS architecture was presented against certain basic network layer breaches such as replay attacks, MITM attacks, and DoS. The system's main functions were to discern between legitimate and malicious packets, as well as to identify the type of attack. The training dataset obtained with Waikato environment for knowledge analysis (WEKA) software was classified using NB, J48, SVM, Zero R, OneR, MLP, and RF. The J48 algorithm proved to be the most prominent of all others. Different current timestamp estimates for the communicating nodes were used in lightweight device authentication and key management mechanisms for edge-based IoT (LDAKM-EIoT), which was proposed in [109]. A comparatively small value of transmission delay factor is employed within every LDAKM-EIoT message transmitted. As a result, an opponent cannot benefit from replaying the previously transmitted communications. As a result, it is possible to achieve security against replay attacks. But the problem with LDAKM-EIoT is, it cannot consider original message replay, in case of delay due to signal problems. Each communication message in the method suggested in [79] has included encrypted nonce, making it impossible for the attacker to change them. The receiver detects the replayed message by checking the nonce once the message is replayed. As a consequence, the suggested approach is resilient to replay attacks. Hu-Fu is an authentication mechanism proposed in [99] that can withstand replay attacks. Even though an adversary can eavesdrop across all communications of an authorized tag, Hu-Fu cannot pass any of these signals. In a non-coupling condition, the attacker can collect the signal of the authorized tag, but Hu-Fu cannot be passed. It can also capture coupling signals, but replaying these will not function since the randomized signal will be distinct the next time. The authentication technique which was stated in [80] successfully prevents replay attacks. It is easy to identify an intruder who attempts to validate himself as a valid user by replaying the communication directed by the legitimate user and gateway without modifying the earliest. This is because during information transmission, a pre-calculated timestamp, the only authorized time to allow any reply, is utilized to defend against replay attacks. To prevent replay attacks, the paper [89] suggested a user authentication system based on biometrics and PUF. A heterogeneous sign-cryption method discussed in [102] can be used to prevent replay attacks very effectively. From the literature, one can say that replay attacks can be prevented by using encryption-based authentication mechanisms.

Routing Attack: In [110], the authors presented a compression header analyzer intrusion detection system (CHA-IDS), which examines 6LoWPAN compression header data in order to minimize individual and combined routing assaults. CHA-IDS is a multi-agent system framework for data collecting, analysis, and operation that captures and manages raw data. To confirm the system's resilience, the CHA-IDS was tested using three types of combinations of attacks: a sinkhole, hello flood, and wormhole. Upon comparing the results of CHA-IDS with SVELTE and Pongle's IDS [111], the results revealed that CHA-IDS outperformed the other approaches, with a 99% positive rate and decreased energy overhead and memory use. The authors of [112] created deep neural network models with excellent precision, accuracy, and recall rates using the IoT routing attack dataset. Based on the F1-score and area under the curve test (AUC) score, this algorithm achieved performance values of up to 99%.

Sinkhole: To prevent sinkhole attacks, the authors [109] developed a real-time IDS. CHA-IDS algorithm introduced in [110] is used to detect sinkhole attacks with a 99% positive rate in comparison with the SVELTE and Pongle's IDS [111]. The probe route-based defense sinkhole attack (PRDSA) technique was introduced in [113] to withstand sinkhole attacks and ensure IoT security, and this is the first study that can identify, bypass, and address the sinkhole all at once. The PRDSA method suggested a routing strategy that combines minimum hop routing, equal-hop routing, and far-sink reverse routing to efficiently avoid sinkhole attacks and discover a secure route to the actual sink, allowing the method to detect sinkholes more accurately.

Blackhole: In [114], the authors suggested the intelligent blackhole attack detection system (IDBA) adapted to automated and connected vehicles. The IDBA utilized four major parameters, with a hop count and sequence number being two of them. The black hole takes advantage of these two to compromise the network's integrity and availability. The other two parameters are network performance outputs that have been impacted as a result of the attack on the first two. So, by combining these four characteristics and pre-calculating the thresholds for future black hole activities, the authors were able to identify the attack quickly using the algorithm.

Wormhole: The advanced hybrid intrusion detection system (AHIDS) presented in [115] automatically identifies WSN assaults. The AHIDS employs a cluster-based framework with an improved LEACH protocol, with the goal of lowering sensor node energy utilization. AHIDS uses the multilayer perceptron neural network in conjunction with fuzzy rule sets to detect anomalies and misuse. Both the back-propagation neural network and the feed forward neural network are utilized to highlight the various sorts of attackers and incorporate the diagnostic results. The wormhole resistant hybrid technique is used to identify wormhole attacks with a detection rate of 99.2%. CHA-IDS algorithm introduced in [110] is used to detect wormhole attacks with a 99% positive rate. In software-defined heterogeneous IoT [116], the wormhole attack analysis using the neighbor discovery (WAND) approach was suggested as a potential wormhole protective measure. WAND found worm-holes without requesting any specific location data and without creating substantial coordination and communication complexity since it identified the resemblance of neighbor counts at a centralized controller. WAND additionally guards against the misleading scenarios created by discovery protocol vulnerability.

Hello flood: The signal strength and distance in AHIDS presented in [115] are used to detect hello flood attacks with a detection rate of 98.2%. To forecast and counteract hello flooding assaults on the RPL protocol in IoT networks, a gated recurrent unit (GRU) network model based on deep learning was developed [117]. Total energy consumption and various power states of the nodes were considered and tested. The suggested model was analyzed using logistic regression and support vector machine (SVM) approaches. With respect to source efficiency and IoT security, the findings validated the model claimed the expected outcomes.

Sybil: The advanced Sybil attack detection algorithm was introduced to detect Sybil attacks with a 99.4% of detection rate [115]. IoT devices are preloaded with a unique ID [79], and the adversary cannot infer the identification from communications sent over the network. Additionally, prior to transmitting or receiving data, each node goes through the authentication mechanism. If the attacker does not have the preloaded credentials, he or she cannot be validated. As a result, the suggested method can withstand Sybil attacks. In [118], the authors presented a game-theory strategy for defending against Sybil attacks. The method, as per its parameters, establishes a global trust barrier in order to preserve node durability in the network while determining its trustworthiness, rendering the attack expensive.

Traffic Analysis: In [119], the authors suggested a traffic morphing strategy to reduce the privacy consequences of traffic analysis threats. This technique alters network data, making it much harder to recognize IoT devices and associated activities. The suggested method protects smart home clients from traffic analysis threats and prevents privacy leaks. A privacy protection solution based on secure network coding was developed in [120] to tackle the issues of personal information privacy. This approach employs a hybrid coding technique to efficiently withstand traffic analysis assaults, maintaining node information privacy.

Sniffing Attack: The authors of [121] tailored the well-known Knapsack crypto-graphic techniques to the communication history-based authentication method. Since it is easy to use and offer strict protection for various applications. They tested the authentication system without and with RSSI data and found that employing RSSI associated with data packet history can provide the advantages of increased entropy confidential bit creation and greater security from packet sniffing threats.

Solutions for MAC Layer Attacks

Collision Attack: The ciphertext-policy attribute-based encryption (CP-ABE) was used in the scheme [122] to ensure the shared key's security (session key). For every encrypted information, CP-ABE gives an access structure, and decryption needs only one subset of the attributes. CP-ABE can fight against collision attacks since the secret key includes a distinct random number for every attribute in the access control mechanism. As a result, illegitimate users will be unable to get the exchange shared key through collision. To ease the detection procedure and eliminate collisions between transmitted packets at the cluster head, the model proposed in [107] depends on the TDMA protocol.

Denial of sleep: The authors developed a denial-of-sleep immune MAC layer for IoT devices communicating through IEEE 802.15.4 connections in [123]. This MAC layer has two key components inside. The first major component is a denial-of-sleep-resistant mechanism for creating session keys across IEEE 802.15.4 nodes in close proximity. The acquired session keys have two purposes: They provide basic wireless security, and they supplement the second primary component's denial-of-sleep measures. A denial-of-sleep-resistant MAC protocol is the second major component. This MAC protocol includes state-of-the-art procedures for attaining high throughput, low energy dissipation, and high delivery ratios.

6LowPAN exploit: In [124], the authors presented 6LowPSec, a security protocol that provides a good end-to-end security solution against the 6LoWPAN exploit. The 6LowPSec makes use of the MAC security sublayer's current device security features.

Solutions for Transport Layer Attacks

SYN flooding: SoftThings, a software-defined network (SDN)-based dynamic threat detection system for IoT networks, was developed by the authors of [125]. In the SDN controller, linear and non-linear SVM classifiers are utilized to identify and prevent flooding and DDoS attacks with 98% accuracy. SDN-based SYN proxy (SSP), a collaboration of the SDN Openflow switch and SDN controller to function as a stateless SYN proxy counter to the TCP SYN flood based on the fundamental packet processing technique of SDN Openflow, was devised by the authors in [126]. The SYN flooding was modeled employing Bayesian inference in [127], and an effective method for identifying flooding-based attacks adaptable to many kinds of wireless ad hoc networks was devised by modifying the mean of the Beta distribution. The suggested technique ensures that mobile nodes get enough sleep time, allowing for more efficient use of battery resources. Without a dedicated hosting node, it can identify request flooding attacks, Hello flooding attacks, data flooding attacks, SYN flooding attacks, and UDP flooding attacks with better precision and TPR.

Session Hijacking: To defend IoT networks from session hijacking attacks, the authors suggested the private key-based binding update for the CoAP-based mobility management protocol in the paper [128]. Prior to executing any binding update activity, clients must confirm the address provenance and legality of mobile nodes. To counteract a session hijacking assault, the gateway must validate the sensor's authenticity continually during the session while providing data like the location, token, and battery, according to [15].

Data Integrity Attack: The authors of [129] presented a decentralized technique known as bubbles of conviction based on the public blockchain that performs smart contracts to ensure data integrity. They imagined a system with a vast number of disparate smart things, each of which can exclusively interact with gadgets in its zone. Data integrity can be provided by a diverse sign-cryption scheme which was introduced in [102].

Solutions for Middleware Layer Attacks

Signature Wrapping: To identify malicious wrapping attacks, the authors of [130] devised a side-channel-based identification system that counts the occurrence of every node in a requested service. The authors of [131] advocated utilizing "positional token" as a reference to the original XML elements getting signed, while signing the Web service requests, to identify XML signature wrapping attacks on signed Web service requests. By contrasting the estimated hash of the Web service request with the hashed information obtained from the signed content in the Web service request, XML wrapping attacks can be identified during validation.

SQL Injection Attack: An application context pattern-driven corpus was published by the authors in [132] to train a supervised learning method. To protect against SQL injection attacks, the model was trained using Microsoft Azure machine learning (MAML) studio using ML techniques such as two-class logistic regression (TC LR) and two-class support vector machine (TC SVM). The authors of [133] investigated different machine learning approaches and demonstrated that they can identify SQL injection attacks with extreme accuracy. The study [134] offered an effective SQL injection attack detection system. The emphasis was on enhancing the data environment and computational information to achieve optimum dependability of SQL injection attack prediction in IoT-driven platforms.

Impersonation Attack: The authors of the paper [109] presented LDAKM-EIoT, designed to defend against various impersonation assaults. The authors presented a lightweight key exchange protocol technique for healthcare IoT depending on hash functions [78]. Wearable devices, a server, and a user device make up their network. Before transmitting the wearable device's health information, it must first verify the consumer device with a cryptographic hash function scheme. The intended protocol's authenticity is assessed with BAN-logic and the AVISPA tool. Impersonation attacks can be effectively mitigated by using this algorithm since it is using a hash function-based password authentication system. An adversary cannot replay the messages sent by CHs, according to [79]. Furthermore, every cluster member (CM) should verify the assailant before relaying the sensory information, even if the adversary purports to be a CH. Since the generator is preinstalled on all nodes, the attacker is unable to calculate the session key; hence, it cannot be validated. As a result, the suggested system can withstand a cluster head (CH) impersonation attack. The method [122] avoids a hijacked mobile device attack using a local user verification procedure and access control. So, the adversary cannot alter the user. As a result, the suggested algorithm avoids impersonation threats because a secure session key agreement is maintained. The authors of [80] developed a three-factor user authentication and session key establishment technique to safeguard IoT networks from impersonation attacks. To achieve mutual authentication and session key establishment, the proposed approach uses ECC, cryptographic algorithm, hash functions, and symmetric encryption.

Privileged Insider Attack: A lightweight authentication algorithm designed in [101] was able to mitigate insider attacks by the mutual authentication algorithm. The authors of [135] suggested a secure authentication strategy for remote medical sensor networks in IoT technology. The system relies on a cryptographic hash function and uses two aspects, namely user identification and passcode. The suggested method accomplishes secure consensual authentication and session key understanding, according to comprehensive security verification. Since the password is protected by a hash function with a mixed random number, a privileged insider attack can be mitigated. LDAKM-EIoT is a technique planned in [109] that is resistant to privileged insider attacks. The insider attack in the RPL IoT environment was handled in [136]. The authors suggested a methodology to prevent this attack by conducting trials with the Contiki tool which is a low-power designed tool for resource-constrained systems. The authors of [137] suggested a big data analytics-based insider attack detection mechanism for IoT systems.

Solutions for Application Layer Attacks

Path-based DoS: A one-way hash chain was designed in the paper [66] to safeguard end-to-end connections in WSNs from PDoS attacks. The suggested technique is simple to deploy in recent WSNs, and it is lightweight.

CoAP Exploit: The authors described a technique for screening out unauthentic and replayed CoAP communications "en-route" on 6LoWPAN edge routers to counteract the CoAP exploit attack [138].

Virus/Malware: The authors of [139] introduced EnDroid, a malware detection approach that uses an orchestra learning algorithm trained on the AndroZoo and Debrin datasets to track extensive and dynamic harmful behavior, such as sensitive information leaks with a 98.2% accuracy rate. The authors of [140] used fuzzy and fast fuzzy patterns to detect malware and showed that these techniques are responsible for accurately detecting malware and can contend with rival different classifiers like SVM, random forests, KNN, and decision trees. An IoT and IoBT malware detection strategy based on the class-wise choice of OpCodes sequence as a characteristic for classification tasks was reported in the work [141]. The approach has robustness in malware detection, with an overall accuracy of 98.37% and a precision rate of 98.59%.

Message Forging: As mentioned in [122] earlier in the collision attack, CP-ABE is able to defend against collision attack. To forge a legitimate signature of a legitimate user, an adversary must have access to the user's private key. An attacker, on the other hand, cannot deduce the private key. On the other side, the adversary will be unable to generate a new, acceptable ciphertext, and signature from the ciphertext and signature of another user. The receiver can check that the ciphertext of the shared key is illegal if the adversary updates it. A cryptographic hash function-based authentication protocol was designed in [135] can be able to prevent message forging attacks. Since it uses the secret key to send or edit messages and the guessing of the secret key is limited to only 3 times. To provide the feature of unforgeability, which prevents message forging attacks, the authors proposed a heterogeneous sign-cryption scheme [102].

Service Interruption Attack: According to [142], installing an effective IDS is a necessary step for running IoT-related services without interruption and protecting networked customers' personal information. The effectiveness of an IDS is determined by the effectiveness of the classifier employed within the IDS to detect attacks. To become an effective IDS, it must be able to identify network intrusions as quickly as possible, allowing it to identify attacks in real-time. The paper [143] proposed a strategy for creating a trustworthy distributed network for delivering IoT services based on edge computing to solve service interruption and validate the processing of information. The central idea behind the system is that IoT devices are validated by a certification authority (CA) via an approved dealer. Those IoT devices that have been already permitted to connect to the CA will use it.

Service Logging Failure: The authors introduced an IoT-based subsystem that maintains the heating and cooling element [144]. The subsystem may identify failures in the heating and cooling equipment of HVAC and alert administrators to critical system failures that occur as a result of these occurrences. The subsystem was installed to discover that it identifies system problems and can communicate this data to operators to act quickly. As a result, system performance is improved. Furthermore, the subsystem can record HVAC equipment along with critical equipment resource consumption in order to develop utilization and help administrators see significant trends and future problems. Authentication, authorization, and audit log services in healthcare IoT were standardized by fuzzy inference systems using Hyperledger Fabric to provide decentralized fuzziness, trust, and the removal of single points of failure [145].

Solutions for Multi-layer Attacks

DoS: The secure version of OpenPLC [146] is resistant to a wide range of attacks, including packet injection, DoS, and MITM attacks, which is the combination of the embedded encryption module and the ML-based intrusion prevention system. A three-tier IDS architecture using various ML algorithms was introduced in [108] to detect DoS attacks in IoT networks. The CH in the method [79] collects only the information recorded from authorized nodes before forwarding it to the base station (BS). Only information from such an authorized CH is accepted later. Moreover, replayed information is rejected by the CH or BS. As a result, DoS assaults are deterred by the MAKA scheme. Low-rate denial-of-service (LDoS) is a genus of DoS attacks. The routing technique is sensitive to an LDoS attack. Due to its small-signal properties, it is hard to detect using traditional methods. Existing intrusion detection algorithms in IoT face a significant hurdle in detecting LDoS attacks. The authors in [147] developed a unique LDoS exploit detection method integrating trust evaluation and Hilbert-Huang transformation. The suggested scheme [80] restricts the number of user authentication requests to the gateway/device to a maximum of three in order to safeguard IoT devices from DoS attacks. After three tries, if the user still does not enter his credentials, login and authentication for that user will be suspended for a long time. As a result, the suggested approach effectively protects against DoS attacks.

DDoS: An SDN-based attack detection system for IoT networks called SoftThings was developed to detect DDoS attacks [125]. The suggested Naive Bayesian classification technique in the multi-agent system-enriched intrusion detection system Naïve Bayes classification method in the multi-agent system-enriched intrusion detection system (NBC-MAIDS) method [148] is a sophisticated and quick network intrusion detection system. When compared to typical intrusion detection systems, the NBC-MAIDS method with the training of several agents for DDoS attacks detection delivers better results. The authors of [149] used the revised CICIDS2017 dataset to train CNN, MLP, long short-term memory (LSTM), and CNN + LSTM deep learning (DL) algorithms for detecting DDoS attacks. With a 97.16% overall accuracy, the CNN + LSTM classifier outperforms the competition. The author also contrasted the DL algorithms to a standard ML method and found that the DL algorithms performed better. To counteract DDoS attacks, the authors in [150] suggested an adaptive system that combines DL techniques and the fog computing concept. A unique source-based defensive system has been presented by the authors. The defender is hosted by the SDN controller to fight against DDoS attacks focused on the network/transport layer in the strategy [150]. LSTM is a DL algorithm that the authors used in this approach. The authors employed SDN to design an MTD system that enhances unpredictability due to the rapidly changing attack vectors in the article [151]. They also used SDN-based honeypots to fool adversaries into thinking that they were IoT devices. Finally, test results demonstrated that using a mixture of SDN-based honeypots and MTD to conceal network assets from adversaries and protect from assaults in the IoT effectively. A DDoS mitigation solution based on a DL approach has been suggested by Ko et al. [152]. They provided a solution that might be adopted in the ISP domain, which is the network juncture that links consumers to the Internet. They have presented a self-organizing map (SOM) that is stacked. The authors used three layered SOM to develop their technique. The method is depending on a DL algorithm that is unsupervised. To mitigate the severity of the high-volume attack, they deployed Apache Spark for decentralized big data processing.

Side-Channel Attack: The authors of [153] suggested a side-channel resilience MTD technique for side-channel assaults based on the quantity of traces. The suggested system combines masking and fresh rekeying techniques, with the highest number of side-channel outflow traces essential to a successful assault driving these methods. The scalability of the aforementioned approach was tested by using an ML-based threat model that greatly decreases the number of traces necessary for a victorious attack. A lightweight encryption method with security from side-channel attacks was suggested in the publication [154]. The advanced encryption standard (AES) algorithm is used in this lightweight encryption algorithm. Sub-algorithms are used in the lightweight encryption algorithm to randomize the initialization vector and generate randomizing keys for each ciphertext. In [155], it was suggested to use dual-rail digital logic architecture to camouflage side-channel information like power usage waveforms and EM waves radiated by encryption processors. By equilibrating the bit transitions of the circuit's terminals, the dual rail makes it impossible for the ciphertext to leak from the power usage.

MITM Attack: The security mechanism called OpenPLC proposed in [146] is intended to protect IoT applications from MITM attacks. A three-tier IDS architecture using various ML algorithms was introduced in [108] to detect MITM attacks in IoT networks. To prevent MITM outbreaks, the authors in [109] projected a technique called LDAKM-EIoT. In [156], an IDS was created using several ML classification algorithms such as SVM, NB, DT, and AdaBoost to identify MITM attacks, which were carried out in the suggested network via address resolution protocol (ARP) poisoning. The study suggested using a high-quality training dataset to improve the performance of ML systems. The authors presented a three-factor-based user authentication technique in [80], in which the user's identification, passcode, and biometrics are utilized to validate the legality of the user via the user's device. Consequently, the gateway verifies the legality of the IoT node's three components, such as identification, pre-secret key, and randomized point during the authentication step. Furthermore, the symmetric session key is used to encode and transmit. As a result, an attacker in the midway cannot forge the contributing part of the session key. Hence, MITM attacks can be prevented.

Node Capture: Node capture attacks can be prevented by using a cryptographic hash function-dependent authentication scheme developed in [135]. If an attacker captures one node, then also the attacker cannot control other nodes. Because each node contains different session keys or secret keys. For the IoT paradigm [157], the authors developed a lightweight three-factor unknown user authentication scheme using PUF. The suggested protocol is immune to node capture attempts, according to informal and formal security investigations using the real-or-random paradigm.

False Data Injection Attack: OpenPLC is a secure mechanism [146] designed to protect IoT devices from injection attacks. An IoT and IoBT malware detection strategy proposed in [141] effectively counters false data injection attacks. The suggested security model [158] employs a range-based behavior filtering strategy and is dependent on a Markov stochastic process, which is used to watch the behavior of every network device. The suggested architecture and model are efficient in finding and forecasting false data injection and DDoS assaults in the 5G-enabled IoT, according to experimental findings. A malicious node detection approach that relies on a correlation concept that avoids fault data injection attacks was presented in [159]. First, time correlation is used to discover anomalies among similar kinds of sensor data. The second step is to identify malicious nodes using spatial correlation. Finally, event correlation is used to verify the malicious nodes that have been found. The authors of [160] investigated the identification and forecasting of false data injection assaults in IoT systems. To observe the behavior of IoT devices and make predictions of false data injection assaults, they initially launched SecIoT, a false data injection attack detection mechanism. Second, a decentralized trust management mechanism was presented, which uses a set of weighted votes to create trust throughout devices. A bandwidth optimization issue was developed to properly distribute the bandwidth of trusted devices to protect against false data injection assaults in communication channels. When it comes to protecting against false data injection assaults, SecIoT has maximum accuracy. The MTD using the IoT-based data replication (MTD-IDR) paradigm was developed in the paper [161]. MTD-IDR formulates an optimization problem for finding the best number of copies for every transmitted signal in the system using linear-matrix inequalities. It effectively diminished the consequence of successful false data injection attempts.

Malicious Code Injection Attack: The authors of [162] propose a new watermarking system based on reinforcement learning that captures stochastic aspects of the generated signal to identify malicious code injection in IoT devices. For comprehensive information, the IoT gateway employed fictitious play (FP) learning. While for partial information, it used LSTM learning. However, a game-theoretic method based on lightweight mixed-strategy Nash equilibrium (MSNE) is being studied to improve the decision-making process of IoT gateways. The suggested framework's message delivery reliability was approximately 100% underneath a one-second attack detection delay, according to the findings. The authors made both information flash and RAM non-executable in [163] to protect against code injection attacks. In the non-secure world, they verified this by initiating two buffer overflow-based code injection assaults that divert control flows to code snippets located in either non-secure information flash or non-secure RAM. Whenever control flows are redirected, defect exceptions are produced in both circumstances. Non-secure and secure flash are non-writable throughout application execution to prevent application code from being tampered.

Spoofing: A deep neural networks (DNN)-based sophisticated authenticating method was designed in [164]. By extracting channel state information (CSI) signals, the author used the device-free approach to reach 92.34%. The authors of [165] devised a PHY-layer authentication scheme that employs multiple MIMO monuments to measure signal RSSI and employs LR to prevent being bound by a known radio channel design. To minimize computational complexity, an incremental aggregated gradient (IAG)-based identification is provided. The suggested authentication method can increase spoofing detection capability, according to simulations and experiments. To prevent spoofing attacks, a lightweight mutual authentication algorithm was developed [101]. IoT devices are secured from IoT vulnerabilities using two security elements in the suggested security system [166]. The challenge-response protocol allows IoT servers to certify the legitimacy of IoT ledgers by attesting their Merkel trees; and the acquisition of the IoT devices' group key and private key quadratic fractions by IoT servers. An adversary must provide proper responses to each and every randomized challenge created by the IoT device to masquerade as a trusted entity in the network. In this strategy, a single missing calculated answer given by the adversary will cause the node underneath assault to notify a trusted IoT activity monitoring system about the presence of a fraudulent IoT authenticator.

Desynchronization: Desynchronization attacks can be prevented by using the algorithm based on a cryptographic hash function suggested in [135]. But the algorithm failed to prevent a desynchronization attack. In such a case, the gateway would not update the pair (time interval of data and data of a particular user) if the final approval message is rejected or lost due to time latency. Such information will be contradictory across gateway and user. To increase security mechanisms and address the user authentication issue, the article [89] presented a user authentication scheme based on biometrics and PUF. The potential alternative mitigated the desynchronization attack based on an initial study.

To protect IoT devices from attacks, different kinds of mechanisms can be employed as mentioned in this section. The kind of mechanism can be encryption, IDS, SDN, MTD, lightweight authentication, ML, DL, or DNN. Depending on the attack model and feasibility of the application, the researcher will be able to design a solution mechanism.

Mechanisms for Combating IoT/IIoT Security Attacks

IoT devices, in reality, barely have any built-in security, making them a prime target for hackers. Most IoT devices are connected to one another, which makes it possible for one compromised device to impact the security of many others. To take advantage of IoT's benefits, especially in the case of Industry 4.0, we must apply different strategies to mitigate security attacks. The techniques used to defend against IoT/IIoT security attacks are listed below.

The primary and most crucial process for ensuring confidentiality throughout transmission is *encryption*. It entails converting the plaintext to ciphertext using the hash function, which can be easily reversed with the knowledge of a secret key.

*PUF*s can be used to increase the security of hardware. The primary idea behind PUF is to use minor variances created during the chip fabrication process to create a unique identity for every device.

It is crucial to be able to identify persistent attacks in addition to blocking them. Due to cost and energy limits, sophisticated anti-virus programs and network analyzers cannot be employed in IoT devices. The *IDS* mechanism can be utilized for this purpose.

Blockchain, a decentralized, distributed, and shared ledger, is the cornerstone of data security. It is a decentralized ledger. The blockchain records are both time-stamped and chronological. Employing cryptographic hash keys, each record in the logbook is firmly tied to the prior record.

SDN's goal is to create an atmosphere where more flexible system solutions can be developed, and network resources can be managed more easily by utilizing a centralized SDN controller.

MTD's purpose is to switch between numerous deployments in a cyber-system on a regular basis, such as altering open network ports, software, network configuration, and so on. As a result, the attacker faces more ambiguity. As a result, the benefit of observation that an adversary has over traditional defense methods is reduced.

In recent times, the field of **_artificial intelligence_** has gained a lot of attention. Many areas employ AI techniques like ML, DL, ANN, and DNN for advancement, and it is also being applied for IoT/IIoT security. By taking a different approach to guarding against assaults than other standard methods, AI seems to be a potential solution for protecting IoT devices from cyberattacks.

Conclusion

Designing IoT applications or Industry 4.0 security is challenging because the IoT is characterized by a vast number of diverse devices that are vulnerable to a variety of threats. The heterogeneity of resources in IoT has also hampered efforts to provide a viable mechanism for protecting the IoT layers. We discussed the major IoT security concerns in this study. The paper summarized 46 kinds of attacks and mentioned various kinds of solutions for each attack. It is noticed that from all the layers in IoT architecture, the network layer is more vulnerable to different types of attacks. It also concluded that DDoS attacks are considerably vulnerable comparable to other attacks. It is observed that different types of mechanisms can be employed for various kinds of attacks. From all the mechanisms, AI, SDN, MTD, and encryption techniques have their own importance according to the feasibility of applications. In preventing or at least detecting the attacks, AI is playing a vital role.

References

1. Velliangiri, S., Kumar, S.A., Karthikeyan, P. (Eds.): Internet of Things: Integration and security challenges. CRC Press (2020)
2. Sundmaeker, H., Guillemin, P., Friess, P., Woelf_é, S.: Vision and challenges for realising the Internet of Things. Cluster Eur. Res. Projects Internet Things, Eur. Commision 3(3), 34–36 (2010)
3. Amaral, L.A., Hessel, F.P., Bezerra, E.A., Corrêa, J.C., Longhi, O.B., Dias, T.F.O.: eCloudRFID_A mobile software framework architecture for pervasive RFID-based applications. J. Netw. Comput. Appl. **34**(3), 972–979 (2011). https://doi.org/10.1016/j.jnca.2010.04.005
4. Ashton, K.: That 'internet of things' thing. RFID J. **22**(7), 97–114 (2009)
5. Yan, Z., Zhang, P., Vasilakos, A.V.: A survey on trust management for Internet of Things. J. Netw. Comput. Appl. **42**(2014), 120–134. ISSN 1084–8045. https://doi.org/10.1016/j.jnca.2014.01.014
6. Abomhara, M., Køien, G.M.:Security and privacy in the Internet of Things: Current status and open issues. In: 2014 International Conference on Privacy and Security in Mobile Systems (PRISMS), pp. 1–8 (2014). https://doi.org/10.1109/PRISMS.2014.6970594
7. Butun, I., Osterberg, P., Song, H.: Security of the Internet of Things: vulnerabilities, attacks, and countermeasures. IEEE Commun. Surv. Tutorials **22**(1), 616–644 (2020). https://doi.org/10.1109/COMST.2019.2953364

8. Internet of Things (IoT) total annual revenue worldwide from 2019 to 2030. Statista Report. https://www.statista.com/statistics/1194709/iot-revenue-worldwide/
9. Waidner, M., Kasper, M.: Security in industrie 4.0-challenges and solutions for the fourth industrial revolution. In: Design, Automation & Test in Europe Conference & Exhibition (DATE), 2016, pp. 1303–1308. IEEE (2016)
10. Hassija, V., Chamola, V., Saxena, V., Jain, D., Goyal, P., Sikdar, B.: A survey on IoT security: Application areas, security threats, and solution architectures. IEEE Access 7, 82721–82743 (2019). https://doi.org/10.1109/ACCESS.2019.2924045
11. Velliangiri, S., Gunasekaran, M., Karthikeyan, P.: Secure Communication for 5G and IoT Networks. Springer International Publishing AG (2021)
12. Hosseinzadeh, S., Hyrynsalmi, S., Leppänen, V.: Chapter 14—Obfuscation and diversification for securing the internet of things (IoT). In: Buyya, R., Dastjerdi, A.V. (eds.), Internet of Things, Morgan Kaufmann, pp. 259–274 (2016). ISBN 9780128053959, https://doi.org/10.1016/B978-0-12-805395-9.00014-9
13. Kouicem, D.E., Bouabdallah, A., Lakhlef, H.: Internet of things security: A top-down survey. Comput. Netw. 141, 199–221 (2018). ISSN 1389-1286. https://doi.org/10.1016/j.comnet.2018.03.012
14. Khanam, S., Ahmedy, I.B., Idna Idris, M.Y., Jaward, M.H., Bin Md Sabri, A.Q: A survey of security challenges, attacks taxonomy and advanced countermeasures in the Internet of Things. In: IEEE Access 8, 219709–219743 (2020). https://doi.org/10.1109/ACCESS.2020.3037359
15. Badhib, A., Alshehri, S., Cherif, A.: a robust device-to-device continuous authentication protocol for the Internet of Things. IEEE Access 9, 124768–124792 (2021). https://doi.org/10.1109/ACCESS.2021.3110707
16. Whitman, M.E., Mattord, H.J.: Principles of information security. Cengage Learning (2021)
17. Ashraf, Q.M., Habaebi, M.H.: Autonomic schemes for threat mitigation in Internet of Things. J. Netw. Comput. Appl. 49, 112–127 (2015). https://doi.org/10.1016/j.jnca.2014.11.011
18. Zou, Y., Zhu, J., Wang, X., Hanzo, L.: A survey on wireless security: technical challenges, recent advances, and future trends. Proc. IEEE 104(9), 1727–1765 (2016). https://doi.org/10.1109/JPROC.2016.2558521
19. Brickerbot Permanent Denial-of-Service Attack (Update A), U.S. Dept. Homeland Security, Washington, DC, USA, 2017. https://ics-cert.us-cert.gov/alerts/ICS-ALERT-17-102-01A
20. Ahmed, A.H., Omar, N.M., Ibrahim, H.M.: Modern IoT architectures review: A security perspective. In: Proceedings of 8th Annual International Conference in ICT: Big Data, Cloud Security, pp. 73–81 (2017). https://doi.org/10.5176/2251-2136
21. Ur, B., Jung, J., Schechter, S.: The current state of access control for smart devices in homes. In: Workshop on Home Usable Privacy and Security (HUPS), vol. 29, pp. 209–218. HUPS 2014 (2013). http://cups.cs.cmu.edu/soups/2013/HUPS/HUPS13-BlaseUR.pdf
22. Nandy, T., Idris, M.Y.I.B., Md Noor, R., Mat Kiah, L., Lun, L.S., Annuar Juma'at, N.B., Ahmedy, I., Abdul Ghani, N., Bhattacharyya, S.: Review on security of Internet of Things authentication mechanism. IEEE Access 7, 151054–151089 (2019). https://doi.org/10.1109/ACCESS.2019.2947723
23. Petrov, V.,. Edelev, S, Komar, M., Koucheryavy, Y.: Towards the era of wireless keys: How the IoT can change authentication paradigm. In: 2014 IEEE World Forum on Internet of Things (WF-IoT), pp. 51–56 (2014). https://doi.org/10.1109/WF-IoT.2014.6803116
24. He, D., Chen, C., Chan, S., Bu, J., Vasilakos, A.V.: ReTrust: Attack-resistant and lightweight trust management for medical sensor networks. IEEE Trans. Inf Technol. Biomed. 16(4), 623–632 (2012). https://doi.org/10.1109/TITB.2012.2194788
25. Kortesniemi, Y., Lagutin, D., Elo, T., Fotiou, N.: Improving the privacy of IoT with decentralised identi_ers (DIDs). J. Comput. Netw. Commun. 2019, 1–10 (2019). https://doi.org/10.1155/2019/8706760
26. Vijayakumar, P., Obaidat, M.S., Azees, M., Islam, S.H., Kumar, N.: Efficient and secure anonymous authentication with location privacy for IoT-based WBANs. IEEE Trans. Ind. Informat. 16(4), 2603–2611 (2020). https://doi.org/10.1109/TII.2019.2925071

27. Garms, L., Lehmann, A.: Group signatures with selective linkability. In: Lin, D., Sako, K. (eds.) Public-Key Cryptography—PKC 2019. PKC 2019. Lecture Notes in Computer Science, vol 11442. Springer, Cham (2019). https://doi.org/10.1007/978-3-030-17253-4_7
28. Tomić, I., McCann, J.A.: A survey of potential security issues in existing wireless sensor network protocols. IEEE Internet Things J. **4**(6), 1910–1923 (2017). https://doi.org/10.1109/JIOT.2017.2749883
29. Calihman, A.: Architectures in the IoT Civilization [Online]. https://www.netburner.com/learn/architectural-frameworks-in-the-iot-civilization/
30. Eldrandaly, K.A., Abdel-Basset, M., Shawky, L.A.: Internet of spatial things: A new reference model with insight analysis. IEEE Access **7**, 19653–19669 (2019). https://doi.org/10.1109/ACCESS.2019.2897012
31. Silva, J.D.C., Rodrigues, J.J.P.C., Saleem, K., Kozlov, S.A., Rabêlo, R.A.L.: M4DN.IoT-A networks and devices management platform for Internet of Things. IEEE Access **7**, 53305–53313 (2019). https://doi.org/10.1109/ACCESS.2019.2909436
32. Bing, F.: The research of IOT of agriculture based on three layers architecture. In: 2016 2nd International Conference on Cloud Computing and Internet of Things (CCIOT), pp. 162–165 (2016). https://doi.org/10.1109/CCIOT.2016.7868325
33. Wu, M., Lu, T.-J., Ling, F.-Y., Sun, J., Du, H.-Y.: Research on the architecture of Internet of Things. In: 2010 3rd International Conference on Advanced Computer Theory and Engineering (ICACTE), pp. V5-484–V5-487 (2010). https://doi.org/10.1109/ICACTE.2010.5579493
34. Kaur, N., Sood, S.K.: An energy-efficient architecture for the Internet of Things (IoT). IEEE Syst. J. **11**(2), 796–805 (2017). https://doi.org/10.1109/JSYST.2015.2469676
35. Navani, D., Jain, S., Nehra, M.S.: The Internet of Things (IoT): A study of architectural elements. In: 2017 13th International Conference on Signal-Image Technology & Internet-Based Systems (SITIS), pp. 473–478 (2017). https://doi.org/10.1109/SITIS.2017.83
36. Virat, M.S., Bindu, S.M., Aishwarya, B., Dhanush, B.N., Kounte, M.R.: Security and privacy challenges in Internet of Things. In: 2018 2nd International Conference on Trends in Electronics and Informatics (ICOEI), pp. 454–460 (2018). https://doi.org/10.1109/ICOEI.2018.8553919
37. Atzori, L., Iera, A., Morabito, G.: The internet of things: a survey. Comput. Netw. **54**(15), 2787–2805 (2010)
38. Babun, L., Denney, K., Berkay Celik, Z., McDaniel, P., Selcuk Uluagac, A.: A survey on IoT platforms: Communication, security, and privacy perspectives. Comput. Netw. **192**, 108040 (2021). ISSN 1389-1286
39. Cirani, S., et al.: A scalable and self-configuring architecture for service discovery in the Internet of Things. IEEE Internet Things J. **1**(5), 508–521 (2014). https://doi.org/10.1109/JIOT.2014.2358296
40. Tsai, CW., Lai, CF., Vasilakos, A.V.: Future Internet of Things: open issues and challenges. Wirel. Netw. **20**, 2201–2217 (2014). https://doi.org/10.1007/s11276-014-0731-0
41. Mahbub, M.: Progressive researches on IoT security: An exhaustive analysis from the perspective of protocols, vulnerabilities, and preemptive architectonics. J. Netw. Comput. Appl. **168**, 102761 (2020). ISSN 1084-8045. https://doi.org/10.1016/j.jnca.2020.102761
42. Ma, G., Li, X., Pei, Q., Li, Z.:A security routing protocol for Internet of Things based on RPL. In: 2017 International Conference on Networking and Network Applications (NaNA), pp. 209–213 (2017). https://doi.org/10.1109/NaNA.2017.28
43. Sethi, P., Sarangi, S.R.: Internet of Things: Architectures, protocols, and applications. J. Electr. Comput. Eng. **2017**(9324035), 25 pages (2017). https://doi.org/10.1155/2017/9324035
44. Mukherjee, A.: Physical-layer security in the Internet of Things: sensing and communication confidentiality under resource constraints. Proc. IEEE **103**(10), 1747–1761 (2015). https://doi.org/10.1109/JPROC.2015.2466548
45. Farris, I., Taleb, T., Khettab, Y., Song, J.: A survey on emerging SDN and NFV security mechanisms for IoT systems. In: IEEE Communications Surveys & Tutorials, vol. 21, no. 1, pp. 812–837, Firstquarter (2019). https://doi.org/10.1109/COMST.2018.2862350

46. Peris-Lopez, P., Hernandez-Castro, J.C., Estevez-Tapiador, J.M., Ribagorda, A.: RFID systems: A survey on security threats and proposed solutions. In: Cuenca, P., Orozco-Barbosa, L. (eds.) Personal Wireless Communications. PWC 2006. Lecture Notes in Computer Science, vol. 4217. Springer, Berlin, Heidelberg (2006). https://doi.org/10.1007/11872153_14
47. Mosenia, A., Jha, N.K.: A comprehensive study of security of Internet-of-Things. In: IEEE Transactions on Emerging Topics in Computing, vol. 5, no. 4, pp. 586–602, 1 Oct.–Dec. 2017. https://doi.org/10.1109/TETC.2016.2606384
48. Zhang, W., Qu, B.: Security architecture of the Internet of Things oriented to perceptual layer. Int. J. Comput. Consum. Control (IJ3C) 2(2), 37–45 (2013). http://ij3c.ncuteecs.org/volume/paper_le/2-2/IJ3C_5.pdf
49. Khairi, A., Farooq, M., Waseem, M., Mazhar, S.: A critical analysis on the security concerns of Internet of Things (IoT). Perception 111 (2015). https://doi.org/10.2136/sssaj1987.036159 95005100060002x
50. Yousuf, O., Mir, R.N.: A survey on the Internet of Things security: State-of-art, architecture, issues and countermeasures. Inf. Comput. Secur. 27(2), 292–323 (2019). https://doi.org/10.1108/ICS-07-2018-0084
51. Liu, W., et al.: Various detection techniques and platforms for monitoring interference condition in a wireless testbed. In: Fàbrega, L., Vilà, P., Careglio, D., Papadimitriou, D. (eds.) Measurement Methodology and Tools. Lecture Notes in Computer Science, vol. 7586. Springer, Berlin, Heidelberg (2013). https://doi.org/10.1007/978-3-642-41296-7_4
52. Deogirikar, J., Vidhate, A.: Security attacks in IoT: A survey. In: 2017 International Conference on I-SMAC (IoT in Social, Mobile, Analytics and Cloud) (I-SMAC), pp. 32–37 (2017). https://doi.org/10.1109/I-SMAC.2017.8058363
53. Mathur, A., Newe, T., Rao, M.: Defence against Black Hole and Selective Forwarding Attacks for Medical WSNs in the IoT. Sensors 16(1), 118 (2016). https://doi.org/10.3390/s16010118
54. Bysani, L.K., Turuk, A.K.: A survey on selective forwarding attack in wireless sensor networks. In: 2011 International Conference on Devices and Communications (ICDeCom), pp. 1–5 (2011). https://doi.org/10.1109/ICDECOM.2011.5738547
55. Mitrokotsa, A., Rieback, M.R., Tanenbaum, A.S.: Classification of RFID attacks. G. E. N. 15693(14443), 14 (2010). https://doi.org/10.5220/0001738800730086
56. Hu, Y., Perrig, A., Johnson, D.B.: Packet leashes: A defense against wormhole attacks in wireless networks. In: IEEE INFOCOM 2003. Twenty-second Annual Joint Conference of the IEEE Computer and Communications Societies (IEEE Cat. No.03CH37428), vol. 3, pp. 1976–1986 (2003). https://doi.org/10.1109/INFCOM.2003.1209219
57. Anu, P., Vimala, S.: A survey on sniffing attacks on computer networks. In: 2017 International Conference on Intelligent Computing and Control (I2C2), pp. 1–5 (2017). https://doi.org/10.1109/I2C2.2017.8321914
58. Borgohain, T., Kumar, U., Sanyal, S.: Survey of Security and Privacy Issues of Internet of Things (2015). arXiv preprint arXiv:1501.02211
59. Stajano, F., Anderson, R.: The Resurrecting Duckling: security issues for ubiquitous computing. Computer 35(4), supl22–supl26 (2002). https://doi.org/10.1109/MC.2002.1012427
60. Wood, A.D., Stankovic, J.A.: Denial of service in sensor networks. Computer 35(10), 54–62 (2002). https://doi.org/10.1109/MC.2002.1039518
61. Znaidi, W., Minier, M., Babau, J.-P.: An Ontology for Attacks in Wireless Sensor Networks. [Research Report] RR-6704, INRIA (2008) (inria-00333591)
62. Nasridinov, A., Byun, J.-Y., Park, Y.-H.: A study on detection techniques of XML rewriting attacks in web services. Int. J. Control Autom. 7(1), 391–400 (2014)
63. WS-Attacks. Attack Subtypes. Accessed: Feb. 9, 2019. https://www.ws-attacks.org/XML_Signature_Wrapping
64. Dorai, R., Kannan, V.: SQL injection-database attack revolution and prevention. J. Int. Commercial Law Technol. 6(4), 224 (2011)
65. Razzaque, M.A., Milojevic-Jevric, M., Palade, A., Clarke, S.: Middleware for Internet of Things: A survey. IEEE Internet Things J. 3(1), 70–95 (2016). https://doi.org/10.1109/JIOT.2015.2498900

66. Deng, J., Han, R., Mishra, S.: Defending against path-based DoS attacks in wireless sensor networks. In: Proceedings of the 3rd ACM workshop on Security of ad hoc and sensor networks. Association for Computing Machinery, New York, NY, USA, pp 89–96 (2005). https://doi.org/10.1145/1102219.1102235

67. Frank, B., Shelby, Z., Hartke, K., Bormann, C.: Constrained Application Protocol (COAP), IETF, Fremont, CA, USA (2011). https://www.ietf.org/archive/id/draft-ietf-core-coap-08.html

68. Rahman, R.A., Shah, B.: Security analysis of IoT protocols: a focus in CoAP. In: 2016 3rd MEC International Conference on Big Data and Smart City (ICBDSC), pp. 1–7 (2016). https://doi.org/10.1109/ICBDSC.2016.7460363

69. Tahir, R., V. U. o. P. Department of Computer Science: A study on malware and malware detection techniques. Int. J. Edu. Manage. Eng. **8**(2), 20–30 (2018). https://doi.org/10.5815/ijeme.2018.02.03

70. Grover, J., Laxmi, V., Gaur, M.S.: Attack models and infrastructure supported detection mechanisms for position forging attacks in vehicular ad hoc networks. CSI Trans. ICT **1**(3), 261–279 (2013). https://doi.org/10.1007/s40012-013-0025-1

71. Feng, Z., Hu, G.: Secure cooperative event-triggered control of linear multiagent systems under DoS attacks. IEEE Trans. Control Syst. Technol. **28**(3), 741–752 (2020). https://doi.org/10.1109/TCST.2019.2892032

72. Grobauer, B., Walloschek, T., Stocker, E.: Understanding cloud computing vulnerabilities. In: IEEE Security & Privacy, vol. 9, no. 2, pp. 50–57, March–April 2011. https://doi.org/10.1109/MSP.2010.115

73. Kumar, S., Sahoo, S., Mahapatra, A., Swain, A.K., Mahapatra, K.K.: Security enhancements to system on chip devices for IoT perception layer. In: 2017 IEEE International Symposium on Nanoelectronic and Information Systems (iNIS), pp. 151–156 (2017). https://doi.org/10.1109/iNIS.2017.39

74. Shabana, K., Fida, N., Khan, F., Jan, S.R., Rehman, M.U.: Security issues and attacks in wireless sensor networks. Int. J. Adv. Res. Comput. Sci. Electron. Eng. **5**(7), 81 (2016)

75. Sajjad, S.M., Yousaf, M.: Security analysis of IEEE 802.15.4 MAC in the context of Internet of Things (IoT). In: 2014 Conference on Information Assurance and Cyber Security (CIACS), pp. 9–14 (2014). https://doi.org/10.1109/CIACS.2014.6861324

76. Pankov, N.: Protect networked IoT devices or protect the network from IoT devices? https://www.kaspersky.co.in/blog/rsa2021-dangerous-iot/22927/

77. Cyberwar in Ukraine leads to all-time-high levels of DDoS attacks Kaspersky. Kaspersky report. https://www.kaspersky.com/about/press-releases/2022_cyberwar-in-ukraine-leads-to-all-time-high-levels-of-ddos-attacks

78. Gupta, A., Tripathi, M., Shaikh, T.J., Sharma, A.: A lightweight anonymous user authentication and key establishment scheme for wearable devices. Comput. Netw. **149**, 29–42 (2019). ISSN 1389-1286. https://doi.org/10.1016/j.comnet.2018.11.021

79. Harbi, Y., Aliouat, Z., Refoufi, A., Harous, S., Bentaleb, A.: Enhanced authentication and key management scheme for securing data transmission in the internet of things. Ad Hoc Netw. **94**, 101948 (2019). ISSN 1570-8705. https://doi.org/10.1016/j.adhoc.2019.101948

80. Sadhukhan, D., Ray, S., Biswas, G.P., et al.: A lightweight remote user authentication scheme for IoT communication using elliptic curve cryptography. J. Supercomput. **77**, 1114–1151 (2021). https://doi.org/10.1007/s11227-020-03318-7

81. Anajemba, J.H., Iwendi, C., Razzak, M., Ansere, J.A., Okpalaoguchi, M.I.: A counter-eavesdropping technique for optimized privacy of wireless industrial IoT communications. In: IEEE Transactions on Industrial Informatics. https://doi.org/10.1109/TII.2021.3140109

82. Kwon, S., Park, S., Cho, H., et al.: Towards 5G-based IoT security analysis against Vo5G eavesdropping. Computing **103**, 425–447 (2021). https://doi.org/10.1007/s00607-020-00855-0

83. Han, Q.: Data-Driven Analysis and Characterization of Modern Android Malware. PhD diss., Dartmouth College (2021)

84. Aman, M.N., Basheer, M.H., Sikdar, B.: A lightweight protocol for secure data provenance in the Internet of Things using wireless fingerprints. IEEE Syst. J. **15**(2), 2948–2958 (2021). https://doi.org/10.1109/JSYST.2020.3000269
85. Amelian, A., Etemadi Borujeni, S.: A side-channel analysis for hardware Trojan detection based on path delay measurement. J Circ. Syst. Comp. **27**, 1850138 (2018). https://doi.org/10.1142/S0218126618501384
86. Nguyen, N.L.N.: New side-channel and techniques for hardware trojan detection. PhD diss., Georgia Institute of Technology (2020)
87. Hazari, N.A.: Design and Analysis of Assured and Trusted ICs using Machine Learning and Blockchain Technology. PhD diss., University of Toledo (2021)
88. Bu, K., Xu, M., Liu, X., Luo, J., Zhang, S., Weng, M.: Deterministic detection of cloning attacks for anonymous RFID systems. IEEE Trans. Ind. Informat. **11**(6), 1255–1266 (2015)
89. Nurkifli, E.H.: The Resilience of DoS Attacks in User-Authentication to Preserving The Availability (2021)
90. Aref, M.A., Jayaweera, S.K., Machuzak, S.:Multi-agent reinforcement learning based cognitive anti-jamming. In: 2017 IEEE Wireless Communications and Networking Conference (WCNC), pp. 1–6 (2017). https://doi.org/10.1109/WCNC.2017.7925694
91. Han, G., Xiao, L., Poor, H.V.: Two-dimensional anti-jamming communication based on deep reinforcement learning. In: 2017 IEEE International Conference on Acoustics, Speech and Signal Processing (ICASSP), pp. 2087–2091 (2017). https://doi.org/10.1109/ICASSP.2017.7952524
92. Upadhyaya, B., Sun, S., Sikdar, B.:Machine learning-based jamming detection in wireless IoT networks. In: 2019 IEEE VTS Asia Pacific Wireless Communications Symposium (APWCS), pp. 1–5 (2019). https://doi.org/10.1109/VTS-APWCS.2019.8851633
93. Jaber, M.M., Elameer, A.S., Mohammed Ali, S.: IoT network security using autoencoder deep neural network and channel access algorithm. J. Intell. Syst. **31**(1) (2021)
94. Zou, Y., Yu, D., Hu, P., Yu, J., Cheng, X., Mohapatra, P.: Jamming-resilient message dissemination in wireless networks. In: IEEE Transactions on Mobile Computing. https://doi.org/10.1109/TMC.2021.3108004
95. Sharma, K., Ghose, M.K.: Wireless sensor networks: An overview on its security threats. IJCA, Special Issue on "Mobile Ad-hoc Networks" MANETs 1495, pp. 42–45 (2010)
96. Alladi, T., Chamola, V., Naren: HARCI: A two-way authentication protocol for three entity healthcare IoT networks. IEEE J Sel. Areas Commun. **39**(2), 361–369 (2021). https://doi.org/10.1109/JSAC.2020.3020605
97. Meidan, Y., Bohadana, M., Shabtai, A., Ochoa, M., Tippenhauer, N.O., Guarnizo, J.D., Elovici, Y.: Detection of Unauthorized IoT Devices Using Machine Learning Techniques (2017). arXiv:1709.04647. http://arxiv.org/abs/1709.04647
98. Khatun, M.A., Chowdhury, N., Uddin, M.N.: Malicious nodes detection based on artificial neural network in IoT environments. In: 2019 22nd International Conference on Computer and Information Technology (ICCIT), pp. 1–6 (2019). https://doi.org/10.1109/ICCIT48885.2019.9038563.
99. Wang, G., et al.: Hu-Fu: replay-resilient RFID authentication. IEEE/ACM Trans. Netw. **28**(2), 547–560 (2020). https://doi.org/10.1109/TNET.2020.2964290
100. Li, Q., Chen, F., Kang, J., Wang, P., Su, J., Huang, F., Li, M., Zhang, J.: Intrinsic random optical features of the electronic packages as physical unclonable functions for Internet of Things security. Adv. Photonics Res. 2100207 (2021)
101. Sharma, G., Kalra, S.: A lightweight multi-factor secure smart card based remote user authentication scheme for cloud-IoT applications. J. Inf. Secur. Appl. **42**, 95–106 (2018). ISSN 2214-2126. https://doi.org/10.1016/j.jisa.2018.08.003
102. Ullah, I., Zahid, H., Algarni, F., Asghar Khan, M.: An access control scheme using heterogeneous signcryption for IoT environments. CMC Comput. Mater. Continua **70**(3), 4307–4321 (2022)
103. Yi, P., Guan, Y., Zou, F., Yao, Y., Wang, W., Zhu, T.: Web phishing detection using a deep learning framework. Wirel. Commun. Mob. Comput. **2018**(4678746), 9 pages (2018). https://doi.org/10.1155/2018/4678746

104. Mao, J., Bian, J., Tian, W., et al.: Phishing page detection via learning classifiers from page layout feature. J. Wirel. Com. Netw. **2019**, 43 (2019). https://doi.org/10.1186/s13638-019-1361-0
105. Lee, J., Ye, P., Liu, R., Divakaran, D.M., Chan, M.C.: Building robust phishing detection system: an empirical analysis. NDSS MADWeb (2020)
106. Bostani, H., Sheikhan, M.: Hybrid of anomaly-based and specification-based IDS for Internet of Things using unsupervised OPF based on MapReduce approach. Comput. Commun. **98**, 52–71 (2017). ISSN 0140-3664
107. Abdalzaher, M.S., Muta, O.: A game-theoretic approach for enhancing security and data trustworthiness in IoT applications. IEEE Internet Things J. **7**(11), 11250–11261 (2020). https://doi.org/10.1109/JIOT.2020.2996671
108. Anthi, E., Williams, L., Słowińska, M., Theodorakopoulos, G., Burnap, P.: A supervised intrusion detection system for smart home IoT devices. IEEE Internet Things J. **6**(5), 9042–9053 (2019). https://doi.org/10.1109/JIOT.2019.2926365
109. Wazid, M., Das, A.K., Shetty, S., Rodrigues, J.J.P.C., Park, Y.: LDAKM-EIoT: lightweight device authentication and key management mechanism for edge-based IoT deployment. Sensors **19**(24), 5539 (2019). https://doi.org/10.3390/s19245539
110. Napiah, M.N., Bin Idris, M.Y.I., Ramli, R., Ahmedy, I.: Compression header analyzer intrusion detection system (CHA–IDS) for 6LoWPAN communication protocol. IEEE Access **6**, 16623–16638 (2018). https://doi.org/10.1109/ACCESS.2018.2798626
111. Raza, S., Wallgren, L., Voigt, T.: SVELTE: Real-time intrusion detection in the Internet of Things. Ad Hoc Netw. **11**(8), 2661–2674 (2013)
112. Yavuz, F.Y., Devrim, Ü.N.A.L., Ensar, G.Ü.L.: Deep learning for detection of routing attacks in the internet of things. Int. J. Comput. Intell. Syst. **12**(1), 39 (2018)
113. Liu, Y., Ma, M., Liu, X., Xiong, N.N., Liu, A., Zhu, Y.: Design and analysis of probing route to defense sink-hole attacks for Internet of Things security. IEEE Trans. Netw. Sci. Eng. **7**(1), 356–372, 1 January-March 2020. https://doi.org/10.1109/TNSE.2018.2881152
114. Hassan, Z., Mehmood, A., Maple, C., Khan, M.A., Aldegheishem, A.: Intelligent detection of black hole attacks for secure communication in autonomous and connected vehicles. IEEE Access **8**, 199618–199628 (2020). https://doi.org/10.1109/ACCESS.2020.3034327
115. Singh, R., Singh, J., Singh, R.: Fuzzy based advanced hybrid intrusion detection system to detect malicious nodes in wireless sensor networks. Wirel. Commun. Mob. Comput. **2017**(3548607), 14 pages (2017)
116. Alenezi, F.A.F., Song, S., Choi, B.-Y.:WAND: wormhole attack analysis using the neighbor discovery for software-defined heterogeneous Internet of Things. In: 2021 IEEE International Conference on Communications Workshops (ICC Workshops), pp. 1–6 (2021). https://doi.org/10.1109/ICCWorkshops50388.2021.9473770
117. Cakir, S., Toklu, S., Yalcin, N.: RPL attack detection and prevention in the Internet of Things networks using a GRU based deep learning. IEEE Access **8**, 183678–183689 (2020). https://doi.org/10.1109/ACCESS.2020.3029191
118. Kumar, B., Bhuyan, B.: Game theoretical defense mechanism against reputation based sybil attacks. Procedia Comput. Sci. **167**, 2465–2477 (2020)
119. Hafeez, M.A., Tarkoma, S.:Protecting IoT-environments against Traffic Analysis Attacks with Traffic Morphing. In: 2019 IEEE International Conference on Pervasive Computing and Communications Workshops (PerCom Workshops), pp. 196–201 (2019). https://doi.org/10.1109/PERCOMW.2019.8730787
120. Zeng, Z., Zhang, J.: Based on the role of Internet of Things security in the management of enterprise human resource information leakage. Wirel. Commun. Mob. Comput. **2021**(5936390), 12 pages (2021)
121. Lee, C., Shen, S., Sahin, G., Choi, K., Choi, H.-A.: A novel and scalable communication-history-based Knapsack authentication framework for IEEE 802.11 networks. In: 2015 IEEE Conference on Communications and Network Security (CNS), pp. 44–52 (2015). https://doi.org/10.1109/CNS.2015.7346809

122. Alrawais, A., Alhothaily, A., Hu, C., Xing, X., Cheng, X.: An attribute-based encryption scheme to secure fog communications. IEEE Access **5**, 9131–9138 (2017). https://doi.org/10. 1109/ACCESS.2017.2705076
123. Krentz, K.-F.:A denial-of-sleep-resilient medium access control layer for IEEE 802.15. 4 networks. PhD diss., Universität Potsdam (2019)
124. Glissa, G., Meddeb, A.: 6LowPSec: An end-to-end security protocol for 6LoWPAN. Ad Hoc Netw. **82**, 100–112 (2019). ISSN 1570-8705. https://doi.org/10.1016/j.adhoc.2018.01.013
125. Bhunia, S.S., Gurusamy, M.: Dynamic attack detection and mitigation in IoT using SDN. In: 2017 27th International Telecommunication Networks and Applications Conference (ITNAC), pp. 1–6 (2017). https://doi.org/10.1109/ATNAC.2017.8215418
126. Dang, V.T., et al.: SDN-based SYN Proxy—a solution to enhance performance of attack mitigation under TCP SYN flood. Comput. J. **62**(4), 518–534 (2019). https://doi.org/10.1093/ comjnl/bxy117
127. Nishanth, N., Mujeeb, A.: modeling and detection of flooding-based denial-of-service attack in wireless Ad Hoc network using Bayesian inference. IEEE Syst. J. **15**(1), 17–26 (2021). https://doi.org/10.1109/JSYST.2020.2984797
128. Oryema, B., Lee, B., Park, J.: Secure mobility management using CoAP in the Internet of Things. In: 2018 IEEE 5th International Congress on Information Science and Technology (CiSt), pp. 514–524 (2018). https://doi.org/10.1109/CIST.2018.8596598
129. Hammi, M.T., Hammi, B., Bellot, P., Serhrouchni, A.: Bubbles of trust: A decentralized blockchain-based authentication system for IoT. Comput. Secur. **78**, 126–142 (2018). ISSN 0167-4048. https://doi.org/10.1016/j.cose.2018.06.004
130. Gupta, A.N., Thilagam, P.S.: Detection of XML signature wrapping attack using node counting. In: Proceedings of 3rd International Symposium on Big Data and Cloud Computing Challenges (ISBCC), pp. 57–63. Springer (2016)
131. Kumar, J., Rajendran, B., Bindhumadhava, B.S., Chandra Babu, N.S.: XML wrapping attack mitigation using positional token. In: 2017 International Conference on Public Key Infrastructure and its Applications (PKIA), pp. 36–42 (2017). https://doi.org/10.1109/PKIA.2017. 8278958
132. Uwagbole, S.O., Buchanan, W.J., Fan, L.:An applied pattern-driven corpus to predictive analytics in mitigating SQL injection attack. In: 2017 Seventh International Conference on Emerging Security Technologies (EST), pp. 12–17 (2017). https://doi.org/10.1109/EST.2017. 8090392
133. Ross, K., Moh, M., Moh, T.-S., Yao, J.: Multi-source data analysis and evaluation of machine learning techniques for SQL injection detection. In: Proceedings of ACMSE 2018 Conference, New York, NY, USA, pp. 1:1–1:8 (2018). https://doi.org/10.1145/3190645.3190670
134. Gowtham, M., Pramod, H.B.: Semantic query-featured ensemble learning model for SQL-injection attack detection in IoT-ecosystems. In: IEEE Transactions on Reliability. https://doi. org/10.1109/TR.2021.3124331
135. Wu, F., Li, X., Sangaiah, A.R., Xu, L., Kumari, S., Wu, L., Shen, J.: A lightweight and robust two-factor authentication scheme for personalized healthcare systems using wireless medical sensor networks. Future Gener. Comput. Syst. **82**, 727–737 (2018). ISSN 0167-739X. https:// doi.org/10.1016/j.future.2017.08.042
136. Li, C., Li, P., Zhou, D., Yang, Z., Wu, M., Yang, G., Xu, W., Long, F., Yao, A.C.-C.: A decentralized blockchain with high throughput and fast confirmation. In: 2020 {USENIX} Annual Technical Conference ({USENIX}{ATC} 20), pp. 515–528 (2020)
137. Khan, A.Y., Latif, R., Latif, S., Tahir, S., Batool, G., Saba, T.: Malicious insider attack detection in IoTs using data analytics. IEEE Access **8**, 11743–11753 (2020). https://doi.org/10.1109/ ACCESS.2019.2959047
138. Seitz, K., Serth, S., Krentz, K.-F., Meinel, C.: Enabling en-route filtering for end-to-end encrypted coap messages. In: Proceedings of the 15th ACM Conference on Embedded Network Sensor Systems, p. 33. ACM (2017)
139. Feng, P., Ma, J., Sun, C., Xu, X., Ma, Y.: A novel dynamic android malware detection system with ensemble learning. IEEE Access **6**, 30996–31011 (2018). https://doi.org/10.1109/ACC ESS.2018.2844349

140. Dovom, E.M., Azmoodeh, A., Dehghantanha, A., Newton, D.E., Parizi, R.M., Karimipour, H.: Fuzzy pattern tree for edge malware detection and categorization in IoT. J. Syst. Arch. **97**, 1–7 (2019). ISSN 1383-7621. https://doi.org/10.1016/j.sysarc.2019.01.017

141. Azmoodeh, A., Dehghantanha, A., Choo, K.R.: Robust malware detection for internet of (battlefield) things devices using deep eigenspace learning. In: IEEE Transactions on Sustainable Computing, vol. 4, no. 1, pp. 88–95, 1 January–March 2019. https://doi.org/10.1109/TSUSC.2018.2809665

142. Chauhan, P., Atulkar, M.:Selection of tree based ensemble classifier for detecting network attacks in IoT. In: 2021 International Conference on Emerging Smart Computing and Informatics (ESCI), pp. 770–775 (2021). https://doi.org/10.1109/ESCI50559.2021.9397033

143. Lukaj, V., Martella, F., Fazio, M., Celesti, A., Villari, M.:Trusted ecosystem for IoT service provisioning based on brokering. In: 2021 IEEE/ACM 21st International Symposium on Cluster, Cloud and Internet Computing (CCGrid), pp. 746–753 (2021). https://doi.org/10.1109/CCGrid51090.2021.00090

144. Derun Karabeyoğlu, E., Karalar, T.C.: IoT module improves smart environment reliability. In: 2017 IEEE 3rd International Forum on Research and Technologies for Society and Industry (RTSI), pp. 1–5 (2017). https://doi.org/10.1109/RTSI.2017.8065883

145. Zulkifl, Z., et al.: FBASHI: Fuzzy and blockchain-based adaptive security for healthcare IoTs. IEEE Access **10**, 15644–15656 (2022). https://doi.org/10.1109/ACCESS.2022.3149046

146. Alves, T., Das, R., Morris, T.: Embedding encryption and machine learning intrusion prevention systems on programmable logic controllers. IEEE Embed. Syst. Lett. **10**(3), 99–102 (2018). https://doi.org/10.1109/LES.2018.2823906

147. Chen, H., Meng, C., Shan, Z., Fu, Z., Bhargava, B.K.: A novel low-rate denial of service attack detection approach in ZigBee wireless sensor network by combining Hilbert-Huang transformation and trust evaluation. IEEE Access **7**, 32853–32866 (2019). https://doi.org/10.1109/ACCESS.2019.2903816

148. Mehmood, A., Mukherjee, M., Ahmed, S.H., Song, H., Malik, K.M.: NBC-MAIDS: Naïve Bayesian classification technique in multi-agent system-enriched IDS for securing IoT against DDoS attacks. J. Supercomput. **74**(10), 5156–5170 (2018). https://doi.org/10.1007/s11227-018-2413-7

149. Roopak, M., Yun Tian, G., Chambers, J.: Deep learning models for cyber security in IoT networks. In: 2019 IEEE 9th Annual Computing and Communication Workshop and Conference (CCWC), pp. 0452–0457 (2019). https://doi.org/10.1109/CCWC.2019.8666588

150. Priyadarshini, R., Barik, R.K.: A deep learning based intelligent framework to mitigate DDoS attack in fog environment. J. King Saud Univ. Comput. Inf. Sci. (2019). https://doi.org/10.1016/j.jksuci.2019.04.010

151. Luo, X., Yan, Q., Wang, M., Huang, W.:Using MTD and SDN-based honeypots to defend DDoS attacks in IoT. In: 2019 Computing, Communications and IoT Applications (ComComAp), pp. 392–395 (2019). https://doi.org/10.1109/ComComAp46287.2019.9018775

152. Ko, I., Chambers, D., Barrett, E.: Feature dynamic deep learning approach for DDoS mitigation within the ISP domain. Int. J. Inf. Secur. **19**(1), 53–70 (2020). https://doi.org/10.1007/s10207-019-00453-y

153. Vuppala, S., Mady, A.E.-D., Kuenzi, A.:Rekeying-based Moving target defence mechanism for side-channel attacks. In: 2019 Global IoT Summit (GIoTS), pp. 1–5 (2019). https://doi.org/10.1109/GIOTS.2019.8766426

154. Hussain Pirzada, S.J., Xu, T., Jianwei, L.: Lightweight encryption algorithm implementation for Internet of Thing application. In: 2020 International Conference on Cyber Warfare and Security (ICCWS), pp. 1–6 (2020). https://doi.org/10.1109/ICCWS48432.2020.9292373

155. Ukezono, T.: Resistance for side-channel attack by virtual dual-rail effect. In: 2021 International Conference on Electrical, Communication, and Computer Engineering (ICECCE), pp. 1–6 (2021). https://doi.org/10.1109/ICECCE52056.2021.9514176

156. Sai Kiran, K.V.V.N.L., Kamakshi Devisetty, R.N., Pavan Kalyan, N., Mukundini, K., Karthi, R.: Building a intrusion detection system for IoT environment using machine learning techniques. Procedia Comput. Sci. **171**, 2372–2379 (2020). ISSN 1877-0509. https://doi.org/10.1016/j.procs.2020.04.257

157. Liu, Z., Guo, C., Wang, B.: A physically secure, lightweight three-factor and anonymous user authentication protocol for IoT. IEEE Access **8**, 195914–195928 (2020). https://doi.org/10.1109/ACCESS.2020.3034219

158. Moudoud, H., Khoukhi, L., Cherkaoui, S.: Prediction and detection of FDIA and DDoS attacks in 5G enabled IoT. IEEE Network **35**(2), 194–201 (2021). https://doi.org/10.1109/MNET.011.2000449

159. Lai, Y., Tong, L., Liu, J., Wang, Y., Tang, T., Zhao, Z., Qin, H.: Identifying malicious nodes in wireless sensor networks based on correlation detection. Comput. Secur. **113**, 102540 (2022). ISSN 0167-4048

160. Moudoud, H., Mlika, Z., Khoukhi, L., Cherkaoui, S.: Detection and prediction of FDI attacks in IoT systems via hidden markov model. In: IEEE Transactions on Network Science and Engineering. https://doi.org/10.1109/TNSE.2022.3161479

161. Giraldo, J., El Hariri, M., Parvania, M.: Moving target defense for cyber-physical systems using IoT-enabled data replication. IEEE Internet Things J. https://doi.org/10.1109/JIOT.2022.3144937

162. Ferdowsi, A., Saad, W.: Deep learning for signal authentication and security in massive Internet-of-Things systems. IEEE Trans. Commun. **67**(2), 1371–1387 (2019). https://doi.org/10.1109/TCOMM.2018.2878025

163. Luo, L., et al.: On Security of TrustZone-M based IoT systems. IEEE Internet Things J. https://doi.org/10.1109/JIOT.2022.3144405

164. Shi, C., Liu, J., Liu, H., Chen, Y.: Smart user authentication through actuation of daily activities leveraging WiFi-enabled IoT. In: Proceedings of the 18th ACM International Symposium on Mobile Ad Hoc Networking and Computing—Mobihoc'17 (2017). https://doi.org/10.1145/3084041.3084061

165. Xiao, L., Wan, X., Han, Z.: PHY-layer authentication with multiple landmarks with reduced overhead. IEEE Trans. Wirel. Commun. **17**(3), 1676–1687 (2018). https://doi.org/10.1109/TWC.2017.2784431

166. Rasheed, A.A., Mahapatra, R.N., Varol, C., Karpoor, N.: Exploiting zero knowledge proof and blockchain towards the enforcement of anonymity, data integrity and privacy (ADIP) on IoT. In: IEEE Transactions on Emerging Topics in Computing. https://doi.org/10.1109/TETC.2021.3099701

Chapter 5
Adopting Artificial Intelligence in ITIL for Information Security Management—Way Forward in Industry 4.0

Manikandan Rajagopal and **S. Ramkumar**

Introduction

The topic of artificial intelligence (AI) is fascinating because of how quickly it is developing. Combining AI with IT service management (ITSM) has the potential to revolutionize service management and usher in a new era of technological advancement [1]. The existing techniques in automation and information exchange in industrial terms are called as Industry 4.0. It has brought in the next generation of networks to ensure more responsiveness and efficiency [2]. Many notable researches have been carried out in bringing the communication gap closer. Algorithms such as TPSO [3] have been formulated to assure optimal data transmission which is vital in Industry 4.0. This chapter describes how far AI has come, how it is now being used in IT service management, and what the future holds for AI in this field. The field of computer science known as "artificial intelligence" (AI) focuses on the study and development of computational methodologies and techniques that give machines the ability to carry out tasks that would typically require human intelligence.

M. Rajagopal (✉)
Lean Operations and Systems, School of Business and Management, CHRIST (Deemed to Be University), Bangalore 560029, India
e-mail: mani4gift@gmail.com

S. Ramkumar
Department of Computer Science, School of Sciences, CHRIST (Deemed to Be University), Bangalore 560029, India
e-mail: ramkumar.drl2013@gmail.com

© The Author(s), under exclusive license to Springer Nature Singapore Pte Ltd. 2023   113
V. Sarveshwaran et al. (eds.), *Artificial Intelligence and Cyber Security in Industry 4.0*, Advanced Technologies and Societal Change,
https://doi.org/10.1007/978-981-99-2115-7_5

Definition of AI

A subfield of computer science, artificial intelligence aims to create software capable of tasks that were once reserved for people. Speech, vision, decision-making, interpretation, and even movement tasks similar to those performed by humans are all topics of study. Artificial intelligence has made great strides in areas such as image identification, machine vision, text-to-speech and speech-to-text translation, machine learning, natural language processing, robotics, and expert systems [4]. Scientific breakthroughs in machine learning, natural language processing, and speech recognition have had the greatest impact on IT service management.

The concept of AI has been around for some time. The study of AI has been ongoing since the middle of the last century. The current explosion in this field can be attributed in large part to the decreasing price of technology and data storage, the increased speed of processing, the scalability made possible by cloud services, and the availability of large masses of data collected through a variety of channels.

As a result of the exciting possibilities presented by artificial intelligence, businesses in a wide variety of industries are making strategic investments in the area. Phones, tablets, and other portable electronics, as well as everyday appliances, all make use of AI [5]. The fact that they can answer our questions and aid in our time management reflects the intelligence of these seemingly mundane technologies. Unfortunately, this intelligence is largely restricted to pre-programed answers to frequently asked inquiries or the application of collected data to the delivery of relevant and helpful information to consumers.

As time goes on, more and more data will be collected by these gadgets, fueling the learning process and inspiring fresh approaches to incorporating AI into our everyday lives.

Definition of AI

There are four distinct categories of artificial intelligence that can be defined by how closely they approach human intelligence [6, 7].

Artificial Narrow Intelligence (ANI)

Also known as "weak AI" or "narrow AI," these methods are only useful for tackling narrow problems within the confines of their intended functionality. Narrow AI excels at performing routine activities, often outperforming human beings at them. Technologies like Siri, Google Translate, and IBM's Watson are just a few examples.

Artificial Broad Intelligence (ABI)

It is also termed as broad AI, and it is a technology which integrates one or more ANIs or its functions to make decisions effectively. Self-driving cars, banking systems' analysis of customers' investment strategies, oil rig maintenance software, and so on can all benefit from enterprise-specific data training.

Artificial General Intelligence (AGI)

These are methods that enable machines to execute intellectual activities at the same level as a human, often known as "strong AI" or "deep AI." The capabilities of general artificial intelligence (AI) include theory of mind, self-awareness, the ability to grasp human beliefs, thoughts, emotions, and expectations, and social interaction and comprehension. General AI has the cognitive abilities of a human being, including the capacity to reason, to strategize, and to make plans based on emotional intelligence and prior knowledge. General AI may be aware of itself in a theoretical sense, but it cannot feel anything. However, at the present time, such progress in AI development has not been accomplished.

Artificial Super-Intelligence (ASI)

Methods that, in theory, is more capable and intelligent than people. There is a substantial effort required to achieve this state but various scientists and investors are inching toward the same. This can be a reality in short time than expected.

Application of AI in IT Service Management

As an IT service management professional, I see significant potential in incorporating AI into service management procedures [8, 9]. We can use the information we gather about our services, our customers' needs, and end-user problems and their solutions to fuel the development of AI-powered systems that will improve service design, delivery, and support. When AI is applied to IT service management, service providers can use automated decision-making tools to analyze data from a variety of sources and make informed choices.

Artificial intelligence (AI) has enormous potential, allowing service providers to enhance their customers' experiences, deliver better services that matter to their businesses, and free up their IT staff to focus on more creative and original tasks. ITIL 4 helps businesses create a comprehensive plan for integrating AI into all facets of IT service management [10, 11]. Artificial intelligence (AI) has the potential to significantly improve and expedite all IT Infrastructure Library (ITIL) processes, but these are specifically used in the domains such as IS security management, insights and knowledge management, measurements and reporting, change management, deployment management, and almost in every aspect of IT service management.

Artificial Intelligence for IT Operations

According to Gartner, "Artificial Intelligence for IT Operations" (AIOps) is the proactive, personalized, and dynamic knowledge gained via the use of big data, AI/ML, and other technologies to support all core IT operation functions. Companies now have easy access to the processing power necessary to run AI/ML algorithms because to the widespread use of cloud computing [10]. Network monitoring, project monitoring, diagnosis of events, incident reporting, analysis of performance, etc., are some of the examples of IT operational activities which can be governed through AIOPs.

Improved customer experience, service quality, and delivery agility are all things that digital technologies are helping to foster, and this is a major factor in the development of AIOps. With a well-thought-out AIOps strategy, businesses may boost output, get rid of risks, and gain a market edge thanks to faster product iteration times.

Change Management

The goals of change management are to aid in the development of change implementation techniques, the exercise of change control, and the facilitation of change adaptation. Effective planning and refined control systems will push change management forward. Organizational performance will be largely dependent on the vendors' ability to adhere to established procedures. As change management becomes strong, effective, and streamlined, it will be crucial to establish a high level of coordination with delivery teams. In addition, businesses should adopt a method predicated on the use of instruments for change management (i.e., preparation, evaluation, prevention, execution, and tracking) [12]. Tools also allow for simulations to be run, which can assist businesses plan for and prepare for any unforeseen consequences that may arise during actual deployment. Improving the success rate of the change and preventing human errors or unauthorized changes can also be accomplished by raising awareness of the importance of the change implementation activity and by ensuring readiness. Therefore, businesses should make sure their top brass is invested in the change management process so they can set the right example and inspire the necessary buy-in and cooperation which was shown in Fig. 5.1

Incident and Problem Management

Organizations can speed-up the process of resolving recurring events and restore service while still adhering to SLAs with the use of standard incident management

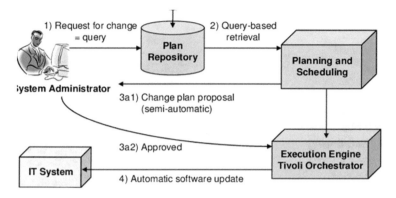

Fig. 5.1 Change management process [13]

models. Zero-touch incident management, in which human intervention is minimized or eliminated altogether, is an important development in the field of incident management [14]. As a result, businesses will be able to deploy automated problem detection, event correlation, ticket tracking, fault remediation, solution validation, and incident clean up to better serve their customers. The management of issues is another area where automation might be useful. During root-cause analysis, for instance, the gathered data can be parsed for insights and used to build appropriate action plans for avoiding a repetition of the incident which was shown in Fig. 5.2

Fig. 5.2 Incident management process [15]

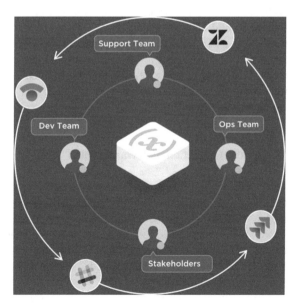

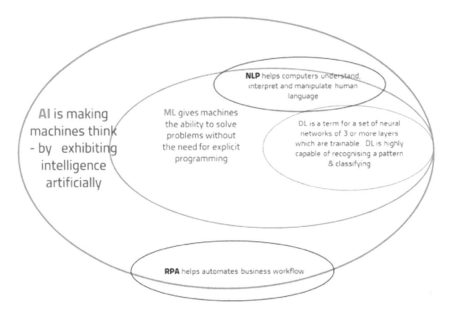

Fig. 5.3 IT asset management process [17]

IT Asset Management

As the digital revolution continues to shake up industries, it is more crucial than ever to have a solid IT asset management (ITAM) plan in place. From procurement to retirement, an IT asset management strategy can help organizations keep their IT assets turning over, keep track of their asset data, and meet their contractual commitments [16]. Robust processes for executing tasks, activities to ensure continuous improvement, and IT automation are all examples of IT asset management principle which was shown in Fig. 5.3

IT Service Integration and Management

Today, businesses often employ the services of several different companies to meet their various IT demands. The Service Integration and Management (SIAM) method [18] is used to coordinate and integrate service delivery among these suppliers. Thanks to the rise of cloud computing and other digital technologies, SIAM is playing an even more important role nowadays. To a large extent, the breadth and nature of a service determine which SIAM model a business employs. The four main SIAM models are as follows:

Client retained as SIAM, with retained firm overseeing and coordinating SIAM's supplier management and vendor relationships. Service Integration and Management (SIAM) that is "co-sourced" means that the client and one of their service providers share the workload. The primary provider who is responsible for the SIAM layer is the lead integrator. The primary integrator is accountable for both service management and SIAM layer management across all service providers. A third-party SIAM provider would handle the SIAM layer, while other service providers handled IT service delivery. There are benefits and drawbacks to each of the aforementioned models. Businesses must assess their management needs in order to determine which services they can handle in-house and which can be outsourced.

There are benefits and drawbacks to each of the aforementioned models. Businesses must assess their management needs in order to determine which services they can handle in-house and which can be outsourced. For the needs of the modern digital era, the old ways of providing IT services are no longer sufficient. Businesses that adopt the latest IT service management (ITSM) principles will see their IT departments become more strategic partners to the company, allowing them to gain a competitive edge and expand to new heights.

Artificial Intelligence for IT Operations

ITIL, and by extension IT service management, includes IT security management. Organizational security incidents can be avoided, found, and fixed with the use of AI. There are less risks associated with untrained or unmotivated personnel when AI is integrated into information security management to verify and predict potential threats and automate responses [19, 20]. When developing new services, it is important to keep AI in mind, as it can greatly improve the system's overall security administration. Imagine, for instance, a service that can compare abnormal readings from its own sensors with baseline data in order to identify and avert potential security threats. For any company or risk manager, information security management is mission vital. The use of AI to spot outliers in IT service management (ITSM) data, spot security holes, and rank problems in need of human attention is just the beginning. In order to better forecast, prevent, detect, and learn from security breaches, businesses must begin integrating AI throughout the whole value chain. This is because improving security is not limited to employing AI for detection and recovery from security-related incidents.

Artificial intelligence (AI) algorithms are used to improve the performance of many modern technologies, such as security systems and devices. Simply put, an algorithm is a set of instructions for a computer or system to follow in order to complete a given task. Artificial intelligence (AI) algorithms can perform a variety of tasks for security technologies, including the detection and identification of images or materials, the classification and matching of images (e.g., using computer vision to differentiate between a person and an animal), and the detection of anomalies in patterns of behavior (e.g., behavioral analysis in surveillance systems) (such as contraband or compounds in X-ray scanners).

Increased detection likelihood and quickness, decreased operator burden and tiredness, and better focus of security staff on critical areas are just few of the ways in which AI in security technology can improve operational security. Artificial intelligence (AI) has the potential to help managers save money, better allocate resources, make better decisions, and even head off potential internal threats by providing possibilities for early intervention. Basic forms of AI are used in many different kinds of security technology [21]. AI-powered devices that can perform multiple independent tasks at once may give the impression that they are more intelligent than they actually are. This does not necessarily indicate a greater level of AI, however, as intelligence tends to go along with the complex and combined decisions instead of different discrete programs a system shall execute.

The Security Technology-Artificial Intelligence Cycle

Artificial intelligence (AI) is used in all phases of the technology cycle used by security systems and devices, from receiving an input to making a decision and then acting on that decision. Figure 5.2 shows the STAIL cycle which conveys the congruence between the sensing, processing, deciding, and acting phases of security technology's life cycle and the input, computation, rule checking, and output phases of the AI life cycle.

In order to explain the security technology-artificial intelligence linguistic cycle, we can look at some real-world applications of security technologies. For example, the electromagnetic energy that is radiated by humans shall be used to create a change in thermal radiation (input/sensing) that can be detected by a passive infrared (PIR) sensor. An alarm state is generated (output/acting) when the rate of change is determined (computational technique/processing) to be greater than a pre-set value (rules/deciding). Whatever AI paradigm or approach is used, this fundamental cycle applies to it.

To the contrary, the statistical AI paradigm is where you will find a facial recognition-based biometric access control system. Facial recognition technology, for instance, relies on extracting and storing a person's facial features from a scanned image, albeit there are various methods by which this can be done [22]. When attempting to identify a person based on a scan of their face, the system first stores a template of that person's face and then uses computational techniques to determine the likelihood that a given scan would match the existing template. By releasing the door's locking mechanism (an example of an output/acting) if the individual's probabilistic matching score is above a certain threshold (rules/deciding), entrance is given. So, while the principals of operation for biometric facial recognition and PIR sensors are different and they operate inside distinct AI paradigms, they nonetheless have a common technical life cycle which was shown in Fig. 5.4

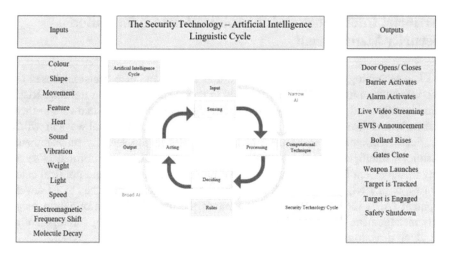

Fig. 5.4 STAIL cycle [23]

Operations Understanding Current Use of Artificial Intelligence in Security Technologies

Observe, detect, control, and response are the four main types of security technology, as defined by ASIS International (20,158). The following is a summary of the current state of artificial intelligence in security technologies across these areas:

- The vast majority of existing security technologies are rooted on the symbolic and AI statistics and high use of ANI approach.
- Machine learning in security technologies is in its infancy, with examples in areas such as video analysis over the network, biometric devices, acoustic systems, and RPAs all of which depend on a confined set of input which cannot interpret the inputs which are not known
- The accuracy and dependability of security technologies can be impacted by the fact that AI can make mistakes without providing any indication that a mistake has been made.
- The development of ASI technology in the IT security has not seen blooming in researchers
- Reviewing how AI is employed for observing, responding, and controlling through technologies can help to clarify as how its applications may vary between security technology categories and types.

Artificial Intelligence and Its Application to Information Security Management

AI is a technology which can self-learn, understand, and take decisions based on the information they acquire and respond to environment based on training. Risks are identified and prioritized by AI, allowing IT to quickly detect malware in their networks and devise a plan to respond to incidents. There are several ways in which artificial intelligence can be used to improve information security management. Artificial intelligence (AI) has challenges in its application to information security management that can be broken down into digital, physical, and political categories, with ANI, AGI, and AIS being the most common approaches to AI development (ASI) [24, 25]. This study examines the use of AI in information security management, weighing the pros and downsides, and outlining some potential research directions.

During the month of November 2018, Amazon had a data problem due to a breach of customer names and email addresses. As reported by the media, Amazon has admitted that a technical fault in one of its systems was to blame for the leak of customer data. A generic email was sent to users whose data had been compromised, informing them that the system had been secured and that they did not need to reset their passwords. Many users, however, incorrectly concluded the email was a phishing attempt. Experts disagree on whether or not the data breach was legitimate, despite Amazon's best efforts to conceal details about it. According to the Identity Management Institute (2020), mistakes are more likely to occur and hackers are given more opportunity to break into a system and produce a breach like the one Amazon encountered when it granted access to information beyond what any individual user actually needed. Businesses have invested in system upgrades and new technology in an effort to better protect themselves from the endless risks they confront as they go about their everyday operations [26]. Artificial intelligence is one such instrument that has aided several businesses (AI).

The past few years have seen tremendous advancements in artificial intelligence (AI). The progress made in this area has led to numerous useful applications. More and more questions are being raised regarding its security as it becomes more widely used in fields including as health care and government administration. Artificial intelligence, like any other technology, has vulnerabilities that can be exploited, posing a threat to the security of enterprises' data management systems.

Cyber-attacks have become all too regular in the global business world, thus many companies are incorporating artificial intelligence (AI) into their information security systems to lessen the danger posed by such attacks. The growing sophistication of data collection, data storage, and data processing within systems organizations has expanded the scope of the application. The way AI works is by teaching itself the strategies and patterns employed by corporate managers and hackers. As stated in [26, 27], the technology can identify any kind of deviation that may occur in the information management systems. Their crucial task is, therefore, detection and prevention.

Fascinatingly, AI aids machine learning by enhancing access to specific data management systems via learning from the data made available to it. AI has the potential to identify both current and future dangers that the company has yet to learn about through traditional means of information gathering. This is yet another way in which AI aids in decision-making. It aids in the development of resistant systems that can adapt to environmental shifts by supplying a number of variables that strengthen the overall deficiencies of the management structure. In the following sections, we will look at the role of artificial intelligence (AI) in information security management, analyzing its potential benefits and drawbacks and outlining potential research avenues.

AI Basics and Early Adopters

AI refers to computer systems that can learn new tasks, reason about new data, and take appropriate action based on that analysis. A modern application of AI can be broken down into three distinct types of operation. The most popular approach to artificial intelligence is the first one that was developed, called aided intelligence. Next, as a developing subset of AI, augmented intelligence (AI) has opened up previously unimaginable opportunities for businesses and individuals alike. Last but not least, autonomous intelligence is currently being developed and will be widely used in the near future [28]. However, in some countries, some aspects of autonomous intelligence are already being implemented. The use of self-driving vehicles, an application of autonomous intelligence technology, has risen in China and other industrialized countries recently.

Companies like Google, IBM, Juniper Networks, and Balbix were early adopters of AI and make heavy use of it in information security management today. For instance, Google Gmail employs AI to filter messages and recommend responses. IBM used AI on its cognitive learning platform to collect and organize data for use in identifying security threats. Last but not least, Balbix employs an AI-powered approach to risk prediction and vulnerability management via its Breach Control Platform. Balbix has developed a preventative measure in response to certain system intrusions. With this newfound capability, the company's cyber security staff can better manage the company's security frameworks, protecting it from threats like ransomware and other external attacks.

Methods AI is Used in Information Security Management

For the purposes of information security management, AI can be broken down into three distinct subfields: artificial narrow intelligence (ANI), artificial general intelligence (AGI), and artificial super intelligence (ASI) which was shown in Fig. 5.5.

Fig. 5.5 Classification of AI
[29]

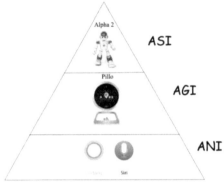

Forms of Artifical Intelligence

Artificial Intelligence (AI) Security Threats

Cyber-attacks can be divided into integrity, confidentiality, authenticity, and non-repudiation issues. Concerns about the safety of AI can arise from three main sources, depending on the specifics of the situation which was shown in Fig. 5.6

- One sort of advanced cyber-attack is known as "espionage," and it occurs when a person or company secretly gathers knowledge about a competitor's computer network in order to launch an assault on that organization [30]. For instance, a hacker can utilize an AI-based engine to go deeply into an IM system and learn more about it via internals like the datasheet.

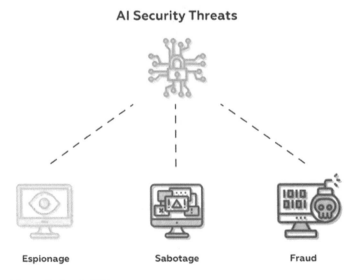

Fig. 5.6 AI security threats [29]

- Sabotage entails intentionally causing harm to an AI system, usually by corrupting its models or overwhelming it with requests it cannot fulfill.
- Fraud—It means mis-classifying roles through data poisoning.
- To affect a system's decision-making, fraudsters may potentially introduce false data or initiate a false interaction during the training phase.
- Malicious usage of artificial intelligence (AI) threatens information security management in many ways, classified under digital, physical, and political security.

AI Basics and Early Adopters

Threat occurs through social engineering are depicted in Fig. 5.7

- **Automated social engineering attacks**: With the help of natural language processing tools, it is simple to copy a victim's writing pattern. This opens the door for AI systems to gather personal information online and use it to send the victim harmful links, emails, and websites.
- **Vulnerability discoveries**: By analyzing historical trends, AI is able to find openings that hackers can exploit without the target's knowledge.
- **Complicated hacking**: AI has the power to automate the election of victim through prioritizing the vulnerabilities which allows the hackers to fasten the process of hacking.

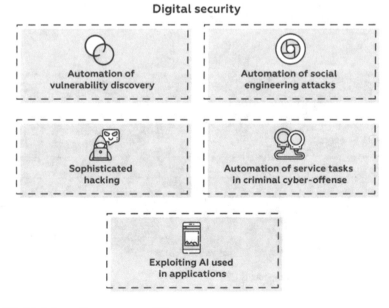

Fig. 5.7 Digital security components [29]

- **AI Automate Target**: AI can automate target victims' election by prioritizing their vulnerabilities, thereby allowing attackers to speed-up their hacking process.
- **Service Tasks for cyber—offenses**—Artificial intelligence (AI) can automate processes that disrupt the data pipeline, such as transaction processing [31].

Physical Security

These occur through advanced technique such as weaponized hard drive which can affect the systems physically which was illustrated in Fig. 5.8.

- **Fake news**—Artificial intelligence enables the production of visual content using methods similar to those used in the natural language processing of text. Since it is not always possible to determine the origin, it may also contain potentially inflammatory material.
- **Automated surveillance**—Without people's knowledge or consent, state and federal governments can use video and audio processing AI platforms to collect and use data about businesses and individuals.
- **Personalized misinformation**—The use of AI in social networks allows for the identification of important negative influencers or those who provide targeted misinformation through said networks.
- **Data manipulation and controlled behaviors**—Using artificial intelligence, sophisticated algorithms can alter data to persuade consumers to do a specific action. For instance, bot-driven large-scale denial-of-information assaults can be used against businesses or individuals to clog up their communication lines and delay the acquisition of genuine data.

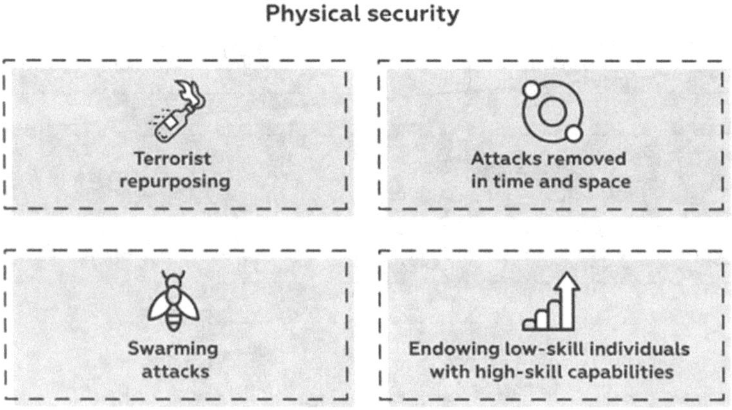

Physical security

Terrorist repurposing

Attacks removed in time and space

Swarming attacks

Endowing low-skill individuals with high-skill capabilities

Fig. 5.8 Physical security components [29]

Benefits/Areas of AI Applications in Information Security Management

Information technology (IT) asset inventory artificial intelligence (AI) is a method for gaining complete and accurate inventories of the people, machines, and software with access to a company's or person's network of computers and other electronic resources [32]. Furthermore, AI helps in quantifying and classifying businesses to facilitate simple stock management.

Threat Exposure

Systems based on AI can deliver the most recent data on both general and company-specific security risks. Companies can use this data to prioritize decisions that leave them vulnerable to attacks and make the necessary system adjustments.

Effectiveness Control

Understanding the interplay between the various security tools and processes in use is essential for organizations to keep their security steady. Here, artificial intelligence (AI) can help by developing a reliable InfoSec program that flags any existing holes.

Prediction of Breach Risks

AI can foretell how, when, and where a corporation will be breached in addition to managing IT asset inventories, assessing risk, and optimizing control. Artificial intelligence (AI) can foresee when a weakness will arise, allowing businesses to allocate tools and resources in advance to either mitigate the effect of the weakness or prepare for it in advance. Prescriptive insights from AI analysis, as stated in "using artificial intelligence in cyber security," can aid an IT security department in setting up and improving its control and processes. Over time, the business has strengthened its defenses against cyber-attacks.

Response to Incidences

When it comes to responding to or prioritizing security alerts, artificial intelligence (AI) systems can provide more helpful contexts. You can rely on them to set up a system for swiftly responding to emergencies and to aid in tracing the origins of accidents so that the same mistakes are not repeated in the future.

Explainability

Keys that can be used to augment human InfoSec, and hence explain analysis and suggestion, can be tapped by AI systems [33]. The procedure is crucial, as stated in "using artificial intelligence in cyber security." Management can use this opportunity to educate employees and external stakeholders on the processes, procedures, and technologies that go into keeping sensitive information safe.

Analysis

As time goes on, it becomes more and more difficult to detect security threats using only traditional security protocols and keep information systems safe. Many companies have documented incidents in which cybercriminals were able to bypass their firewalls in the recent past; these criminals are constantly looking for new ways to exploit vulnerabilities in computer systems [34, 35]. As a result, businesses need to take a more strategic approach to thwarting hackers.

With its wide variety of proven security technologies, artificial intelligence (AI) has emerged as a leading contender in the field of data management. Through the use of machine learning, AI systems are able to identify and report on irregularities in data. Machine learning is a subfield of AI that involves automatically identifying patterns in data and then extrapolating future outcomes based on the system's existing knowledge and experience. In some ways, the results produced by AI systems can be compared to those produced by humans in terms of reasoning and general cognitive ability.

Isolating the compromised data is a task that artificial intelligence (AI) can perform. These separations are crucial because they enable users or organizations to fix vulnerabilities and protect themselves from future malware attacks. Experts in the field of security have benefited from AI because it has helped them eliminate distracting background noise from their data. Because it has a firm grasp of the cyberspace, the technology can study the data presented and use that knowledge to spot unusual behavior. As a result, businesses may now leverage AI in three distinct areas of their information security management to bolster the effectiveness of their existing cyber security infrastructure and procedures:

Level 1: Prevention and Mitigation

Artificial intelligence (AI)-enabled systems are sophisticated can contain hidden protections that allow them to prevent data errors, losses, or access by unauthorized people. The systems use adaptable algorithms to aid administration in making decisions that better the IT department as a whole. When a system is brand new and has not been breached, it is easier to prevent attacks and detect them.

Level 2: Detection

It is common practice to use AI systems as reference points in order to learn how typical operations are carried out. Because signatures vary depending on the system under scrutiny, they are frequently used for detection. Signature recognition and regular updates are crucial to the operation of every system. Artificial intelligence [36, 37] believes that AI systems have built-in and networked sensors that can spot discrepancies in data during storage and transmission. When it comes to protecting users' privacy, data integrity, confidentiality, and authentication, monitoring software is indispensable.

Level 3: Response

This is the most time-consuming phase because it decides how effectively the AI is employed to counter threats. Using log files to conduct data searches is an example of the kind of human labor that can be intelligently replaced by AI. Robotics and AI make it simple to refocus efforts and develop new value-added endeavors founded on pooled expertise. By doing so, it can improve the intelligence community's ability to respond to threats from within or without the facility's walls.

In response, AI has been put to good use by the development of "intelligent traps" known as "honey potters," which generate a replica of the target environment in order to entice attackers, track down security flaws, and then rewrite the code to close them. Some networks can use AI-enabled dynamic segregation to divert attackers away from particularly vulnerable nodes. In a nutshell, AI systems boost information security management effectiveness by sifting through data to find high-likelihood signals and then concentrate on plugging the gaps where the system is most vulnerable. Because of its flexibility in data management and security, artificial intelligence (AI) is finding more and more applications in fields including education, health care, and industry, as reported in [38, 39]. Because AI is powered by machine learning, it holds the promise of enabling the development of error-free cyber centers in the future.

Conclusions

In recent years, artificial intelligence (AI) has emerged as a critical tool for supplementing human efforts in the management of information security. Artificial intelligence (AI) provides an excellent framework for analyzing, protecting, mitigating, detecting, and responding to gaps in security protocols that organizations use now that firewalls and other traditional technologies can no longer protect individuals' and businesses' dynamic information systems. Due to AI's ability to rank threats, IT departments will soon be able to detect malware in their networks in real time and devise effective countermeasures. Response to incidents, breach prediction, efficiency control, and inventory management are just a few of the many uses of AI in IS management that have been considered. Information security management may be made more stable, however, thanks to AI, which enables cyber security teams to create a robust and human-like machine that exceeds the bounds of their understanding of data management and security regulations. We need to give AI the ability to intelligently and automatically recover from possible risk, not just detect it. In this chapter, we will look at how artificial intelligence (AI) may be used to better anticipate, detect, and learn from security issues, as well as how ITIL might include AI more broadly into its IT security management practices.

References

1. White Paper on Artificial Intelligence: a European approach to excellence and trust. 1st edn. European Commission (2020)
2. Velliangiri, S., Manoharn, R., Ramachandran, S., Krishnasamy, V., Karthikeyan, V.R.P., Kumar, P., Abishek, K., Dhanabalan, S.S.: An efficient lightweight privacy preserving mechanism for industry 4.0 based on elliptic curve cryptography. IEEE Trans. Ind. Inform. **18**(9), 6494–6502 (2021)
3. Sangeetha Francelin, V.F., Daniel, J., Velliangiri, S.: Intelligent agent and optimization-based deep residual network to secure communication in UAV network. Int. J. Intell. Syst. **37**(9), 5508–5529 (2022)
4. Waguie, F.T., Al-Turjman, F.: Artificial intelligence for edge computing security: A survey. In: International Conference on Artificial Intelligence in Everything (AIE), pp. 446–450 (2022)
5. Azan Basallo, Y., Estrada Senti, V., Martinez Sanchez, N.: Artificial intelligence techniques for information security risk assessment. IEEE Latin Am. Trans. **16**(3), 897–901 (2018)
6. Zhang, Z., Hamadi, H.A., Damiani, E., Yeun, C.Y., Taher, F.: Explainable artificial intelligence applications in cyber security: State-of-the-art in research. IEEE Access **10**, 93104–93139 (2022)
7. Al-Suqri, M.N., Gillani, M.: A comparative analysis of information and artificial intelligence toward national security. IEEE Access **10**, 64420–64434 (2022)
8. Rizvi, S., Scanlon, M., Mcgibney, J., Sheppard, J.: Application of artificial intelligence to network forensics: Survey, challenges and future directions. IEEE Access **10**, 110362–110384 (2022)
9. Wu, H., Han, H., Wang, X., Sun, S.: Research on artificial intelligence enhancing internet of things security: A survey. IEEE Access **8**, 153826–153848 (2020)
10. Kim, H., Lee, Y., Lee, E., Lee, T.: Cost-effective valuable data detection based on the reliability of artificial intelligence. IEEE Access **9**, 108959–108974 (2021)

11. Gupta, B.B., Tewari, A., Cvitić, I.: Artificial intelligence empowered emails classifier for Internet of Things based systems in industry 4.0. Wirel. Netw. **28**, 493–503 (2023)
12. Radanliev, P., De Roure, D., Nicolescu, R.: Digital twins: artificial intelligence and the IoT cyber-physical systems in Industry 4.0. Int. J. Intell. Robotics Appl. **6**, 171–185 (2022)
13. Vogt, J.: Where is the human got to go? Artificial intelligence, machine learning, big data, digitalisation, and human–robot interaction in Industry 4.0 and 5.0. AI & SOCIETY **36**, 1083–1087 (2021)
14. Radanliev, P., De Roure, D., Page, K.: Cyber risk at the edge: current and future trends on cyber risk analytics and artificial intelligence in the industrial internet of things and industry 4.0 supply chains. Cybersecurity **3** (2020)
15. https://www.dnsstuff.com/change-management-process. Last accessed on 26 January 2021
16. Keleko, A.T., Kamsu-Foguem, B., Ngouna, R.H.: Artificial intelligence and real-time predictive maintenance in industry 4.0: a bibliometric analysis. AI Ethics **2**, 553–577 (2022)
17. https://www.bmc.com/blogs/itil-v3-incident-management/. Last accessed on 2 January 2021
18. Becue, A., Praça, I., Gama, J.: Artificial intelligence, cyber-threats and Industry 4.0: challenges and opportunities. Artif. Intell. Rev. **54**(5), 3849–3886 (2021)
19. https://www.itil-docs.com/en-in/blogs/asset-management/it-asset-management-best-practices. Last accessed on 20 January 2021
20. Binder, C., Neureiter, C., Lüder.: Towards a domain-specific information architecture enabling the investigation and optimization of flexible production systems by utilizing artificial intelligence. Int. J. Adv. Manuf. Technol. **123**, 49–81 (2022)
21. Kaupp, L.; Nazemi, K.; Humm, B: Evaluation of the flourish dashboard for context-aware fault diagnosis in industry 4.0 smart factories. Electronics **11**(23), 3942 (2022)
22. AKhan, I.U., Aslam, N., AlShedayed, R., AlFrayan, D., AlEssa, R., AlShuail, N.A.: A proactive attack detection for heating, ventilation, and air conditioning (HVAC) system using explainable extreme gradient boosting model (XGBoost). Sensors **22**(23), 9235 (2022)
23. Stavroulakis, G.E., Charalambidi, B.G.: Review of computational mechanics, optimization, and machine learning tools for digital twins applied to infrastructures. Appl. Sci. **12**(23), 11997 (2022)
24. Shahbazi, Z., Byun, Y.-C.: Analysis of the security and reliability of cryptocurrency systems using knowledge discovery and machine learning methods. Sensors **22**(23), 9083 (2022)
25. https://www.cimcor.com/blog/cybersecurity-lifecycle. Last accessed 12 January 2021
26. Yin, J., Wu, J., Gao, C., Jiang, Z.: LIFRNet: A novel lightweight individual fish recognition method based on deformable convolution and edge feature learning. Agriculture **12**(12), 1972 (2022)
27. Koblah, D.S., Acharya, R.Y., Capecci, D., Dizon-Paradis, O.P., Tajik, S., Ganji, F., Woodard, D.L., Forte, D.: ACM Transactions on Design Automation of Electronic Systems (2022)
28. Lin, H., Yu, Z., Peng, S., Bian, B.: Security issues in commercial application of artificial intelligence. In: 2021 3rd International Conference on Artificial Intelligence and Advanced Manufacture (AIAM2021), pp. 1174–1177. ACM, New York (2022)
29. Trifonov, R., Manolov, S., Tsochev, G., Pavlova, G.: Recommendations concerning the selection of artificial intelligence methods for increasing of cyber-security. In: Proceedings of the 21st International Conference on Computer Systems and Technologies (CompSysTech '20), pp. 51–55. ACM, New York (2020)
30. Iwendi, C., Ur Rehman, S., Javed, A.R., Khan, S., Srivastava, G.: Sustainable security for the internet of things using artificial intelligence architectures. ACM Trans. Internet Technol. **21**(3S), 01–22 (2022)
31. Patel, K., Sheth, K., Mehta, D., Tanwar, S., Florea, B.C., Taralunga, D.D., Altameem, A., Altameem, T., Sharma, R.: RanKer: An AI-based employee-performance classification scheme to rank and identify low performers. Mathematics **10**(19), 3714 (2022)
32. https://www.wipro.com/cybersecurity/how-to-make-artificial-intelligence-core-to-your-cybersecurity-strategy/. Last accessed 12 January 2021
33. https://www.ibm.com/thought-leadership/institute-business-value/en-us/report/ai-cybersecurity. Last accessed 15 January 2021

34. Tikhonov, A.I., Sazonov, A.A., Kuzmina-Merlino, I.: Digital production and artificial intelligence in the aircraft industry. Russ. Eng. Res. **42**, 412–415 (2022)
35. Oprach, S., Bolduan, T., Steuer, D.: Building the future of the construction industry through artificial intelligence and platform thinking. Digitale Welt **3**, 40–44 (2019)
36. Yu, J.Y., Kim, Y., Kim, G.: Intelligent video data security: A survey and open challenges. IEEE Access **9**, 26948–26967 (2021)
37. Chakkaravarthy Sethuraman, S., Mitra, A., Li, K.C., Ghosh, A., Gopinath, M., Sukhija, N.: Loki: A physical security key compatible IOT based lock for protecting physical assets. IEEE Access **10**, 112721–112730 (2022)
38. Chen, H., Zhang, Y., Cao, Y., Xie, J.: Security issues and defensive approaches in deep learning frameworks. Tsinghua Sci. Technol. **26**(6), 894–905 (2021)
39. Kassekert, R., Grabowski, N., Lorenz, D.: Industry perspective on artificial intelligence/machine learning in pharmacovigilance. Drug Saf. **45**, 439–448 (2022)

Chapter 6
Intelligent Autonomous Drones in Industry 4.0

Kriti Dwivedi, Priyanka Govindarajan, Deepika Srinivasan, A. Keerthi Sanjana, Ramani Selvanambi, and Marimuthu Karuppiah ⓘ

Introduction

The drone sector is one that is expanding quickly. Devices that were initially made for military use have changed and now have a wide range of commercial uses. Adaptable computer systems can execute various data processing and pattern matching algorithms, and a drone is currently a smart, automatic guided, as well as a partially self-aware device that could function both independently and in a swarm. They are currently available for purchase at consumer electronics stores and are widely used and distributed throughout the world. Around the world, drones are used in a variety of fields. A wide range of industries, including farming, disaster response, border security, asset monitoring, traffic monitoring, and many others use drones extensively. Their use is now soaring in popularity. This is mainly because of the

K. Dwivedi · P. Govindarajan · D. Srinivasan · A. Keerthi Sanjana · R. Selvanambi
School of Computer Science and Engineering, Vellore Institute of Technology, Vellore, Tamil Nadu 632014, India
e-mail: kriti.2020@vitstudent.ac.in

P. Govindarajan
e-mail: priyanka.govindarajan2020@vitstudent.ac.in

D. Srinivasan
e-mail: thirumani.deepika2020@vitstudent.ac.in

A. Keerthi Sanjana
e-mail: keerthisanjana.a2020@vitstudent.ac.in

R. Selvanambi
e-mail: ramani.s@vit.ac.in

M. Karuppiah (✉)
School of Computer Science and Engineering and Information Science, Presidency University, Bengaluru, Karnataka 560064, India
e-mail: marimuthume@gmail.com

quick developments in mobile embedded computing, which enable the integration of different sensors and controllers on drone systems, allowing them perceive but also comprehend both internal condition and the environment around them. The most recent use of drones was for riot control, infection surveillance, face detection, and logistic purposes during the global pandemic [1].

Drone technology is quickly developing and holds the ability to both interrupt and elevate our standard of living. The majority of drones have two different flying modes. By configuring the flight schedule and linking this to the drone's computer prior to take off, a drone can fly in the first mode entirely autonomously without any human input. Consider a case where a drone finds an internal pipeline leak. It will give the operator a head start. Drones technologies are increasingly used for public security tasks, including rescue missions and disaster zone management. In comparison with manned aircraft, drones have many benefits for rescue missions, including faster deployment, increased maneuverability in confined locations, superior sensing technology, and ease of usage. Drones are less rigid and more affordable than utilizing manned planes. Drones are also utilized in disaster mitigation since they can swiftly and easily cover large region, giving interested parties like first responders, law enforcement, and cleanup personnel a highly detailed birds-eye picture of the area that will allow them to proceed with greater knowledge and efficiency. Over the past few years, drones have become more useful in people's daily lives. Due to the spread of the disease and the development of new technologies, there is now a larger need to create new means of accessing goods and services.

Drones are gradually becoming a part of everyday life. They have also helped with attempts to fight fires, find the missing, and perhaps even preserve the environment over the past couple of years. It is realistic to assume that a large range of prospective advancements and innovative features will lead to a wide range of future uses for drones. In the next ten years, drones, a game-changing technology, will definitely evolve in ways you never imagined conceivable. Since drone technology is constantly developing and they are already used in many different industries, the future of drones is highly promising.

Scope of Study

This chapter "intelligent autonomous drones in Industry 4.0" focuses on design of drones, reactive control, obstacle detection, drones application, security issues, and legal uncertainties; finally, the case studies of real time are included to give in depth details on intelligent drones.

Design of Drones

Drones rely heavily on sensors. With the help of these sensor technologies, drones can perceive and comprehend their surroundings. The environment they are flying in may allow them to take evasive and proactive activities. In order to guide the drone's mobility in relation to its environment, including its spatial location, speed, as well as direction, sensors that help the drone fly are largely employed to measure the drone's condition [1]. Numerous sensors are needed for the completely autonomous operation to monitor and understand the atmosphere, with a collection of procedures to analyze the sensor data. Effectively resolving these problems presents significant hurdles, primarily as a result of the rapid expansion of technology.

Sensor Technologies

Accelerometer: To locate and orient the drone while in flight, accelerometers are used.

Gyroscope: By detecting angular velocity in 3 dimensions, the gyroscope helps calculate the amount at which angles in pitching, rolling, and yaw change.

Electronic indicator: Drones can navigate using the compass to determine its location. It offers three-dimensional information about magnetic fields.

Barometer: The barometer measures altitude and aids in drone navigation.

Global Positioning System (GPS): Users that have GPS devices have accessibility to geostationary satellite position data.

Drone Platform

A variety of payloads can be carried by a drone platform, ranging from relatively complex payloads (such as a video camera system having adaptive zoom support) to simple payloads (like a fixed camera with fixed lenses having a mass as less as 200 g) as shown in Fig. 6.1.

Payload: The qualities and requirements of the drones can change depending on the varied drone payloads utilized for each application. Drones may possess a very small payload capacity because the payload weight may severely influence the duration of the flight. This restriction is largely determined by the power of the combination of the motors and propeller. The drone's primary battery's capacity also becomes significant when the drone fuels the payload, cutting down on total flight duration.

Fig. 6.1 Drones in action

Another option to be considered is the drone's level of autonomy, which ranges from fully unmanned operation to complete remote pilot control [1].

Radar: Radar systems use the echoes that return from radio waves sent out to build topographical maps or navigate around obstructions. They can be made to scan huge regions quickly and are less susceptible to weather, smoke, and dust than LiDAR and cameras.

Telecommunication sensors: Drones with telecommunication hardware, including Wi-Fi antennas, may be able to provide information about a person's whereabouts.

Electronic sniffers: Tools that sample the air and report on any harmful, radioactive, or other compounds present in the region to track pollution.

Flight Concept

Flight requires a lot of energy, especially as the size of the device is reduced. This is frequently the outcome of practical problems that occur when a vehicle's size is reduced, such as a decline in electromagnetic motors' power density, a reduction in efficiency of transmission brought on by an increase in friction due to gears and bearings, and an increase in viscous losses brought on by a decrease in Reynolds numbers. At the most fundamental level of independence, scaling problems create barriers that make it difficult to simply maintain flight for long enough to complete higher-level mission tasks. In order to overcome these challenges, drone designers must balance the trade-offs occurring due to choosing the components of the vehicle. The majority of small aircraft are rotorcraft or fixed-wing, propeller-driven models. The latter has categories such as quadcopter variations which have lately gained popularity, coaxial twin rotors, and traditional primary and tail rotor layouts. For each of these layouts, as well as for bigger propeller-driven aircraft, there have been relatively minor incremental advancements in the design of the propeller over the past few decades.

Electromagnetic motors are frequently used in rotorcraft and fixed-wing drones with propeller propulsion. Few aircrafts having flapping wings also make use of electromagnetic motors, but a linkage system is required in order to translate the rotating action of the motor to the flapping action of the aircraft's wings. As the scale gets smaller, conventional motors lose efficiency, become extremely difficult to build, and require enormous gearing in order to achieve the proper wing or propeller velocity. Therefore, extra activation mechanisms are needed for vehicles weighing less than a few grams. In addition to the issues posed by the selection and production of actuators as the vehicle's size decreases, there are also concerns regarding the vehicle's overall production.

Conventional production methods including composite material injection molding, additive printing and subtractive machining, are used at scales where electromagnetic actuation and bearing-based rotary joints are feasible. Combining aerial as well as terrestrial talents can be useful in various situations, including search and rescue, parcel delivery especially in constrained spaces, and environmental sensing.

Reactive Control of Autonomous Drones

The concept of reactionary control, which outperforms the time-triggered method, is presented by making use of the characteristics of current control circuitry and the hardware on which it operates. Reactive control bases control decisions on variations in the navigation sensors that have been noticed and is only used when necessary [2]. Hence, depending on the situation, the execution rate varies. This enables us to achieve more timely control decisions compared to time-triggered control, better hardware utilization, and reduce the requirement for excessive control rate provisioning. There are two elements involved. At a ground-control station (GCS), specialised software allows users to set mission parameters, including the coordinates to cover with waypoint navigation and the action to take at each waypoint. The measures to be taken at every waypoint are set by users at a ground-control station (GCS) using specialized software. The GCS is typically just a standard computer connected to the drone by a low-bandwidth and long-range radio [3].

The drone's autopilot software is in charge of providing the low-level control necessary to move the drone toward the next coordinate on its own. A range of sensor inputs, including accelerations and GPS coordinates, is fed into the control loop. It processes them so that actuators, such as electrical motors, can be operated to establish the drone's attitude, which is another word for the drone's three-dimensional orientation. Autopilots rely on resource-constrained embedded electronics because of their small size, high prices, and high energy consumption.

Most drone PID controllers tweak the proportional component to dominate. Balanced weights reduce the derivative component [3]. Navigation sensor calibration can minimize the integral component. As a result, executions are produced in which: (i) modest differences in inputs of the sensor end up matching small variations in settings of the actuator; and (ii) as long as sensor inputs do not vary, settings of the actuator stay essentially unchanged. Autopilot software works on phoneslike devices, especially sensing equipment. Drone technology may not have existed without cell phones' effort to improve sensors. The sensing equipment of drones

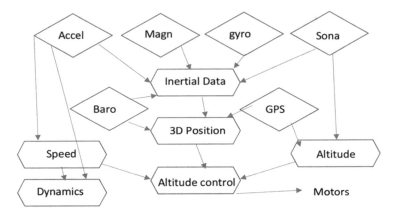

Fig. 6.2 ArduPilot control loop

are designed to be energy-efficient and of high-frequency. We regularly check the navigation sensors and only perform the control logic when needed. Reactive control dynamically adjusts control rate. Control runs frequently when sensor inputs vary, maybe more frequently than a time-triggered solution. Control execution slows when sensor inputs are stable, releasing resources for other uses. Drone platforms depend on autopilots. Their implementation generally requires dedicated hardware as shown in Fig. 6.2.

Example:
ArduPilot is a robust open-source autopilot system for drones and ground robots. Embedded hardware runs ArduPilot. They offer interrupt-driven modes for energy-efficient high-frequency sampling. Pulse-width modulation encodes autopilot actuator commands (PWM). Electronic speed controllers turn them into motor current (ESC). Since autopilots must constantly alter attitude, we chose aerial drones. For example, a ground robot might not conduct a control loop once it has achieved its goal because it usually keeps its location. The drones have a Waspmote that runs on its own power and samples Hall-effect sensors at control loop speed. The hexacopter is more resilient due to its larger frame and stronger engines. As a result, the control logic makes more decisions that are the same as the quadcopter.

Issues and Solutions:
Identifying sensor input changes that affect control decisions is the challenge. We record a sensor trigger when a sensor input changes and the control decisions change. Finding a trigger in sensor readings is difficult to predict. These circumstances depend on sensor precision and calibration, drone characteristics, control logic, and actuator output granularity.

To solve this problem, an online lightweight approach that automatically adjusts itself to numerous cases of the characteristics listed above is selected. This strategy does not assume any expertise of any of the aforementioned aspects, such as control logic. Because of this, our solution is applicable to a wide variety of implementations.

PID controllers for autopilots make the assumption that sensor readings are taken in parallel at a constant rate. The control problem is impacted when assumptions are removed, which necessitates developing a new theoretical model.

Legal Uncertainties

A hypothetical but plausible situation featuring an autonomous drone that falls from the sky and exposing persons on the ground in danger is investigated from the points of view of both German and English private law. The issue of legal ambiguity is something that developers have to deal with, and this real-world example is used to demonstrate the problem.

RAPID Project

The risk-aware autonomous port inspection drones (RAPID) project will, in accordance with a pan-European research effort on autonomous drones in particular, give an outline of a number of the unpredictability encountered during the procedure of autonomous drones in overall. This project is part of the risk-aware autonomous port inspection drones (RAPID) initiative. This project is being used as an example to show the legal problems that developers of autonomous drone technology confront, with a specific emphasis on civil accountability. When it comes to the widespread usage of autonomous drones, one of the most important requirements is safety. As a consequence of this, unmanned aerial vehicles (UAVs) must be capable of detecting and evading (D&A) any potential obstacle in their path of flight, and this must take place independently of the intervention of any operator.

Successful Deployment of ANN/DLNN

In order to successfully deploy an ANN or DLNN in the field, it is essential for the network to be able to correctly classify both known inputs and inputs that have not been identified before and that are present in the execution environment. One example of this would be a bird flying close to the drone. This skill can be described using a variety of terminology, including underfitting and overfitting. A neural network is said to be overfit if it is able to perform its operations with a fair level of accuracy on the dataset it was trained on, but it performs poorly when it is given fresh input.

Regulations

The legal gray area that surrounds the use of drones is not the result of a deficiency in legislation; rather, it is the result of confusion regarding how to apply already-existing laws to this new technology as well as disparities in the legal systems of individual nations. There are a number of additional aspects that give rise to worries, including the possibility for drones to make decisions that appear to be made independently and how the law treats this capability.

Operators of drones and distant pilots should take precautions to ensure that they have received adequate training on the relevant union and governmental rules related to the proposed operations. This education should focus on topics such as safety, confidentiality, data security, legal responsibility, insurance, security, and the environment. In the event that an autonomous drone was to fall from the sky, it is helpful to examine who would be accountable for damages caused by drones, as well as the conditions and the breadth of this obligation. Doing so can help show the issues that operators of drones are likely to confront. In any event, it seems that the company would be responsible for any injuries caused regardless of how the drone was handled at this time when an employee is flying one on behalf of the company when an accident occurs when the employee is operating the drone. In these kinds of situations, the only way for the owner to avoid complete responsibility is to provide evidence that the person who was hurt was partially responsible for the disaster.

English and German Law

The law in England that governs drones that fly independently is still in its infant stages at this point. Because of the potential for legal repercussions in the event of irresponsible drone operation, it may be required in some situations for the owner of a drone, or in some cases, the operator of the drone to provide evidence that the drone was not used in a negligent manner. The Civil Aviation Act of 1982 makes it clear that stringent accountability must be demonstrated in cases involving third-party liability. After looking at the situation from the points of view of English and German private law, it is clear that under English law, culpability in this case is primarily based on the law of torts. This can be seen now that the situation has been examined from the perspectives of English and German private law. If a tort has been committed, the owner of the property has the legal right to submit a claim for damages against the person who perpetrated the tort.

And if we consider things from the German point of view, According to German law, the two regimes of civil liability that specifically apply to the scenario of a drone falling from the sky are fault-based liability under Section 823 and subsequent sections of the Bürgerliches Gesetzbuch, and strict liability under the Luftverkehrs-gesetz (LuftVG; German Air Traffic Act), which establishes a special liability regime for aircraft (including drones). The Luftverkehrsgesetz (LuftVG; German Air Traffic Act) also establishes a special liability regime for aircraft.

These countries were selected to illustrate the geographical diversity of the RAPID project's Member States. However, they also serve to emphasize the dramatic dispar-ities in legal systems that exist among countries whose attitudes to law and order are generally similar to one another.

Drone Obstacle Avoidance

The objective should always be to keep the drone from getting itself into precarious circumstances such as corners and getting trapped. In continuation of the previous point, it is typically necessary to have a skilled pilot in order to operate these drones in more complicated conditions such as inside. On the other hand, it is not always possible or desirable to have a human pilot in control of a drone. Due to the constantly growing demand for drones, it is essential that they be able to fly autonomously. There is a considerable challenge presented by the drone's capacity to automatically and dynamically navigate around obstacles in tough environments.

Methodology:
Only very small and lightweight sensors are used by the drone. These sensors identify impediments and allow the drone to automatically avoid them. It is constructed on a synaptically flexible direct two-neuron looping network. Because of brain dynamics and network synaptic plasticity, the drone is able to successfully adjust its obstacle avoidance behavior to successfully navigate complicated landscapes containing obstructions, curves and bends, and dead ends [4].

It is possible that the network is made up of three primary neurons that operate in discrete time. Signals are transmitted to the device from sensors that are attached to the front of the drone. These sensors are used to detect impediments in the path of the drone. It is possible to set the range of each sensor to around 50 cm. After then, the raw sensory data can be mapped to the range $[-1, 1]$, where $[-1]$ indicates that there are no obstacles in the range, and $[1]$ indicates that there are obstacles nearby (approximately 20 cm). After then, the output of the control network may be repurposed into a yaw instruction that directs the behavior of the drone. If this configuration is used, the drone will be directed to turn right if it detects an obstacle on the left, and it will be directed to turn left if it detects an obstacle on the right. The yaw value will be positive if the drone detects an obstacle on the left and negative if it detects an obstacle on the right.

Parallelization of Local Path Planning for High-Reliable Autonomous Drones

This focuses on the planning of localized paths to avoid barriers and proposes a rapid method to obtain the paths in advance for a variety of scenarios. It is necessary to make a plan on how to navigate around the obstacles in order to avoid them. It is difficult, in terms of the amount of computation time required, to estimate the courses for several different scenarios in advance because obstacles could emerge in a variety of different ways [5]. Parallelization of the state lattice planner, which is one of the approaches for path planning, is something that should be considered in order to prepare for a variety of potential outcomes.

One method for getting started with this is called final state sampling, and it involves taking a sample from each of the sites on the map that are candidates for the next state's placement. In order to build a path for the sampled final states, one can make use of an appropriate algorithm. By analyzing the trajectory of each generated path, it is then possible to assess whether or not the path will come into contact with an obstruction. If it does not, the next step is to select the route that will incur the least amount of damage to the system. When computing the paths in advance for a variety of circumstances, the optimization phase of the path planning algorithm that is being utilized will be the component of the process that moves at the slowest pace. The process of generating the path, which requires time-consuming optimization, needs to be streamlined in order to accomplish the goal of distributing the produced path among the several possible outcomes. The suggested method is responsible for initially creating the sample paths, and after that, the sample paths are assigned in order to arrive at the final states. After ensuring that there are no collisions along the pathways leading to the final states, the evaluation and selection steps can then be carried out. Because of this, it is now possible to build short pathways for a variety of different scenarios.

It is therefore possible for the initial state lattice planner to make a comparison between the time taken and the time taken by the applied algorithm to obtain all paths for various conditions.

Short-Range Telemetry Communication for Autonomous Drone Navigation

We can switch out the traditional telemetry modules used in dedicated telemetry systems with the less expensive HC-05 Bluetooth modules for short-range telemetry systems. The Bluetooth modules can configured as a master–slave pair so that the RPi and flight controller may communicate with one another. Some researchers have utilized autonomous flight and mission planning using traditional Pixhawk boards. The Pixhawk Mini 3DR (PM3DR), a lighter model more suited for small-sized drones, was used for this study. To assure seamless communication, it was necessary to establish protocols and calculate baud rates through try-and-error tests. Additionally, a code for image processing was created that computes the centroid of the target image and returns the location of the coordinates of the target for a secure landing. The 10 m communication range provided by Bluetooth telemetry modules is adequate for testing the drone's software and functionality during autonomous flying [6].

Autonomous Drone Guidance and Landing System Using AR Markers

Visual markers are measurement tools that are low-cost, slim, and light and do not require any power. The ID number can be identified with just one camera that is installed on a drone by using the drone's relative location and attitude. A hybrid marker (HM) that is comprised of an ARToolKit, a large conventional AR marker, and a tiny, extremely accurate marker, in addition to a marker post that has the HM attached on either side of a hexagonal cylinder is proposed. A marker post is typically placed near the geographic center of a landing base [7]. When a large ARToolKit of an HM is spotted in the distance by drones, this serves as a cue for the drones to head in the direction of the marker post. When the drone is getting close to HM and it detects a little amount of LentiMark, it will then carry out landing control based on its position within the HM coordinate system. This technology is suitable with any drone that is equipped with a camera. Due to the fact that the only main piece of equipment that makes up the system is a marker post, it is exceedingly compact and portable. In addition, the approach enables precision landing thanks to LentiMark's positioning inaccuracy, which is only about 10 mm on average.

LentiMark is a visual marker that has a very high degree of accuracy. The most significant drawback associated with standard augmented reality markers is that their pose estimate accuracy suffers when viewed from the front (lenticular angle gages). LentiMark was able to solve yet another challenge associated with AR markers, known as pose ambiguity [7].

Detecting Rogue Drones

In order to monitor and keep track of aircraft, researchers are developing new technology for air traffic management. The rapid development of new technologies offers a primary source of the many challenges that must be surmounted before these issues can be effectively resolved. The majority of current efforts are focused on legislative and technological attempts to regulate operational procedures. As a component of the envisioned technology, which is the primary focus of this article, a variety of land and aerial systems for identifying rogue drones are now being investigated and tested. In order to monitor and keep track of aircraft, researchers are developing new technology for air traffic management.

Drones Application

Artificial Intelligence: The interesting marriage of artificial intelligence (AI) and unmanned aerial vehicles (UAVs), sometimes known as drones, is gradually becoming more widespread. This is especially true when it comes to the logistical issues of transferring products and equipment that must be dispersed in the event of an emergency. This is due to the fact that it is both affordable and maneuverable, and it is also somewhat quick. In addition, leveraging artificial intelligence (AI) capabilities, the drones in Wuhan, China were able to identify individual persons, which made it much simpler to direct the dispersal of the crowd and the isolating of specific areas. Drones are quickly becoming a significant factor in nations, particularly in the development of smart cities. This is a direct effect of the substantial income that they bring in [8].

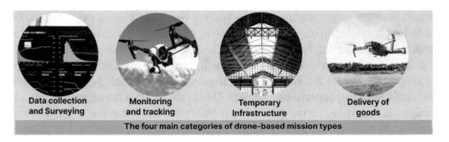

Fig. 6.3 Drone-based mission types

On the other hand, the safety of these technologies has been completely over-looked. This presents a challenge because it is possible for attacks such as wireless threats, hijacking, and man-in-the-middle attacks to be carried out against these devices. There has been a breach in data security, which has led to the exposure of both equipment and medicinal supplies. Moreover, because of this, these devices are susceptible to malicious hacking and the theft of private information such as voice and facial identification data, which might be used for private authentication mechanisms. Because of this, more security is necessary because it will be detrimental to both the general population and governments that make use of AI and drones as shown in Fig. 6.3.

Medical use: We can claim that the roles that autonomous drones have been used in during COVID-19 have been favorable ones because they have interacted with the populace during time-sensitive circumstances. The pandemic's most notable job is transporting emergency medical supplies to abandoned buildings or to persons who cannot go to the designated place to pick up these supplies [9]. Because the UK Government mandated limited movement to prevent the virus from spreading and reduce interaction with other people, the United Kingdom (UK) used these devices as a safer alternative to prevent the sickness from spreading. Researchers in Australia collaborated with Dragonfly to create drones that can measure body temperatures and look for signs of viral diseases like COVID-19. The drones can detect symptoms from a person's location as close as ten meters away and relay this information in real time to the necessary individuals [10].

E-commerce use: Using drones that travel at a speed of about 65 miles per hour, the Google-owned business "Wing" has carried numerous parcels across the USA in thousands of automated trips. When performing logistical duties, these drones have proven essential, especially since they reduce person-to-person contact.

Monitoring the road transportation network: Road network condition monitoring is essential for effective traffic management. Various metrics can be obtained with current sensing technology, including vehicle counts at a particular site, traffic densi-ties along a stretch of road, and speed and position from moving cars. Considering that drone technology is non-intrusive, does not cost anything to deploy, and frequently

Fig. 6.4 Sectors utilizing drones

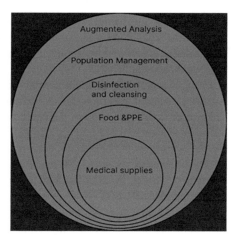

offers spatiotemporal measurements for the entire network, it is developing as a potential solution that can solve these constraints.

Examination of power lines: The ability to deliver electricity without interruption is a basic necessity. However, it is a challenging and expensive undertaking to adequately examine and maintain a nation's extensive, outdated, and frequently inaccessible power line network in order to detect or avoid the emergence of faults. Drones offer an alluring substitute for manual inspection that can boost productivity and cut down on inspection time and expense. By employing visual equipment to patrol the lines and look for problems that need to be rectified, they assist.

Civil defense: Future autonomous first responders for civil defense protection are thought to be drones. Drones currently support firefighting operations, strengthen the capabilities of troops responsible for civil protection, and speed up the reaction time of rescue teams. Drones help with a number of tasks connected to the early detection of anomalous conditions, disaster preparedness for different natural or man-made disasters (such as forest fires, floods, earthquakes, or landslides), and search and rescue operations in this context as shown in Fig. 6.4 [11]. A collection of drones can collectively carry out a task significantly more efficiently by using novel swarm optimization approaches and autonomous-control algorithms [12]. Additionally, there will be countless chances to use machine learning to enhance drone perception abilities by utilizing the vast volumes of data that are becoming accessible for a variety of purposes.

Drones in Manufacturing

Drone technology has been the subject of extensive research about its technical capabilities (such as robotics, control, and computer vision), but there has been far less study regarding the actual use of drones in industry. There is now a disconnection between the technological advancements of drones (i.e., "what drones can do") and the profitable manufacturing applications they can provide (i.e., "what it makes sense for them to accomplish").

From a theoretical standpoint, drone technology can be categorized as an AMT. A computer-based invention utilized in manufacturing processes is called an AMT. The use of drones falls within the category of AMT. A computer-based invention utilized in manufacturing processes is called an AMT.

AMTs are a variety of contemporary manufacturing technologies, including flexible manufacturing systems, CNC machines, and industrial robots (FMS). An aircraft, a remote controller, installed payload (cameras, sensors, carriers, etc.), local navigation support system, energy supply system, and information technology (IT) infrastructure make up a drone system's hardware. The software is made up of the scripts and algorithms that manage the flight and payload operations of the drone and allow it to interact with the controllers, navigation system, and IT systems.

An AMT application might be "stand alone," linked to other technologies, or integrated with them. The majority of today's drone applications are standalone and made to do a single task. Examples include using video and infrared cameras in the oil and gas industry to inspect difficult-to-reach equipment. AMTs that are connected together perform their own jobs while simultaneously coordinating with other technologies. AMTs that are integrated into other technologies must perform their functions. Currently, relatively few drone applications are eligible to be incorporated AMTs. Applications used in inventory cycle counts to automatically update inventory data in a warehouse management system are an example of a linked drone system. AMTs that are integrated into other technologies must perform their functions. Currently, relatively few drone applications are eligible to be incorporated AMTs.

The ability to perform physical tasks is an AMT's physical capability. Readers and sensors for RFID, as an example. We refer to an AMT's capacity for data processing as its analytical capabilities. High analytical capability and low analytical capability can be distinguished by different levels of data processing during use, which is not always related to lengthy programming. For instance, a straightforward special-purpose CNC machine has a high programming setup, but it often just executes its program while in operation, making it an AMT with limited analytical power. For drone systems, it makes logical sense to divide an AMT's analytical capabilities from its physical capabilities. Simple photography or filming is instances of input data that is not processed by a drone with minimal analytical capability. The input data are converted to different forms of data or information by a drone with strong analytical capabilities. For instance, a drone system with a thermal camera collects input data and generates images of temperature radiation in the form of thermal data. A drone with limited physical strength can only fly; it cannot carry out any other physical tasks. A drone with strong physical capabilities may move objects (in addition to the mounted cargo) (such as delivering packages, parts, or tools), as well as perform physical operations in addition to flying (e.g., spraying chemicals or repairing scratches).

Advantages of Using Drones in Manufacturing

Drones can significantly reduce costs, especially in manufacturing facilities in inspection-intensive process industries. Drone inspections minimize the quantity of labor-in depth paintings and reduce scaffolding too. The speedier completion of duties is a related potential advantage. Inspecting drones for difficult-to-reach installations and equipment speeds up operations thanks to its quick setup and greater agility. Drones can help numerous operations reduce dangerous jobs. Drones can take the place of manual human inspection of dangerous and difficult-to-reach equipment in particular. Applications for drones face technical difficulties caused by limitations in current battery technologies. Operational issues then follow. Alert and capable pilots are needed for manual operations. Pilot fatigue from long flights can easily lead to human error. Even experienced pilots find it difficult to maneuver around gates, doors, pillars, ventilation systems, fire suppression systems, cranes, utility gates, and large machines in production environments. The requirement for skilled drone pilots who cannot only fly drones safely but also have a thorough understanding of the duties and objectives involved is one of the organization's problems. The fourth difficulty relates to statutory laws and regulations.

Autonomous Aerial Counter-Drone System

The scientific and industrial communities have shown a significant amount of interest in unmanned aircraft systems (UAS) because of their potential to revolutionize a wide range of application scenarios. However, as UAS capabilities continue to advance with better intelligence and autonomous characteristics, there is a chance that public areas and vital infrastructure could face new dangers. Importantly, there are currently insufficient methods to reliably find, track, and safely intercept rogue drones. Detection methods such as RF signal sniffing, sensors, and computer vision (FAA) are used as shown in Fig. 6.5. The counter-drone system launches with the intention of locating the errant drone and stopping its activity using mobile phone jamming, importantly to keep your bearings.

The pursuer drone uses a cutting-edge method based on indications of opportunity to self-localize itself in space. Stationary CUAS systems have constant access to energy resources, which increases their surveillance range. A portable FH system has

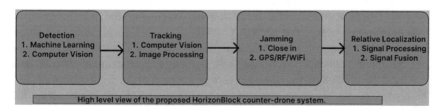

Fig. 6.5 High-level view of the proposed horizon block counter-drone system

the advantage of mobility. However, due to their small size and simple architecture, they cannot autonomously detect and locate UAS. Airborne solutions are installed on UASs and can patrol specific areas to detect and track nearby malicious UASs before neutralization takes place. Moving on to detection and tracking: commercial UASs on the market have flight times of 30 min or less, so scanning an area for potential threats is not very effective.

Steps

Creating a dataset and tagging drone image data are the first steps in the detection process. Currently, a dataset of about 700 drone photos that were captured from various angles and at a distance of up to 30 m is used. After detecting the rogue drone, the counter-drone system's tracking algorithm is engaged. In order to achieve fast tracking, the detection's bounding box is utilized. In this work, a jamming system is implemented using an software-defined radio (SDR) and LabVIEW software. Using LabVIEW software to generate interfering baseband signals, a USRP B200 SDR radio card paired with an omnidirectional antenna transmits these signals to jam the GNSS receiver [6]. When jamming, the jammer emits jamming signals on specific radio bands to disrupt the communication link between the transmitter and receiver. Also, these methods of interference usually aim to send signals from the drone that is being chased all the time. Field tests were conducted to validate the proposed RPS algorithm, using different transmission bands and the proposed RPS-OL. The SDR module aboard was fitted with a broadband antenna, allowing for the collection of SOP signals over a wide frequency range. To put all of these anti-drone features on a single UAS agent, hardware and software solutions are used. For hardware integration, a processing unit that is physically small enough to be implanted on the agent but computationally powerful enough to meet all needs is required in order to incorporate all previously described methods into a single automated system. The processing unit must simultaneously work with all mechanisms and four gigabytes of LPDDR4 RAM.

In addition to the onboard processing unit, the system also requires the necessary hardware to enable communication between the onboard processing unit and the UAS. This allows you to control the UAS, receive telemetry data and, above all, view the video stream. As far as software integration goes, the onboard processing unit runs Ubuntu 18.04 LTS, an operating system compatible with all the necessary libraries and programming languages. Additionally, the connection between the system and the observer is made possible by Ubuntu's compliance with ROS. NVIDIA's CUDA, which offers hardware acceleration for the detector and the tracker, is also compiled alongside OpenCV. Additionally, the SDR's necessary software is loaded, enabling the acquisition of signals that present a possibility for self-localization and signal transmission for jamming implementation. Finally, the manufacturer's SDK for the UAS is implemented on the processing unit, enabling the collection of different data from the drone, including the camera stream and control of the flight controller [13]. The algorithms for detection, tracking, and self-localization are subsequently fed with the data. As a result, the system is self-contained, autonomous, and intelligent and can control the UAS on its own. Due to system restrictions, the suggested system was put to the test in a variety of tracking and interception trials to ensure its viability in practical applications. Important information for upcoming enhancements was gleaned from those experiments. For instance, utilizing the recommended hardware, a maximum image processing speed of 30 frames per second at 720p resolution is attained, which severely hampered our tracking performance, especially at higher drone travel speeds. Performance restrictions apply to the detector and tracking components. The counter-attack range is also constrained by the jammer component.

Internet of Things

The goal of Internet of Things systems is to connect and communicate with real-world objects for a variety of uses. Connected objects can be used as sensors or actuators to send and receive data over the network. Machine-to-machine communication is essential for IoT systems to automate, update, and deliver new services. This study proposes a platform for managing connected drones indoors [14]. The networked drone is controlled using an indoor flight plan specified by the user in her Web application. The application handles all user actions performed during flight plan creation and translates them into automated communication flows between the server and the selected drone. The suggested method enables quick and effective control of a single linked drone or a network of connected drones. In the background, there are a variety of electronic devices and connected objects that are a part of IoT systems. Sensors and small embedded electronics, which typically command sensors and actuators, are the most prevalent connected items. Robotics-related technologies also advance quickly, adding new features, and bringing down costs [15]. The proposed platform can be summarized as shown in Fig. 6.6.

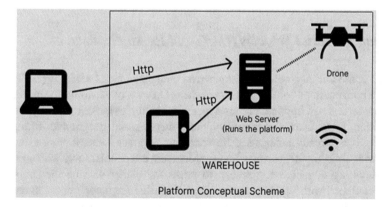

Fig. 6.6 Platform conceptual scheme

Use of Drones in IoT

For managing and monitoring a network of drones, some IoT systems use connected drones with high-level services (such as the ability to move between places or complete tasks). These systems will enable a number of new functions, including the ability to manage many drones simultaneously. These kinds of systems have a few theoretical models, but there are not many functional platforms. This study suggests a brand-new IoT infrastructure for controlling connected drones in enclosed spaces. These programs perform a variety of tasks, including creating flight plans and enabling constant connection with the drones to send and carry out these plans.

Used drones must have Wi-Fi and an open API in order to be controlled by a computer program. Users can remotely access this application using various devices, such as tablets or laptops, thanks to the Web server that runs the platform Web application. The platform was utilized repeatedly. In three flight plans, the platform was used to command and collect data from a connected drone. The platform was set up in a soccer field indoors. The primary Web application was being run on a laptop that was acting as a server. A flight plan had been assigned and submitted to the drone. At the end of the experiment, the drone performed all tasks satisfactorily. The program provided the drone with the necessary movement commands to reach the site, and the drone transmitted platform data from internal sensors and batteries. We have verified that the sensor data displayed in the application is correct and updated in real time while the flight plan is running. The proposed platform was evaluated after being divided into two parts. In the first section, we explored flight plan's online application to see if users could easily control a connected drone. The application initiates a two-way communication process with the drone during flight planning. This means adjusting the movement and location of all drones within the defined flight plan. The second part describes the automated communication process between the application and the drone to execute a series of flight plans.

Unmanned Aerial Vehicles (UAV) and Its Applications

Drones in the military, commonly known as unmanned aerial vehicles (UAVs), are used for both unarmed combat and surveillance. Drone uses outside of the military are developing, and new applications are appearing. Examples include delivery services, journalism, humanitarian aid, archeological surveys, and geographic information systems. It is already being used for mapping, weather forecasting, and wildlife monitoring. In computer science, image processing is a popular topic with numerous algorithms, using such a technology to sense and gather data on the shop floor. UAVs equipped with LiDAR cameras are being used for mapping, the creation of 3D spatial models of buildings, structures, and geographic areas, the execution of surveying measurement activities, the counting and categorization of crops and trees, and a variety of other applications. At the moment, there are businesses that are able to provide these services on demand, and there are even services that act as brokers in order to locate drone companies, pilots, and operators. UAVs equipped with video cameras are used for a variety of purposes, including gathering video footage of public events like concerts or sporting competitions, carrying out surveillance tasks like those performed by the police when they are following suspects or by the lifeguards when they are watching the beaches, and serving recreational purposes like competing in races. There are multiple projects currently underway to utilize unmanned aerial vehicles (UAVs) for transportation and logistics. One well-known illustration of this is Amazon's plan to start delivering items using drones in the not-too-distant future.

Case Study: The Flying Eye System (FES):
The flying eye system (FES) that is here suggested consists of an autonomous drone that can fly inside industrial facilities on a predetermined path, gather visual data, and then interpret the data to create valuable information. In this system, the drone or drones take off from a base station and fly through the building, following the flight path to capture visual data at focal locations before returning to the base station and uploading the information. The visual data are analyzed in order to extract usable information, which is then sent into the ERP and/or manufacturing information system (MIS) software. Operationally, the flight route is a predetermined course that passes over the facility's points of interest. The Z-dimension (the altitude) is properly set to avoid obstructing production processes and people below. At focus points, the drone may drop to a reasonable height, though, to improve the quality of the data it collects there [16].

The drone is autonomous during the flight along the flight path, allowing it to avoid unexpected impediments in its path. With the onboard cameras, data are recorded as either still photos or continuous video. Image processing capabilities on board are not required because the captured data are processed at the ground station. When the drone is in pre-flight mode, it recharges its batteries within the base station.

The drone needs to be equipped with things like probes to connect to charging pads at the base station and indicators to inform the controlling system to change the drone's state to "ready for mission." The mission continues where it left off after the recharge. A powerful computer called the processor is located at the base station. The drone sends the visual data to the CPU as it starts to recharge when it arrives at the base station. A channel for data transfer must be included in the drone's charging apparatus, which has been approved for the upcoming mission. Utilizable managerial information is obtained after data processing, and it is supplied to ERP or MIS [17]. The term "MIS" refers generally to software that may manage or plan. An ideal manufacturing execution system (MES) is capable of gathering data automatically and without human involvement, processing the data, and operating systems in a manufacturing plant as shown in Fig. 6.7.

Fig. 6.7 Flow of drone data collection system

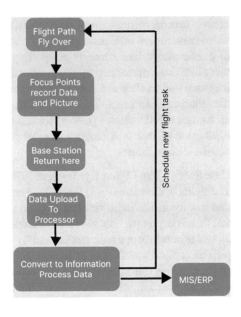

AI-Based Pipeline Inspection by Drone for Oil and Gas Industry in Bahrain

The procedure for inspecting oil and gas platforms is quite important. Such a technique requires taking into account a number of things. Cost, safety, and the environment are thought to be important considerations in the inspection process.

Solution: The proposed solution uses drones equipped with thermal cameras to monitor oil and gas pipelines and find leaks and cracks in dangerous remote locations. The system uses AI-based onboard processing to perform leak detection. Additionally, the device incorporates an onboard high-precision accuracy classifier. In order to attain real-time processing performance and reduce the amount of time needed to notify pipeline leaks, this classifier has been accelerated. As a result, it used hardware implementations of parallel processing. A real-time alarm will be provided by this system with a delay of under 100 ms. This study intends to lower the overall cost of current inspections.

Methodology:
Throughout this study, a machine learning-based algorithm for the proposed system is implemented. The performance of various machine learning (ML) algorithms was evaluated. Comparison, etc. Decision tree algorithms (DT), random forest algorithms (RF), and support vector machines (RF) are examples. The performance comparison measures' evaluation metrics were the F1 score, accuracy, precision, recall, and actual methane (CH4) leakage data from a pipeline were used in the algorithms for this study as shown in Fig. 6.9. GPS was used to pinpoint the exact location

of each leak. In order to further this investigation, a conference was held with the Bapco reliability engineering department. Different pipelines from Bapco include pipes that are overhead, underground, on the ground, and underwater. Depending on the kind of oil or gas it contained, the pipes' size and substance may differ. They also feature water pipelines for cooling and heating. Regarding the inspection procedure, each day at the start of each work shift, Bapco employees undertake a regular visual examination of the oil and gas platforms. Otherwise, they monitor the pressure gages that are attached to the pipes from a control room. If they see any out-of-the-ordinary readings, they visually inspect the pipe and follow it until they locate the leak. When a gas pipe leaks, the inspector will be equipped with a gas detector that can tell whether there is a leak or not. Bapco also creates hydrazine, a gas that is regarded as being extremely toxic. As a result, the hydrazine platform's operators always wear hydrazine detectors.

Disaster Management

Drones excel in three humanitarian assistance scenarios is as shown in Fig. 6.8.

Communications, coordinations, terrain coverage, and search operations are essential for success. In the final scenario, the drone quickly identifies a distributed group of people with disabilities and calculates their ages and movements. Drones also detect the victim's electromagnetic radiation more quickly and accurately. The area of crisis management (CM) and emergency response (ER) is one of the most

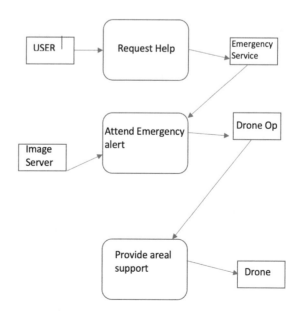

Fig. 6.8 Drones for emergency applications

Fig. 6.9 System block
diagram

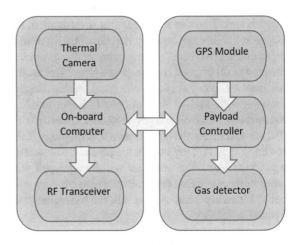

System Block Diagram

important application areas [18]. Autonomous drones can reach faster than human resources and can reach places where land access is considered. Impossible at present.

The deployment of unmanned aerial vehicles (UAVs) that are capable of flying independently cuts down on the likelihood of mishaps happening as a result of a mistake made by a human pilot.

Teams in search and rescue accept drones because of their autonomy. However, autonomy brings with it strict standards that must be met regarding the safety and security of the underlying architecture and procedures. The purpose of the safety guarantees is to prevent additional casualties that may be caused by malfunctions in the drone system. The purpose of the security guarantees is to eliminate the possibility that third parties, in addition to humanitarian organizations in charge of relief operations, could abuse drones, particularly in conflict zones, for activities such as espionage, the unwanted disclosure and publication of pictures of victims in the press, or the deliberate obstruction of relief operations [18].

Methodology:

During the process of developing an embedded system, the safety and security concerns of the system are managed by the SysML-Sec environment. A connected automobile architecture that is both able to withstand network attacks and include safety-important subsystems has been successfully defined with the help of SysML-Sec, which has been effectively applied inside that architecture. With the use of UML and SysML diagrams, SysML-Sec is able to support the following steps of a methodical process: requirements elicitation, attack capture, hardware and software partitioning, and software design [14]. The free and open-source toolkit known as TTool provides support for the SysML-Sec standard.

In addition to diagram modeling, TTool provides formal verification of security and performance related properties from partitioning diagrams and safety and security properties from design diagrams. Surrounding videos were recorded on the

Parrot platform. These videos are transmitted from the UAV to a remote computer, which uses the information to autonomously pilot the drone according to its intended use. Embedded model systems in unmanned aerial vehicles (especially drones that capture video, connect to WIFI networks, and execute remote commands) and how remote control computers handle communications and processing.

Agricultural Drones

Properties:

When a drone is outfitted with flight planning software that enables it to autonomously follow the route and altitude of flight, the user simply needs to draw a perimeter around the area that needs to be mapped in order to complete the mapping process. Photographs will be taken by drones, which will utilize onboard and camera sensors in addition to built-in GPS in order to determine the timing of each picture and the degree to which images will overlap [19].

Crop dusting refers to the use of drones that are able to carry tanks of fertilizer and insecticide and spray crops with a greater level of precision than a tractor. Workers who manually sprayed these crops with pesticides in the past now face less risk of exposure to the chemical as a result of this change, which lowers costs [20]. Current autonomous drones have been used to try and acquire aerial photographic images in limited places, mainly in the Sleman district area of Yogyakarta. This is part of the process of utilizing drones to support GIS, which involves using drones to support GIS. The production of a scenario map required this action to be taken, and it was taken. In addition, hybrid aerial-ground drone operators will have the ability to gather data in addition to performing a wide variety of other jobs. In order for the drone to take flight while operating in autopilot mode, it requires a minimum of eight satellites. This is one of the requirements that must be met before normal operations may begin.

Crowd Density Estimation Challenges and Solution

In this environment, there are a number of challenges, including variations in density and scale, as well as brightness, height from the UAV platform, occlusion, and inefficient pose estimation. In addition, there are height differences between the UAV platform and the target.

Closed-circuit television (CCTV) cameras are currently the most common method used to watch crowds of people, despite the fact that this method is fraught with a variety of issues. These issues include coverage that is restricted to a relatively small area as well as the incessant need for the participation of people in crowd control. Deep learning frameworks show a lot of promise for intelligent crowd analysis from frames of video, despite the fact that there are a variety of challenges to conquer when attempting to recognize people from shaky UAV camera systems.

Methodology:

Even though some of the current recent studies extracted their data using SIFT and SURF in addition to deep learning methods, other studies made use of deep learning methods for human detection in order to get an estimate of the crowd [21]. The gathering of training images of the subject that has to be identified is the first step in the process of developing CNN algorithms. In this instance, the VGG-16 architecture is a well-known one that was utilized for the purpose of training enormous quantities of pictures of a variety of people. An estimate of the crowd's density is one of the key strategies that may be utilized in conjunction with deep learning architectures to solve the problem of crowd flow estimates. This can be accomplished by gathering information about the number of people in the crowd. STNNet is a method that was proposed in a study that was published to jointly manage the problem of density map estimation in crowded scenarios that were captured by drones. STNNet was developed to estimate the number of people in an area using drone footage [22].

Campus Priority System

Being an autonomous operation where human safety is paramount, the multi-rotor drone C carries loads with weight restrictions and deploys to the desired location. An autopilot, ground stations, and a controller or pilot form an autonomous flight system. A companion computer should be considered for development purposes. This prevents modification of the flight control source code, increases processing power, and enables the use of new tools. This process is initiated by an operator on your computer. A companion computer responsible for translating commands from high-level languages into standard drone communication protocols runs the process code. A companion computer and ground station prompt each drone her component to perform the tasks of an air traffic controller as the air traffic controller masquerades as a drone. When the operator submits a launch request, the launch status requests flight conditions (sensor calibration, home location, etc.). The system should prevent the drone from taking off if one of her sensors fails or if the drone is not in its home position.

Autonomous Drone Control Within a Wi-Fi Network

Initially, one autonomous drone is built that can carry out a predetermined flight plan. The development of a single autonomous drone allows the interconnection of other autonomous drones that have been built similarly. Drones can communicate with each other and send and receive flight command codes that tell the receiving drone how to fly a specific flight path.

The best drone is the well-known, readily available Parrot AR2. Through a Wi-Fi communication link, it is controlled by a platform, such as a smartphone or a tablet, on which the desired application is installed. During a drone flight, the AR2 drone also sends a substantial amount of navigation data to the controlling consul. An autonomous drone needs a hardware platform that can run a program that is stored in its memory and that can send flight instructions to the drone over a Wi-Fi communication line (in this case) with two hardware architectures: Notebook control and Wi-Fi NodeMCU control. By sending the correct AT commands to the drone, flight control can be achieved.

Drone Autonomous Navigation on Forest Trails:
The primary goal of the system is to locate routes between trees that can be followed by drones. Deep learning artificial neural networks or convolutional neural networks are the primary tools for solving this problem [23]. The number of layers and dimensions of local networks used in the issue of autonomous drone navigation based on technical targeting systems are demonstrated.

Solution:
A classification and decision-making method implemented using artificial neural networks and fuzzy logic [24]. Deep learning neural networks, impulse neural networks, and neuro-fuzzy inference systems have been proposed to solve the problem of autonomous neural networks. Research shows that increasing the intelligence and autonomy of mobile robotic objects through deep computational parallelization is a promising way to solve these problems. This requires the development of dedicated hardware platforms that ensure the highest possible efficiency in implementing these algorithms and methods. A decision system, a recognition and classification system, a parallel image processing system, and a panoramic video system form the framework of the created system.

Methodology:
Each of the components of the system carries out digital processing of the signals; it receives and transmits the results to the input of the system. The primary task of the system is to filter the acquired optical image and divide it into central and peripheral regions to extract the data needed for the recognition and classification efficiency of the system. The system is a classifier that has learned to identify passable paths in the received video image. It is a deep learning convolutional direct propagation artificial neural network with a complicated setup [25].

At this level of development, the proposed system is trained using an inverse propagation approach for 100 epochs. The training data consist of roughly 20,000 images, including original photos of various forested terrains as well as images that are derived from the original images, which were created by offsetting, rotating, and reducing the originals. At this point in development, a 4-core ARM Cortex-A53 microcontroller and a 128- to 256-core GPU can be used to construct the intelligent control system of each system agent [25].

Improving Image Recognition Accuracy by Contrast Correction in Autonomous Drone Flight

A drone can be controlled to fly to a certain location using attitude control based on a GPS position sensor. But the drone must be able to know exactly where to land when it gets to its destination, which it cannot do with the current level of GPS accuracy.

A learning phase and a recognition phase make up the image-recognition process. In the learning phase, machine learning is used to transform learning photos into feature data and feature data into learning result data. When images are recognized, learning results are transformed into discriminators. In the recognition stage, the photos that were sent in are turned into data about their features, and recognition is done using the rules that were figured out by machine learning.

Susceptible Attack Methods Against Autonomous Drones

Most attacks aim to take advantage of flaws in the communication channel or other aspects of their target that are unique to them. We think there is a big chance that drones will be used wrongly during the epidemic because they were rushed into service without enough security.

Man in the middle:
A "man-in-the-middle" (MiTM) attack is the act of an unauthorized individual or party hacking into a communication path to intercept, intercept, and possibly alter legitimate messages. In a typical interaction, the drone and its transmitter establish her Wi-Fi connection, through which the drone and controller exchange data. In contrast, the attacker's goal in a MiTM attack is to intercept, send, and receive data intended for other network users. ARP cache poisoning is a popular method for accomplishing this. By employing this technique, the attacker can impersonate network devices by sending bogus or fake ARP responses to corrupt ARP tables. Consequently, they force all communications to go through their gadget.

Countermeasures: The basis of MiTM attacks is the idea of illegally monitoring data flows, intercepting data flows, and modifying data variables. The recommended security infrastructure for securing telecommunications data links (VPNs) is the use of virtual private networks. A VPN service can be added to the drone network infrastructure to encrypt the network layer and provide a level of anonymity regarding user credentials, transmitted data, and geographic location. These security measures ensure the confidentiality of data within the drone's security chain.

DoS:
A DoS attack is a one-to-one availability attack, meaning it can be launched from a single system. Distributed denial of service (DDoS) on the other hand is a many-to-one availability attack. Technically, DDoS attacks require the use of other previously compromised systems (computers, drones, etc.) that are currently running zombie

programs. The attacker activates the zombies whenever he is ready to launch a DDoS attack. To generate a denial of service, they next bombard the target with a massive amount of network traffic. Different types of DoS assaults exist. The drone or the remote controller can be the target of a DoS or DDoS attack on drones. The deauthentication flood attack is successful when DoS attacks are made against the remote controller. In this attack, the remote controller is continually sent deauthentication packets, causing it to disengage from the drone. The JSON assault is a successful attack in the event of a DoS attack against the drone. This attack is carried out by pretending to send the drone a request.

Countermeasures against DoS/DDoS attack: In the event of GPS interference, the implementation of adaptive antenna (ADA) in the drone architecture enables uninterrupted workflows and overcomes GPS interference to ensure consistent operation with associated GPS. PNT functionality is guaranteed. ADA uses advanced digital signal processing techniques to protect drones from interference and jammers in one or more layers. The multi-band interference immunity provided by ADA makes this possible.

GPS Spoofing:
The global positioning system (GPS) is a radio navigation system based on satellites that provides geolocation and time information to GPS receivers. Civil GPS signals are intended for public use and are open to all users without restriction, making them susceptible to assaults like signal obstruction, jamming, and spoofing. Civil GPS is used by a lot of devices, especially drones, which depend almost entirely on GPS signals for navigation, sensors, and other functions. The two primary spoofing techniques used in a GPS spoofing assault are overt and covert tactics. "Overt strategy" means that the attacker does not try to hide the fact that they attacked. "Covert strategy" means that the attacker tries to hide the fact that they attacked from both the GPS receiver and the drone's navigation system [25].

Countermeasures: For GPS spoofing, we need a way to ensure that the GPS continues to work as intended. Older GPS signals, known as "legacy GPS signals," contain encrypted binary code. Y-codes (encrypted precision codes), most commonly used by the military, broadcast on the L1 and L2 frequencies. A dynamic sequence changes the encrypted binary code 10.23 million times per second, causing the Y code to randomly and uniquely change its combination key. It is virtually impossible for an attacker to spoof his GPS pointing in the direction of the Y code, because without the encrypted key, he cannot generate the Y code.

Reauthentication Attack:

It enables the station or access point to de-identify users by transmitting packets for deauthentication. The access point does this to conserve resources. Whenever there are instances of high resource demand or whenever there have been inactive users on the network, and with most civilian unmanned aerial vehicles (UAVs) do not use security protocols or have default passwords set by the manufacturer, allowing attackers to send drone reauthentication packets to de-identify their transmitters is easy [26]. The threat actor's motivation is to target these drones for economic reasons when governments face attacks from other pressing concerns. Data confidentiality, integrity, and availability are the three basic building blocks that make up the cyberspace security chain. Integrity protects the originality and authenticity of data, while confidentiality guarantees that data entering and leaving a system can only be read by authorized persons. Availability also makes sure that services or data in a system are always the same and can be accessed.

Measures to Prevent a Reauthentication Attack: The installation of an external firewall with Egress, Ingress, and Address filtering facilitates the examination of data flow exiting the drone's network and traffic flow within it in relation to previously established network administrator policies. Both the types of data streams from your workplace and data flows from external sources can be directly controlled with the help of external firewalls. This makes it less likely that an attacker will launch an attack by first sending a reauthentication packet to the drone or its controller. A strong password is also recommended to protect your data over the network. With recent use of AI and drones, the proposed architecture is a powerful and secure way to protect against malicious actors. Drone deployments have increased significantly as a result of the pandemic, and it is crucial that these systems are properly safeguarded. When it comes to drone utilization, the framework takes important industrial players into consideration [27–32]. Operators, service providers, and manufacturers are the three categories that these playmakers fall under. It is essential that the actions taken by these players follow established policies, laws, security standards, and mitigation plans. The framework takes into account laws governing geographic travel, such as flight limits and bandwidth limits. Drones and those operating in jurisdictions that recognize the importance of meeting these standards will be better covered by the framework [33–38]. Despite all these technical aspects, the framework recognizes the importance of listing laws and safety guidelines that may be relevant to drones, based on case law and reasons for using drones. This is categorized into licensing and geographic legislation, certification, control and mitigation, policy implementation and evaluation, and radar restrictions, which focus solely on drones and their flights.

Conclusion

Drones have a promising future ahead, and they can accomplish tasks that humans cannot. The applications of drones are numerous. Drone technology is developing quickly, as are the supporting technologies that make it possible. Not too long from now, not only will our grocery shopping be delivered by a drone, humans shall commute in a flying autonomous vehicle. Hundreds of other uses for drones are currently being developed as a result of the numerous investments being made in this promising sector. The presence and use of drones will have some kind of impact on almost every element of our life. Delivery drones cannot only help businesses cut costs and save time on the last mile, but they can also meet customers' urgent needs.

Everything that has a benefit also has a drawback. Drones might not be utilized to their full capacity because of various drone flying restrictions. But with the expanding technological sector, all of the current concerns and issues with drones will eventually be resolved.

References

1. Kyrkou, C., Timotheou, S., Kolios, P., Theocharides, T., Panayiotou, C.: Drones: augmenting our quality of life. IEEE Potentials **38**(1), 30–36 (2019)
2. Bregu, E., Casamassima, N., Cantoni, D., Mottola, L., Whitehouse, K.: Reactive control of autonomous drones. In: Proceedings of the 14th Annual International Conference on Mobile Systems, Applications, and Services, pp. 207–219 (2016)
3. Anand, M., Kumar, A., Francis, A.B., Mohan, K., Midhun, R., Samshad, M.: Short range telemetry communication for autonomous drone navigation. In: 2020 IEEE Recent Advances in Intelligent Computational Systems (RAICS), pp. 131–135. IEEE (2020)
4. Devos, A., Ebeid, E., Manoonpong, P.: Development of autonomous drones for adaptive obstacle avoidance in real world environments. In: 2018 21st Euromicro Conference on Digital System Design (DSD), pp. 707–710. IEEE (2018)
5. Hirota, Y., Taniguchi, I., Onoye, T.: Parallelization of local path planning for high reliable autonomous drones. In: 2020 International SoC Design Conference (ISOCC), pp. 67–68. IEEE (2020)
6. Floreano, D., Wood, R.J.: Science, technology and the future of small autonomous drones. Nature **521**(7553), 460–466 (2015)
7. Tanaka, H., Matsumoto, Y.: Autonomous drone guidance and landing system using AR/high-accuracy hybrid markers. In: 2019 IEEE 8th Global Conference on Consumer Electronics (GCCE), pp. 598–599. IEEE (2019)
8. Hartmann, J., Jueptner, E., Matalonga, S., Riordan, J., White, S.: Artificial intelligence, autonomous drones and legal uncertainties. Eur. J. Risk Regul. 1–18 (2022)
9. Claesson, A., Fredman, D., Svensson, L., Ringh, M., Hollenberg, J., Nordberg, P., et al.: Unmanned aerial vehicles (drones) in out-of-hospital-cardiac-arrest. Scand. J. Trauma Resuscitation Emerg. Med. **24**(1), 1–9 (2016)
10. Butt, U.J., Richardson, W., Abbod, M., Agbo, H.M., Eghan, C.: The deployment of autonomous drones during the COVID-19 pandemic. In: Cybersecurity, Privacy and Freedom Protection in the Connected World, pp. 183–220. Springer, Cham (2021)
11. Valsan, A., Parvathy, B., GH, V.D., Unnikrishnan, R.S., Reddy, P.K., Vivek, A.: Unmanned aerial vehicle for search and rescue mission. In 2020 4th International Conference on Trends in Electronics and Informatics (ICOEI) (48184), pp. 684–687. IEEE (2020)
12. Apvrille, L., Roudier, Y., Tanzi, T.J.: Autonomous drones for disasters management: safety and security verifications. In: 2015 1st URSI Atlantic Radio Science Conference (URSI AT-RASC), pp. 1–2. IEEE (2015)
13. Souli, N., Makrigiorgis, R., Anastasiou, A., Zacharia, A., Petrides, P., Lazanas, A., et al.: Horizonblock: implementation of an autonomous counter-drone system. In: 2020 International Conference on Unmanned Aircraft Systems (ICUAS), pp. 398–404. IEEE (2020)
14. Pereira, A.A., Espada, J.P., Crespo, R.G., Aguilar, S.R.: Platform for controlling and getting data from network connected drones in indoor environments. Futur. Gener. Comput. Syst. **92**, 656–662 (2019)
15. Vásárhelyi, G., Virágh, C., Somorjai, G., Nepusz, T., Eiben, A.E., Vicsek, T.: Optimized flocking of autonomous drones in confined environments. Sci. Robot. **3**(20), eaat3536 (2018)
16. Maghazei, O., Netland, T.: Drones in manufacturing: exploring opportunities for research and practice. J. Manuf. Technol. Manag. **31**(6), 1237–1259 (2019)

17. Gunal, M.M.: Data collection inside industrial facilities with autonomous drones. In: Simulation for Industry 4.0, pp. 141–151. Springer, Cham (2019)
18. López, L.B., Manen, N.V., Zee, E.V.D., Bos, S.: DroneAlert: autonomous drones for emergency response. In: Multi-Technology Positioning, pp. 303–321. Springer, Cham (2017)
19. Budiharto, W., Chowanda, A., Gunawan, A.A.S., Irwansyah, E., Suroso, J.S.: A review and progress of research on autonomous drone in agriculture, delivering items and geographical information systems (GIS). In: 2019 2nd World Symposium on Communication Engineering (WSCE), pp. 205–209. IEEE (2019)
20. Bruzzone, A., Longo, F., Massei, M., Nicoletti, L., Agresta, M., Matteo, R.D., et al.: Disasters and emergency management in chemical and industrial plants: drones simulation for education and training. In: International Workshop on Modelling and Simulation for Autonomous Systems, pp. 301–308. Springer, Cham (2016)
21. Kim, D., Oh, P.Y.: Toward lab automation drones for micro-plate delivery in high throughput systems. In: 2018 International Conference on Unmanned Aircraft Systems (ICUAS), pp. 279–284. IEEE (2018)
22. Krishna, S.L., Chaitanya, G.S.R., Reddy, A.S.H., Naidu, A.M., Poorna, S.S., et al.: Autonomous human detection system mounted on a drone. In: 2019 International Conference on Wireless Communications Signal Processing and Networking (WiSPNET), pp. 335–338. IEEE (2019)
23. Zhilenkov, A.A., Epifantsev, I.R.: System of autonomous navigation of the drone in difficult conditions of the forest trails. In: 2018 IEEE Conference of Russian Young Researchers in Electrical and Electronic Engineering (EIConRus), pp. 1036–1039. IEEE (2018)
24. Lan, C.W., Wu, C.J., Shen, H.J., Lin, H.C., Chu, P.J., Hsu, C.F.: Development of the autonomous battery replacement system on the unmanned ground vehicle for the drone endurance. In: 2020 International Conference on Fuzzy Theory and Its Applications (iFUZZY), pp. 1–4. IEEE (2020)
25. Saif, A.F.M., Mahayuddin, Z.R.: Crowd density estimation from autonomous drones using deep learning: challenges and applications. J. Eng. Sci. Res. (2021)
26. Abosuliman, S.S., Almagrabi, A.O.: Routing and scheduling of intelligent autonomous vehicles in industrial logistics systems. Soft. Comput. 25(18), 11975–11988 (2021)
27. Abdel-Basset, M., Chang, V., Nabeeh, N.A.: An intelligent framework using disruptive technologies for COVID-19 analysis. Technol. Forecast. Soc. Chang. 163, 120431 (2021)
28. Kim, D., Oh, P.Y.: Lab automation drones for mobile manipulation in high throughput systems. In: 2018 IEEE International Conference on Consumer Electronics (ICCE), pp. 1–5. IEEE (2018)
29. Karuppiah, M., Saravanan, R.: A secure remote user mutual authentication scheme using smart cards. J. Inf. Secur. Appl. 19(4–5), 282–294 (2014)
30. Karuppiah, M., Saravanan, R.: A secure authentication scheme with user anonymity for roaming service in global mobility networks. Wirel. Pers. Commun. 84(3), 2055–2078 (2015)
31. Karuppiah, M., Kumari, S., Li, X., Wu, F., Das, A.K., Khan, M.K., Basu, S.: A dynamic id-based generic framework for anonymous authentication scheme for roaming service in global mobility networks. Wirel. Pers. Commun. 93(2), 383–407 (2017)
32. Kumari, S., Karuppiah, M., Li, X., Wu, F., Das, A.K., Odelu, V.: An enhanced and secure trust-extended authentication mechanism for vehicular ad-hoc networks. Secur. Commun. Netw. 9(17), 4255–4271 (2016)
33. Karuppiah, M., Kumari, S., Das, A.K., Li, X., Wu, F., Basu, S.: A secure lightweight authentication scheme with user anonymity for roaming service in ubiquitous networks. Secur. Commun. Netw. 9(17), 4192–4209 (2016)
34. Naeem, M., Chaudhry, S.A., Mahmood, K., Karuppiah, M., Kumari, S.: A scalable and secure RFID mutual authentication protocol using ECC for Internet of Things. Int. J. Commun Syst 33(13), e3906 (2020)
35. Karuppiah, M., Das, A.K., Li, X., Kumari, S., Wu, F., Chaudhry, S.A., Niranchana, R.: Secure remote user mutual authentication scheme with key agreement for cloud environment. Mob. Networks Appl. 24(3), 1046–1062 (2019)
36. Maria, A., Pandi, V., Lazarus, J.D., Karuppiah, M., Christo, M.S.: BBAAS: Blockchain-based anonymous authentication scheme for providing secure communication in VANETs. Secur. Commun. Netw. (2021)

37. Pradhan, A., Karuppiah, M., Niranchana, R., Jerlin, M.A., Rajkumar, S.: Design and analysis of smart card-based authentication scheme for secure transactions. Int. J. Internet Technol. Secur. Trans. **8**(4), 494–515 (2018)
38. Li, X., Niu, J., Bhuiyan, M.Z.A., Wu, F., Karuppiah, M., Kumari, S.: A robust ECC-based provable secure authentication protocol with privacy preserving for industrial internet of things. IEEE Trans. Industr. Inf. **14**(8), 3599–3609 (2017)

Chapter 7
A Review on Automatic Generation of Attack Trees and Its Application to Automotive Cybersecurity

Kacper Sowka, Vasile Palade, Hesamaldin Jadidbonab, Paul Wooderson, and Hoang Nguyen

Introduction

Over the past decade, vehicles have been extended with various functionalities and amenities in order to increase usability and implement various quality of life features. In fact, it has been reported that nowadays cars can contain a staggering number of electronic control units (ECUs) (anywhere between 40 and 120 [1–5]) controlling every aspect of the vehicle, from the infotainment system right down to safety critical features such as the breaks [1]. As a result of this explosion in complexity, many vulnerabilities have emerged [1, 5–9] as potentially exploitable by external attackers. Some examples of automotive interfaces found to be vulnerable to external attacks are

- Bluetooth [6, 8]
- Telematics [1, 6, 9, 10]
- Cellular [6, 9]

K. Sowka (✉) · V. Palade · H. Jadidbonab
Coventry University, Coventry, UK
e-mail: sowkak@coventry.ac.uk

V. Palade
e-mail: ab5839@coventry.ac.uk

H. Jadidbonab
e-mail: ad4953@coventry.ac.uk

P. Wooderson
Horiba MIRA, Nuneaton, UK
e-mail: paul.wooderson@horiba-mira.com

H. Nguyen
Swansea University, Swansea, UK
e-mail: h.n.nguyen@swansea.ac.uk

© The Author(s), under exclusive license to Springer Nature Singapore Pte Ltd. 2023   165
V. Sarveshwaran et al. (eds.), *Artificial Intelligence and Cyber Security in Industry 4.0*, Advanced Technologies and Societal Change,
https://doi.org/10.1007/978-981-99-2115-7_7

- Wi-Fi [6, 9]
- ODB-II port [5, 6]
- Radio data system [6, 9, 11]
- Tyre pressure monitoring system [6, 9]
- Remote keyless entry [6, 9].

This variety presents attackers with plenty of opportunities to exploit vulnerabilities within individual interfaces and leverage the connectivity provided by the on-board networks, such as via the controller area network (CAN) bus, to propagate attacks between ECUs [5, 7, 12, 13] and imperil the safety of the vehicle. Therefore, it has become clear that a comprehensive cybersecurity evaluation of automotive cybersecurity is a necessary precondition to commercial distribution of vehicles in much the same fashion as traditional functional safety considerations. This is reflected by the emergence of various initiatives aiming to promote cybersecurity within the automotive industry, such as the ISO/SAE 21434 standard for automotive cybersecurity [14], automotive SPICE [15] and the EVITA project [16]. This effort culminates in new legislation requiring stricter cybersecurity assurance of the modern vehicle, such as the recently adopted regulations for vehicle cybersecurity of the World Forum for the Harmonization of Vehicle Regulations [17], which mandates that vehicle manufacturers demonstrate that their vehicles meet certain requirements in order to obtain the type approval necessary for commercial distribution. For instance, statement 7.3.3 of the new regulations states that "The vehicle manufacturer shall perform, prior to type approval, appropriate and sufficient testing to verify the effectiveness of the security measures implemented" [17], and further, statement 7.3.6 says that "The vehicle manufacturer shall identify the critical elements of the vehicle type and perform an exhaustive risk assessment for the vehicle type and shall treat/manage the identified risks appropriately" [17].

To assess cybersecurity in systems such as automotive networks, human expert-driven practices in the form of risk assessments and penetration testing [14] are necessary and have been rendered increasingly intensive, error prone and subjective procedures due to increasing complexity. In order to aid the experts in their task, attack trees [18] have established themselves as a useful formalism for performing risk assessments in the form of attack path enumeration [14], as they represent threats in a structured and intuitive manner. However, their construction is also proving a challenging and error prone task [19, 20], leading to increasing interest in performing automatic or semi-automatic construction of attack trees. In addition, certain authors have explored utilising attack trees for cybersecurity testing in a systematic fashion within the automotive domain [21–23] in a manner reminiscent of penetration testing, which makes attack trees a potentially promising direction for research into the automation of automotive cybersecurity testing alongside its traditional role in performing risk assessments [14]. Within the context of automatic attack tree generation, an obvious question is the degree to which artificial intelligence (AI) can be utilised in order to produce attack trees, the type of AI one could feasibly utilise to this end, and the domain-specific requirements which would have to be met in order to provide practically feasible automotive cybersecurity assurance [24].

To that end, this chapter explores the state of the art in automatic generation of attack trees and explores how it could act as a means towards facilitating further automation in automotive cybersecurity assurance. Potential challenges towards this will be identified along possible solutions to implementing an AI driven attack tree generation procedure, in which potentially fruitful research directions are identified for subsequent work on automatic attack tree generation. The contributions of this chapter are.

- A comprehensive survey on the state of the art in attack tree generation with respect to the ISO/SAE 21434 [14] standard's suggested risk assessment use case and a potential security testing use case as a further research direction.
- Discussing the applicability of attack tree generation in automotive cybersecurity, bringing together perspectives from both academia and industry with respect to how attack trees are used and generated.
- Identifying challenges and research directions within AI, which may harmonise perspectives between academia and industry to enhance the viability of generated trees for risk assessment and testing purposes.

While there are several surveys on attack trees [19, 20, 25, 26], some with generation as one of the considered topics [19, 20], there is no survey explicitly focused solely on the generation of attack trees. Furthermore, the methods mentioned by these surveys are sparse and focus on a group of the more cited works of Gadyatskaya et al. [27], Ivanova et al. [28], Vigo et al. [29] and Pinchinat et al. (ATSyRA) [30] (see Fig. 7.1). In addition, the existing surveys do not pay much attention to the potential for using attack trees with cybersecurity testing [21–23], which this paper will consider more explicitly within the context of the automotive use case [13, 31]. Further, some advances in machine learning, particularly in fields such as graph neural networks [32] and deep learning, will be discussed as potential avenues towards enhancing attack tree generation. Finally, it should be noted that some of the authors of this work have previously published a paper on the feasibility and obstacles for automatic attack tree generation to find use in the automotive industry [24]. While this work bears many similar themes, the purpose here is to more fully explore the state of the art in attack tree generation and critically examine the specific deficiencies in current approaches with respect to their utility in the automotive sector, whereas the previous paper focused on a more conceptual look at the possible applications and challenges for attack tree generation in the automotive industry.

Automotive Cybersecurity Assurance

Cybersecurity can be described in terms of three main principles: confidentiality, integrity and availability, with some authors further extending this with objectives such as authentication, authorisation and non-repudiation [12, 33, 34].

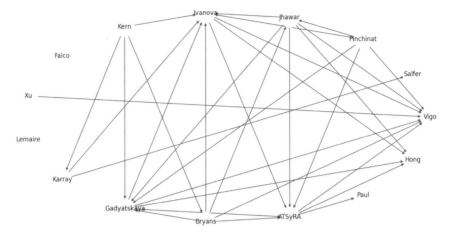

Fig. 7.1 Citation graph summarising the mutual citations between the attack tree generation methods discussed in this paper. An arrow directed from A to B signifies a "A cited B" relation, for instance "Karray cited Salfer". "et al." suffixes were omitted for brevity. The mapping between names and reference in this paper can be found in Table 7.1

　　Confidentiality refers to the requirement that data must only be accessible to those authorised to access it, integrity mandates that information needs to be resilient against corruption and modification, while availability necessitates that data and services within the system are readily available when needed. At first glance, automotive cybersecurity might be viewed from a confidentiality lens much more naturally than in terms of integrity and availability due to the assumption that attackers operate from an economic motivation, seeking to acquire assets (such as personal data) stored within the vehicle to sell for monetary gain [35]. However, the cyber-physical nature of modern vehicles means that attacks disrupting availability or integrity can imperil the safety of not only passengers, but also pedestrians by restricting access to or triggering critical safety systems such as the brakes [1] or airbags [23]. Therefore, it is imperative to verify that all these principles are being upheld, and thus that the system is secure, by utilising a variety of techniques.

　　A vehicle undergoes several phases throughout its life-cycle, from development, through operation and eventual decommissioning, each stage offering different activities for evaluating the cybersecurity of a given vehicle. Accordingly, the ISO/SAE 21434 standard for automotive cybersecurity works with three headings relating to the life-cycle of a vehicle: concept phase, product development phases and post-development phases [14]. Regarding the concept phase, an analysis of the system is performed which entails activities such as attack path analysis in an effort to identify potential threats. Alongside this, the risks associated with the identified threats are ascertained and a decision on the action to be taken regarding the risk is made, which includes options such as attempting to reduce or avoid the risk on one end to simply accepting the risk on the other.

Hence, the goal of an initial risk assessment is to categorise potential threats based on factors such as their potential impact and likelihood, often in the form of a calculation [4, 36], then determining if the risks associated with the threats identified meet the criteria required for commercial distribution and act accordingly in the case that they do not. It is difficult to estimate the exact nature and amount of vulnerabilities any given system may be imbued with, a sentiment particularly relevant in the automotive case due to the large amount of heterogeneous components from various sources [13, 31, 37]. Thus, a risk assessment focuses on the identification of threats that are deemed most likely exploitable by real attackers and most damaging if realised, though this is often a strenuous task given the sheer amount of potential threats. Within the automotive industry, attack trees have seen use as a formalism for performing such risk assessments [4, 14, 36, 38] as they provide an intuitive and structured method for the enumeration of potential faults, which can help mitigate some of the challenges related to complexity and readability of threats, along with a mechanism for computing attributes such as possibility or cost [18], which can act as a useful means of calculating the subsequent risk. In particular, clause 9.4.2 of the ISO/SAE standard for automotive cybersecurity refers to risk assessment using attack trees as a means towards identifying cybersecurity goals [14].

Further along, in the product development phase, verification of the identified cybersecurity requirements and validation of the identified cybersecurity goals (both potentially sourced from the preceding risk assessment activities such as attack tree analysis) must be performed according to requirements RQ-10-12 and RQ-11-01 of the ISO/SAE 21434 standard, respectively [14]. This can be achieved by means of various testing techniques, as outlined in recommendation RC-10-03 and example 1 associated with RQ-10-12 for verification activities. Naturally, the optimal choice of testing technique is situation dependent, which is further decided by factors such as resources available plus the intended scope and objectives of the tests. In order to explore this further, various qualities of vehicular on-board networks should be considered:

- Volume [1–5]: The sheer amount of ECUs present within the vehicle presents a considerable challenge for enumerating not only their individual quirks and properties but also the emergent interactivity between them all.
- Connectivity [1, 5, 7, 12, 36, 39]: ECUs belong to a complex interconnected network, all too often exhibiting a naive trust of adjacent devices in a manner reminiscent of the early Internet.
- Heterogeneity [1, 6, 13, 31, 37, 39]: In terms of both vehicles themselves and the components within them, which tend to be sourced from various different manufacturers with little to no communication due to their proprietary nature.

The key consequence of this is that due to the system being composed of a large number of interconnected components from various sources, precise specifications of the entire system are generally not available to a single entity [31]. Thus, due to this inherent obscurity, a black-box or grey-box approach is often necessitated in order to discover automotive vulnerabilities [13, 31, 37]. Specifically, the black-box/white-box distinction refers to the accessibility to data on the inner workings of the system, such as source code [33], where a total absence of data results in a black-box where the only basis for testing is the input/output behaviour of the system. That being said, this is not a strict dichotomy and varying levels of accessibility to data are often the practical reality, sometimes manifesting in a "grey-box" scenario [31, 33] in which white and black-box techniques are blended.

To that end, penetration testing involves enumerating vulnerabilities by purposefully compromising the system-under-test by a domain expert (or team of experts) [31, 40, 41]. This approach is corroborated by the ISO/SAE 21434 standard with recommendation RC-11-01 pertaining to the cybersecurity validation of a vehicle and recommendation RC-10-03 pertaining to performing tests to search for vulnerabilities [14]. Performing penetration testing gives insight into how an attacker might exploit a system in practice, which is of critical importance when considering how to secure the on-board networks and respond to potential attacks. Moreover, an empirical validation of the system being secure provides additional assurance alongside an analysis based on high-level design specification and risk assessments.

Since an attacker does not follow a strict specification in order to exploit a system but rather works under uncertainty, they find vulnerabilities by analysing the behaviour of the system and adapt their strategy accordingly [40]. Consequently, the attack unfolds in response to how the system behaves given each of the attacker's actions, with each exploit causing a chain reaction that eventually leads to an exploit with the uncertainty being lifted as the attacker probes the system further [40]. This reliance on the input/output behaviour of the system typically places penetration testing in the black-box category, however, it can also function as a grey-box testing technique [33]. However, despite its utility, penetration testing comes with a wide array of caveats in the automotive use case:

- Safety: Naturally, purposefully breaking a car renders it unfit for further use as it could cause latent damage and thus compromise the safety of anybody using it afterwards.
- Generality: Due to their heterogeneous nature, tests performed on a certain vehicle might not generalise well to other vehicles since the results of the test can depend heavily on the specific components making up the system.
- Cost of vehicles: It is very expensive to acquire a modern car to use for testing. Given that generalising the results from one car to another could be difficult, it would be desirable to test on a wide variety of cars.
- Time constraints: It can take a long time to perform full scale attacks on all potentially exploitable interfaces and to verify them in a comprehensive manner.
- Expert knowledge and experience is required: Naturally, this is an activity only an expert would be able to perform as an intricate understanding of not only cybersecurity but also of the domain is required to account for as many threats as possible.

For these reasons, the development of cybersecurity testbeds for vehicles [2, 13, 21, 39] and simulating penetration testing [2, 40, 42] are active areas of research, alongside utilising what information there is to perform a grey-box rather than entirely black-box test. While attack trees are not traditionally considered as a means of performing testing, recent research has highlighted that they could be utilised for systematic and automatic penetration testing of vehicles based on executing the attack paths enumerated within an attack tree.

For instance, Dürrwang et al. [23] establish a methodology for systematically deriving test cases for automotive systems from attack trees and demonstrate the feasibility of this approach on an example testbed where the authors successfully cause the deployment of airbags. Furthermore, Cheah et al. [21] demonstrate that by using scripts associated with the atomic attack actions within an attack tree (the leaf nodes), test case generation can not only be achieved automatically but the execution of these test cases can also be automated in many instances (barring cases such as social engineering) for a case study concerning the automotive Bluetooth interface. Finally, Mahmood et al. [22] demonstrate a similar automated test case generation and execution methodology for automotive over the air updates, in which an attack tree is used as an input to a tool developed by the authors which generates test cases by parsing the attack tree and identifying individual actions within the tree with executable scripts, after which a sequence of actions along with their scripts is produced by the tool which can execute these automatically. Substantially, automation of penetration testing would provide significant benefits for the easing of human burden involved in cybersecurity assurance and coupled with a robust testbed and an automatic attack tree generation methodology, it could provide a significant degree of automation in the testing of vehicles.

Attack Trees

Graph-based attack models are a staple in the literature [2, 25, 40, 43] due to their intuitive and expressive nature, though the world of graphical attack models is subject to a significant amount of variety [40]. A subset of these methods focuses on directed acyclic graphs (DAG) specifically since a directed model can explicitly express causality between actions [25] and the acyclicity mitigates challenges related to complexity [2]. Within the realm of DAG-based attack modelling, there too exists a wide variety of approaches. Kordy et al. [25] establish a comprehensive taxonomy of DAG methods within their survey in terms of 13 different aspects relating to things such as the orientation (attack or defence), purpose, structure et cetera.

Among these methods is the attack tree, which can often be mistaken as analogous to a wide variety of related methods such as fault trees and attack graphs. This is in part due to the fact that there is little in the way of a standardised definition of attack trees [28, 40]. As such, different practitioners might take varying approaches in utilising and distinguishing them from related models. Despite this lack of consistency, various prevailing trends and attributes can be inferred from their common usage in the literature:

- Acyclic directed tree structure [8, 25, 28, 29, 40, 44–46]
- Leaf nodes represent actions while inner nodes represent goals [29, 45, 46]

- Refinement structure requiring goals to be broken into sub-goals with narrower levels of abstraction [19, 27]
- From the attackers perspective, meaning that the tree represents a malicious attacker deliberately exploiting the system rather than benign faults [8, 29, 44]
- Annotations, which include things like probability or cost [29, 40]—logical operators governing refinement:

  – Disjunction (OR) [8, 25, 28, 44–46]
  – Conjunction (AND) [8, 25, 28, 44–46]
  – Sequential conjunction (SAND) [8, 44–46].

The preceding enumeration is far from exhaustive, and an attack tree need not include every characteristic listed above to be considered an attack tree. This makes the task of distinguishing attack trees from related models somewhat of a loaded question, as it presupposes a concrete definition setting it apart from other graph-based methods in the first place. Pragmatically, however, attack trees can generally be distinguished from models such as fault trees by their attacker focused perspective and from attack graphs by their DAG structure.

In terms of their practical utility, the system agnosticism of attack trees makes them well suited for the enumeration of attack vectors in a comprehensive manner. Whereas an expert written description of such faults is loaded with domain-specific artefacts, an attack tree gives a much clearer overview of the vulnerabilities, making them a sort of symbolic language for expressing potential threats. As such, attack trees have seen wide use in academic works concerning various aspects of automotive cybersecurity [4, 14, 21–23, 36, 45].

Despite this distinctive lack of consistency, various attempts have been made at formalising attack trees. In terms of syntax attack, trees are fairly simple, as DAG structures they can be expressed in terms of a straight forward generative grammar. However, there are still differences within the literature [27, 46]. Formal interpretations of attack trees play a key part in the literature on attack tree generation, which the forthcoming review will highlight, as many generation methods predicate the task of generating an attack tree as a formal mathematical process, which generally transforms some formalised information on a system into an attack tree.

Literature Review Methodology

Naturally, an attack tree generation method requires some source of information to drive the creation of new trees, such as the identity of the various nodes within the tree and the relationships between them. Pinchinat et al. [47] identify three possible sources of such information used throughout the literature, which will use the following definitions in this paper:

- Model: A formal model of the system could be a graph representing the physical network or a more abstract model of an enterprise encapsulating actors, assets and physical infrastructure. Includes any formalisation of system or environment entities.

- Library: Encapsulates some form of expert knowledge in a general manner, for example, cybersecurity and exploit databases as well as any form of knowledge base or pre-defined set of patterns or attacks. In general, a library should contain generic patterns rather than actual data from the system.
- Trace: This can include system logs and data on previous attacks or actions, anything which contains information on actions performed on the system. This differs from the library in that a trace contains specific actions performed on the system rather than general patterns.

The way in which each method evaluated was categorised with respect to input is summarised in Table 7.1. As outlined in greater detail in the preceding section, the primary objective of this paper is to evaluate attack tree generation methods on their applicability to the automotive domain, one aspect of which is the black-box use case for vehicles [13, 31, 37], and as such the discussion will consider the applicability of these methods in a black-box setting. Significantly, however, this is not to the exclusion of white or grey-box applications, but rather that given the overabundance of methods utilising models of specific systems (see Table 7.1) the white-box perspective is better explored, and as such the black-box necessitates a more explicit commentary on how it could be accommodated. Note also that although these methods will be evaluated in terms of how they could apply to a testing context in addition to risk assessments, they were not explicitly designed for such a purpose and as such any noted shortcomings are not a criticism of the authors, but an identification of directions in which generation methods would have to proceed in order to be better applicable to a testing context.

The following review is structured in two distinct sections: the first deals with attack tree generation literature in general, while the latter considers attack tree generation for the automotive domain specifically, each organised chronologically from earliest publication to latest. This organisation was chosen in order to maintain focus on the automotive cybersecurity use case; with the general use case being much more numerous, it risked overshadowing the intended focus of this paper. Papers on attack tree generation were collected using two principal sources: search engines and citations. Two different search engines were used in order to locate sources: DBLP and Google Scholar, with search terms such as "attack tree generation", "attack tree generator" and "attack tree construction". Many of these papers cross reference each other, but not uniformly (which can be seen in Fig. 7.1). Therefore, this chapter provides a more comprehensive overview of the methods for attack tree generation that have been found, and reviews them with respect to the criteria outlined above.

Table 7.1 Summary of attack tree generation methods in the literature by their inputs

Author	Input			Comment
	Model	Library	Trace	
Hong et al. [48]	X			Works with logical reduction methods on a model of the physical network infrastructure, e.g. a graph of nodes representing hosts with edges representing network links
Paul [49]	X	X		Relies on a formal language risk assessment study identifying feared events, which a cybersecurity database, both meeting the definitions of a library. In addition, a system architecture acts as a model for this method
Vigo et al. [29]	X			Model takes the form of a process calculus determining the interactions of the system and an adversary
Pinchinat et al. (ATSyRA) [30]	X	X		Semi-automated, with the input library as to the relations between high-level and low-level actions, etc., being interfaced to the user during generation
Ivanova et al. [28]	X			Arguably, the pre-defined manner in which model components are transformed could be considered a library, but these transformations are too high level and general to consider real "expert knowledge" being encapsulated rather than just being an intrinsic part of the generation methodology
Gadyatskaya et al. [27]	X	X	X	Can either use a library for refinement specification + attack traces as semantics or a system model which can provide both
Xu et al. [50]	X	X		Risk assessment and hazard analysis fall under the definition of a "library" outlined above
Falco et al. [51]	X	X		A variety of different "libraries" are used in the form of exiting tools and vulnerability databases

(continued)

Table 7.1 (continued)

Author	Input			Comment
	Model	Library	Trace	
Lemaire et al. [52]	X	X		The templates utilised are analogous to libraries in the same manner as Jhawar et al. and Bryans et al.
Jhawar et al. [53]	X	X		Since the set of assumptions used in this method encodes a priori knowledge of the underlying system, it is being considered a model, albeit a more reasonable one than many others in this category
Pinchinat et al. [47]		X	X	Uniquely, this method does not require a model of the system, relying instead on a library and trace
Salfer et al. [35]	X			A system and attacker model are both utilised, with a lot of properties being modelled for both (such as cost and skills)
Karray et al. [36]	X			Uses a model specifically designed to mimic the automotive network much like Salfer et al.
Bryans et al. [45]	X	X		Uses an abstract model of an automotive network with a library of templates encapsulating general patterns of attack with respect to the system components within the model
Kern et al. [54]	X			Models data dependencies between various functions within a vehicle to enable attack tree generation

Review of Attack Tree Generation Methods

Broadly speaking, literature on automatic attack tree generation has its origins within the work on automatic attack graph generation. Specifically, this typically takes the form of generating attack graphs from a model of the physical system such as the network architecture, via model checking [43]. However, this approach suffers from numerous limitations such as computational complexity resultant from state-space explosion [29, 48], constraining the scope of attacks to the physical network architecture without regard to socio-technical considerations and their unsuitability for black-box scenarios due to reliance on rich models of the underlying system. While attack graph generation methods continue to see new developments and application in the automotive domain [2], attack tree generation has begun to diverge as a distinct topic within the literature due to increased interest in attack trees, a development which can be seen clearly within the automotive domain in particular [8, 14, 16, 35, 36, 45].

An early example of this divergence is in the work of Hong et al. [48], who work with a more general notion of "attack representation models", which encapsulates both attack graphs and attack trees. Inheriting the typical use case of attack graph generation methods [43], this paper is concerned with the efficient generation of attack trees for networked systems and deems the existing methods of attack tree generation inefficient, such as min-cuts from attack graphs or enumerating all attack paths and structuring them into an attack tree. To that end, Hong et al. introduce two logical reduction techniques to create smaller attack trees that maintain the same security analysis potential as their larger counterparts based on a model of the physical network infrastructure. These logical reduction techniques are a full path calculation with similar nodes being grouped together, and an incremental path calculation which recursively expands attack paths to minimise node repetition. While this method has since been eclipsed in relevance by subsequent methodologies, the general approach of inferring attack paths from a model of the system and then structuring them into an attack tree will remain a recurring theme. There are two main issues stemming from relying on a model of the physical network infrastructure: unsuitability for a black-box scenario (regarding both the layout and identity/behaviour of the network and nodes) and the fact that this constrains attack tree generation purely to physical infrastructure with no regard for socio-technical components (which will become an area of interest in later attack tree generation research [28]).

Independently, a feasibility study on the automatic construction of attack trees was performed by Paul [49], which concludes that "draft" trees can be generated in a meaningful manner, however, the "lower parts" of the attack tree are left to manual completion. This study highlights the role of attack trees in early risk assessments and reinforces the notion that their application is shrouded in uncertainty pertaining to the exact architecture of the system, which is congruent with their role within the concept phase of automotive cybersecurity [14]. Additionally, a generation methodology is put forward by the author which makes use of a risk assessment study composed of strictly defined formal language definitions, these definitions identify feared events with respect to system components sourced from a system architecture model and a security knowledge base (which correspond to a model and library, respectively, with regard to the definitions established above). These are then used to generate the attack tree by a series of structuring actions according to specific elements of the models provided. In terms of risk assessment, this work takes the perspective that large scale risk assessments become cumbersome and as such attack trees can be a good means of reducing cognitive load with respect to human analysts, and to that end this paper has established that attack trees generation could be a useful means towards simplifying this process. Regarding a potential testing use case, although some parallels between the perspective presented in this paper and the automotive black-box may be drawn, such as in the inherent uncertainty, these remain two very distinct cases. In particular, this study considers the development of a known system in conjunction with a pre-defined architectural framework, while the automotive cybersecurity use case considered here necessitates a black-box or at least grey-box penetration testing approach [13, 31]. These differ mainly in the fact that the issue is not incomplete or poorly defined architectural data, but the total lack of access to such data, thus necessitating penetration tests in the first place [31].

The trend towards utilising formal methods and models is a recurring one at this stage, which can further be seen in the work of Vigo et al. [29]. The method outlined by Vigo et al. differs from the work of Paul [49] and Hong et al. [48] by representing the system in terms of a process calculus, where an adversarial process is defined which wishes to drive a process representing the system to a compromised state. The attacks are then enumerated in terms of "communication channels" and the knowledge of them necessary to reach this desired state. Channels in this context refer to any interface through which an attacker must gain entry in order to compromise the given system, such as a door or a wireless communication link, where the knowledge of said channel represents the ability to access it such as being in possession of a key or password. Notably, while this method improves upon the model checking based state of the art at the time [43], it still works by essentially extracting attack paths from a given system specification with the major diversion being the use of a process calculus rather than model checking. Therefore, the necessary channels must be explicitly defined within the system model, which is an undesirable property for a black-box scenario and renders the evaluation of unexpected behaviours difficult to implement within this framework. Furthermore, the abstraction of all communication links into channels means that this approach is more abstract than the work of Paul [49] or Hong et al. [48] and is thus more generally applicable, but may lack nuance in domain-specific tasks. As far as utility for risk assessment goes, the authors show that reducing attack trees to a logical formula allows for a quantitative analysis with respect to minimising or maximising some cost metric, and thus a minimum/maximum cost attack path can also be derived using this method, provided that a mapping between individual channels and costs is available.

One way of approaching the task of practical utility of attack tree generation is via a semi-automatic approach in the form of a tool rather than a fully automatic generation scheme. To that end, Pinchinat et al. [30] introduce the methodology utilised in ATSyRA, a tool for synthesising attack trees. This is based on modelling the system using an attack graph, extracting attacks from this graph and structuring them into an attack tree. This seems reminiscent of the approach taken by Hong et al. [48] where an attack graph is used as the initial model which is subsequently transformed into an attack tree, though in this case the attack graph is a model representation of the system itself rather than sourced from a model of the system. A novelty of this approach lies in the fact that this initial attack graph is sourced by automatically interpreting high-level documentation in the form of a hierarchy of high-level and low-level actions, somewhat reminiscent of the "refinement structure" later described by Gadyatskaya et al. [27]. This hierarchy would necessarily be sourced from a library of how high-level actions correspond to low-level actions structured or abstracted in a particular way, similar to the later "refinement specification" of Gadyatskaya et al. [27]. While one can see the appeal of such approaches, it should be noted that this approach is only possible by shifting the responsibility for the identification of relationships between nodes onto the user, who has to provide a library of pre-defined relations between the "low-level" and "high-level" actions. In addition, this methodology fills the need for expert insight by an explicit interface for the user to drive generation and thus can be a practically useful utility for performing risk assessments in a manner similar to the approach taken by Paul [49] where the focus is on easing the cognitive load but does little to advance the question of how to substitute this expert insight within a fully automatic methodology.

Ivanova et al. [28] aim to extend the domain of attack tree generation beyond the purely technical and into the socio-technical sphere, which the authors posit is not feasible with the methods Vigo et al. [29] and Hong et al. [48] introduce. Conceptually, the premise is a familiar one: begin with a model of the system and transform it into an attack tree. The key change being that the modelling framework used by Ivanova et al. allows for the encapsulation of not only physical infrastructure but also more diverse components such as human actors, access rights and policies by predicating the generation of attack trees on policy invalidation. Firstly, a goal asset and attacking actor are established, then a transformation function identifies the potential attack paths for the attacker obtaining this asset by transforming each system component it encounters to a corresponding attack tree. Concurrently, for each missing credential, the transformation function is called recursively to construct a tree for obtaining the given credential. Each type of system component (location, policy, data, items, etc.) has a pre-defined transformation into an attack tree, which means it relies on a rather high level of abstraction regarding each component. Additionally, the claim to extend the domain into human factors is entirely contingent on the model used, leading to limitations regarding unexpected behaviour not being accounted for (which is even more prominent when concerning human agents). Thus, the quality of the threat enumeration provided by this method is entirely dependent on if the right interfaces were explicitly present in the model itself similarly to Vigo et al. [29]. Additionally, the generation methodology could have been made more robust by sourcing patterns of attack from a library more versatile than the inflexible transformation rules utilised, which shift much responsibility onto the model required. While this work does not explicitly consider the subsequent calculations associated with a risk assessment, the authors note that this could be achieved by associating each of the elements with appropriate metrics such as likelihood and cost, which can then be used as part of a risk computation using the resultant attack tree. For the case of testing, an approach such as this one introduces the difficulty of social engineering into the scope, which can be problematic to effectively emulate.

A key limitation of formal methods is that they can suffer from usability issues when put in the hands of non-security experts. To that end, Xu et al. [50] aim to improve the usability of attack tree generation for interoperable medical devices (e.g. pacemakers) by removing the need for formal system models as an input and instead depend on a hazard analysis and operational workflow sourced using a process modelling tool to describe the system. The hazard analysis consists of a description of system states detrimental to the patient's well-being in a representation that can be easily used by clinical staff rather than formal mathematical models, in a manner somewhat reminiscent of the formal language definitions of feared events within the work of Paul [49]. The generation proceeds by taking the operational workflow and the hazard analysis and deriving fault trees, which are here distinguished from attack trees by being limited to a path for a single specific hazard. Afterwards the fault trees are combined into an attack tree, with faults leading to harm "in the exact same way" being grouped under a common root node, and the fault tree descriptions are modified from describing benign faults to specifically adversarial actions. In practice, this approach is reminiscent of the work of Ivanova et al. [28] as its main contribution is a different underlying modelling method being used to make it more accessible to a given domain, though it should be noted that not using mathematical formal modelling tools does not detract from the inherently deterministic nature of the generation process in relation to potential attack vectors having to be explicitly present within the data. That being said, much in the same way as the work of Pinchinat et al. on ATSyRA [30], this can be a valuable improvement upon the usability of attack tree generation for performing risk assessments in the given domain.

A key obstacle in the utilisation of attack trees for easing the cognitive load involved in performing risk assessments is that the resultant trees need to be intuitive and easy to read for human experts. To that end, a refinement aware generation scheme is proposed by Gadyatskaya et al. [27] with the additional caveat that generated trees must adhere to a specific refinement structure, which posits that the transition from root to leaf nodes must be of a lesser level of abstraction. In fact, the lack of such a structure is a critique levelled by the author against previous attack tree generation schemes, such as Vigo et al. [29] and Ivanova et al. [28]. To begin with, the authors formalise their vision of the "attack tree generation problem" as the problem of deriving an attack tree which satisfies a given semantic specification and respects a given refinement structure. In particular, this method employs a formal grammar for the syntax and series–parallel graphs for the semantics [46] of attack trees then reduces the attack tree generation problem into the edge biclique problem, for which a greedy heuristic can be applied. As to the generation methodology, it relies on a semantic and refinement specification as input which can be derived from a system model formulated as a labelled transition system. As an alternative to the system model, a refinement specification could feasibly be sourced from a library of relations between different attacks such as a cybersecurity database, and the traces required for the semantics could be sourced from logs or attack paths, which is indeed the approach explored by the subsequent work of Pinchinat et al. [47]. While the adherence to a refinement structure is certainly a valuable insight, as it would enhance the practical usability of generated attack trees in performing risk assessments within the industry [19], it requires a large amount of additional information pertaining to the nature of the actions which the attacker can take and the way in which the system would respond to them. As far as a potential testing use case goes, since this method requires either attack traces or a system model from which attack traces can be extracted, its utility for producing attack trees to be used for testing via the generation of test cases [21–23] could be feasible if in place of the model a refinement specification with data on practically executable actions along with a relevant trace were used.

In a relatively isolated work within the literature (see Fig. 7.1), Falco et al. [51] apply the problem of attack tree generation to an "industry sector agnostic" Industrial Internet of Things domain, whereby the scope of the outlined methodology can be applied to any "IP networked system" in any industry sector. Firstly, the authors propose a unified master methodology for the construction of attack trees comprised of multiple different established frameworks in order to make it applicable to multiple industrial domains. Specifically, this methodology combines Lockheed Martin's cyber kill chain, OWASP, and various frameworks by MITRE Corporation along with Kali Linux tools to produce a practically viable attack tree strategy. In addition, the hierarchy of the attack tree is given further meaning with each subsequent level symbolising the answer to: when, where, what and how the attack will be executed, which is determined by each of the existing frameworks outlined above. For instance, the cyber kill chain determines the "when" of attack trees, while the "how" is handled by the kali Linux tools. This is in contrast with many of the above methods, as this work explicitly outlines the practicality of how the generated attack trees are to be applied to a given real world domain without abstracting away key questions (such as the precise nature of actions, which in this case are handled with Kali Linux tools). This is significant, as these types of trees can much more naturally be applied to subsequent creation of test cases, similarly to how a procedure with scripts attached to nodes within the tree was utilised for test case generation [8, 21] (albeit in a less comprehensive manner than the work outlined here). As for the generation methodology itself, it requires as input a model of the system and a set of rules encapsulating the way in which the system can be compromised. A drawback of this method is the large state-space resultant from an expressive model, thus limiting the speed at which this method can work, though the authors insist that the intended application is for offline audit tasks rather than real-time analysis and as such speed is not as significant. Additionally, the common issue persists in requiring not only detailed information on the nature of the system but also on its behaviours in response to given attacks. However, a valuable insight by the authors points out that data on such behaviour exists already and has been underutilised by previous methods, such as the frameworks encapsulated within this method like the Common Weakness Enumeration (CWE) and Common Attack Pattern Enumeration and Classification (CAPEC) by MITRE corporation. Although the main intended use case for this generation method is for the enumeration of potential threats in a risk assessment context, the explicit inclusion of penetration testing tools as the low-level basis for actions within the attack tree opens up the potential for a practically viable generation scheme with respect to a testing context [21–23].

A brief paper by Lemaire et al. [52] focuses on attack tree generation for cyber-physical systems, which as a domain is closely applicable to the automotive use case. In addition to generation based on a system model, this method also provides an evaluation step based on a model of the attacker defined by the capabilities and credentials that they possess, thus facilitating a subsequent risk assessment after generation. Generation of attack trees is done using a set of attack goals and templates, which are iterated over and attached to nodes which match the root of said template starting with a root node identified with the attack goal. Adversarial models for the system and attacker are somewhat reminiscent of the method Vigo et al. [29] proposed, with both using this attacker model to find an "optimal" attack in addition to the attack tree generated by quantifying a metric such as cost, which could be extended to other variables relevant in a risk assessment. Thus, this method seems to encapsulate attack tree supported risk assessment as a whole rather than just the construction of attack trees. However, this may be of limited utility in a testing context as the analysis aspect of attack trees would instead be performed via an empirical penetration testing approach through test case generation [21–23]. Interestingly, this paper seems to be the first which explicitly mentions the use of "templates" for the generation of attack trees, an avenue which will subsequently receive more attention [45, 53], but which nevertheless seems isolated as far as citations go (which can be seen in Fig. 7.1).

A semi-automated approach is put forward by Jhawar et al. [53], who propose a methodology in which an automated process is used to enhance an initial tree manually created by a team of experts. Much like the work of Pinchinat et al. with ATSyRA [30], the authors aim to produce a more practically viable methodology by keeping the human experts in the loop throughout the generation process. In essence, generation proceeds by augmenting the initial attack tree using a library of sub-trees in a fashion similar to the templates of Lemaire et al. [52]. In order to facilitate this, a system of annotations for nodes is devised which relates nodes within the tree to sub-trees within the library. As an example, the authors automatically construct a library from the National Vulnerability Database (NVD) storing formalised descriptions of vulnerabilities, which are then applied to high-level attack trees sourced from CAPEC which expand the high-level trees with NVD entries. This is reminiscent of the way Falco et al. [51] utilise existing cybersecurity knowledge bases for attack tree generation, however, in this case, the use is much more limited. In place of a rich system model, Jhawar et al. instead provide system information in terms of facts describing the environment, which could be a better choice for a black-box testing use case as it allows for different degrees of information on the system to be known. However, this can still be a limiting factor as a small amount of known facts could still impede the generation. Thus, while this method is better equipped to express black-box scenarios, it may not be well equipped to actually handle them in a testing context.

As an alternative to the many model-focused attack tree generation methods proposed, Pinchinat et al. [47] present a "model-free" library-based attack tree generation method. Similarly to Gadyatskaya et al. [27], the authors produce a formal definition of the "attack tree synthesis problem", which consists of producing an attack tree given an input library that conforms to an input trace observed in the real system. The library format used by Pinchinat et al. closely resembles the refinement specification of Gadyatskaya et al., while the trace is formalised as a sequence of propositions describing the state of the system at each time step (informally defined as an observed attack). Thus, the trace becomes the way in which the underlying system is ascertained in place of a model. A significant advantage of this is that it is better suited to a black-box testing scenario, as a trace representing an attack on the system is more easily obtainable in a penetration testing context than a full model of the system. However, for this to truly be practically feasible a mapping between practically observed actions on the system and possible leaf nodes within the library must exist, which may necessitate a comprehensive knowledge of the underlying system, mitigating somewhat the utility of the model-free aspect to black-box testing. Thus, while this methodology is significant for its model-free nature, it still requires a lot of manual work in assembling the library to offer practical usability. As far as risk assessment is concerned, a model-free construction process followed by an analysis step in which factors such as likelihood and cost are introduced for a specific system and propagated up the tree is congruent with how attack tree analysis was originally described [18] and as such provides a good first step in exploring how attack tree generation can further distance itself from purely model-based methods.

Attack Tree Generation in the Automotive Domain

While the above section tackled the general issue of attack tree generation, it is useful to focus on methods specifically designed with the automotive domain in mind, to ascertain how they adapt the existing body of work and if they successfully overcome the challenges within this field.

Salfer et al. [35] introduced a method for the automatic generation of attack trees with the automotive use case in mind not long after the work of Vigo et al. [29] and Paul [49] was published. Conceptually, this work bears certain similarities to these methods, in that a high degree of modelling was used to capture the potential interactions between an attacker and the underlying system. Specifically, a system model is defined which encapsulates the software, assets and communication of a highly generalised automotive on-board network. In addition, an attacker model is also defined in the form of "attacker profiles", which represent a human adversary with varying skillsets and motivations who aims to acquire assets found within specific ECUs within the network (e.g. personal data within an infotainment unit) for economic gain, which is similar to the approach Lemaire et al. [52] would apply 4 years later for cyber-physical systems in general. The distinct modelling of the underlying system and an adversary wishing to compromise is reminiscent of the approach Vigo et al. [29] utilised, albeit with the attacker model used by Salfer et al. and its interactions with the system being much more nuanced than the more general "communication channels" in terms of which Vigo et al. understood every interaction. Meanwhile the system model utilised by Salfer et al. is much simpler than the one used by Paul [49], which may be attributed to the more focused domain-specific nature of this work as opposed to the work of Paul, who worked with an existing architectural framework and security knowledge base. In regard to the system model, being sourced from actual engineering and design documents [2, 35], it may be argued that it can provide a seemingly accurate account of the underlying system for a white-box context; however, there are still many assumptions and generalisations made in regards to the skills and motivations of the attacker in reference to the attacker model itself and the way in which it interacts with the system. Particularly, the assumption is made that the attacker approaches the vehicle with an "economic" focus, seeking to extract valuable data assets which can later be sold; however, this fails to factor in attacks of a more directly malicious nature like denial of service, which in the automotive case has significant safety implications via mechanisms such as the breaks [1] and introduces the difficulty of attempting to envisage the "value" of individual assets and contrast them against the true perception and ability that the attacker possesses, especially in the "black market" setting that the authors describe. In addition, vehicles are often made up of many third party components for which precise design specification and behaviours are not available [13, 31], thus mitigating the ability to model the underlying system accurately in terms of the behaviour for these individual components. In terms of performing a risk assessment, this method utilises several metrics such as an attacker's skill, individual resource dependent on factors such as code size and known vulnerabilities. Further, this eventually leads into calculations determining the minimum costs for attacks against a given system model with respect to a given attacker profile. While very expressive, this method depends on a lot of data which, as outlined above, may not always be available or accurate. As far as applicability to testing is concerned, the authors posit that the evaluation resultant from this method can help penetration testers identify potentially heavily attacked targets more precisely, however, this assertion does not hold in a black-box environment due to the dependency on detailed models of the underlying network in regard to not just the architecture but also the behaviour.

Another method explicitly tackling the generation of attack trees for the automotive domain is that of Karray et al. [36], which purports to retain the "general approach" of formal methods for attack tree generation but shuns the modelling approaches used by previous authors, which are replaced with a model deemed more suited for automotive networks. Specifically, the model utilised resembles the approach of Salfer et al. [35] regarding the types of nodes being used to encapsulate aspects of the automotive network, with service, hardware, communication and data nodes utilised by Karray et al. being roughly equivalent to the software, ECU, communication and asset nodes from the model of Salfer et al., respectively. In contrast to Salfer et al., however, there is no explicit attacker model utilised and thus less assumptions are made about the nature of the attacker, although the attributes with which the system model is annotated with resemble the approach of Ivanova et al. [28], such as policy and access rights. Indeed, this approach also resembles Ivanova et al. in that a graph transformation of the model is utilised and thus assumptions on the behaviour of the attacker are shifted into the mechanism of transformation and the way in which attributes such as policy and access rights are interpreted with respect to the attack actions carried out by the attacker on the system. Thus, this approach inherits the shortcomings of Salfer et al. in the difficulties with sourcing precise data on the underlying architecture of the vehicle, which is something alluded to by the authors in the conclusion. This approach also inherits a caveat which can be applied to the work of Ivanova et al., in that the solution to the problem of generating attack trees adequate for a given application domain might not lie in simply altering the modelling approach of a fundamentally constrained generation methodology. Specifically, the use of pre-defined and inflexible rules for graph transformations as the basis for attack tree generation may be a limiting factor, particularly in the case of attempting to account for unexpected behaviours and vulnerabilities within a system which are often the basis of cyber-attacks [33]. Much like Ivanova et al. levels the criticism that the work of Vigo et al. [29] only considers the purely technical system without regard to the socio-technical, Karray et al. accuse the preceding work of not being applicable to the automotive domain due to unsuitable modelling approaches. Regarding the manner in which a risk assessment is to be performed, the authors refer to a calculation function for risk based on the impact and likelihood of a set of attack scenarios. Thus, the author's goal in generating attack trees is to facilitate the enumeration of such attack scenarios and to assess which vulnerability states are attainable for a given system along with which sets of actions can lead to it, which would presumably involve calculating the risk associated with each scenario further down the line.

Another automotive centric method is introduced by Bryans et al. [45], where a "template"-based generation technique is outlined, in which pre-defined templates, which encapsulate known patterns of attack, are used as the building blocks for the attack tree along with a model of the CAN bus. This method bears a resemblance to the methodology of Jhawar et al. [53] and Lemaire et al. [52] with the templates/libraries being sets of sub-trees which are inserted into the main tree in each case. Each template represents an abstract species of attack and, in contrast to the method of Lemaire et al., rather than containing labels of specific components, variables are used which permit an appropriate node to be substituted in their place. Generation is similarly performed via a recursive expansion of leaf node variables present within the tree, in which a valid component from the model is identified and/or a subsequent template is appended onto the chosen leaf node. This use of templates can be contrasted with the transformations used by Karray et al. [36]; rather than using a pre-defined system of transformations, the templates are more versatile and expressive. This is due to the fact that rather than simply mapping each attack to the type of node, an open-ended system of "variables" is used in which different components can be substituted depending on the defined constraints. Much like the work of Salfer et al. [35] and Karray et al. [36], this method relies on a model of the automotive networks. In this case, the model identifies the identity of each ECU, the various sub-nets and entry points for attackers. Unlike Salfer et al., there is no explicit attacker model, and the system model does not encapsulate behaviour or assets, which are instead handled by the various templates within the library, with the identity of each ECU being used to find a suitable template. Arguably, this is a more suitable approach as it lessens the burden of enumerating the behaviour of individual ECUs, with both Karray et al. and Salfer et al. encapsulating software/services explicitly within the model by contrast. While this information still needs to be sourced via the library, it is more generally applicable and, as seen with some other methods, existing cybersecurity databases can be used to populate the library [51, 53]. In terms of utility for performing risk assessments, the main goal of this method is the enumeration of potential threats in terms of a library of pre-defined patterns, which can then be annotated with relevant risk related values to ascertain the level of risk present in a given model. For the case of testing, since the model utilised in this method does not attempt to encapsulate specific behaviours like Salfer et al. [35] and allows for a more dynamic library to the strictly defined transformations of Karray et al. [36], it could be more easily transposed into a black-box setting. Further, as Falco et al. showed [51], it is possible to supplement generation with penetration testing tools, such as the various tools which come with Kali Linux, which can be implemented as the leaf nodes within the template library in a manner similar to how certain authors associated leaf nodes of attack trees with self-contained scripts [21, 22].

Finally, the most recent work exploring attack tree generation for automotive cybersecurity (at the time of writing) is that of Kern et al. [54]. This work is highly focused on relevance to the ISO/SAE 21434 standard [14] and Regulation 155 [17] through the lens of risk assessments. This method, like many others in this review, focuses on a model-based approach to generating attack trees from a model of the system architecture. Specifically, this model proceeds in a source-sink fashion taking data elements through the modelled architecture. Conceptually, the aim of modelling functionality of the various ECUs within the network is to reduce each operation into an atomic operation with a single output which can then be structured in a hierarchical manner to achieve emergent behaviours more closely related to the actual ECU behaviour. However, the authors concede that this process may be difficult to scale and so there is no strict requirement that the modelled functionality be this granular. These functions are the interconnected with directional dataflows between each other, with additional information on the relationship between inputs into each function (e.g. OR/AND refinement). Before generation can proceed, further manual steps need to be taken such as assigning damage scenarios to various key assets within the network and assigning threat agents to specific damage scenarios. Thus, with all the key elements and connections between them defined, feasible attack paths can be calculated with respect to each threat agent by determining which agents have access to which interfaces (e.g. OBD-II port) and what threat scenarios they wish to execute. Scenarios which cannot be executed directly must follow the data dependencies defined in the model. Much like the previous approaches explored in this review, the utility of this method is constrained by access to information, and the requirement to accurately model not just the vehicle but also the behaviour and interests of threat agents.

Challenges and New Directions in Attack Tree Generation for the Automotive Domain

Trends in Research

Attack tree generation has previously been described as a model transformation [19, 28], and many of the works explored in this survey seem to treat is as such; calling into question the extent to which such methods can truly be called "automatic generation" rather than "automatic transformation". In many cases, the assumption behind such utilisation of models is that it is easier to create a good model of the system rather than a good attack tree [19], which may hold in many cases, however, this neglects to consider cases in which information availability does not permit for expressive models of the system. Furthermore, constraining attack tree generation to a model dependent perspective can have adverse effects in the form of shifting the attention of researchers away from attack trees which are practically viable [19], such as in the case of many attack trees being generated with no regard to human readability [19, 27]. This is particularly problematic given that a significant motivation for automatic generation of attack trees in the first place is making security assurance easier in the practical domain by easing the "cognitive load" involved in performing them [19, 49].

And so the issue of reliance on formal models of the underlying system continually resurges, which renders these methods unsuitable for scenarios such as black-box security testing, increasingly relied upon in the automotive domain [13, 31, 37]. Furthermore, over-reliance on faults perceptible to the system model constrains the generated attack tree to well explored and understood aspects of systems, since accurately modelling parts of the system necessitates a good understanding of their behaviour, placing the enumeration of unexpected faults out of bounds, despite these sorts of behaviours playing a key part in real world cyber threats [33]. In addition, a model-based approach makes it difficult to scale the enumeration of possible attack paths with newly discovered vulnerabilities, as new possibilities emerge, it is vital to respond as fast as possible and infer what possible attack may be enabled using this new vulnerability, which would have to be properly contextualised within the model (such as what component is affected and potentially which functionality it exploits), which is not always clear at first. An alternative is to instead shift the source of information towards libraries, as they can more easily be extended as new vulnerabilities are discovered and can be applied to a wide variety of systems, in contrast to a model of a single system. Several promising steps in this direction have already been taken recently, which can be seen in works such as that of Falco et al. [51], who extensively use existing cybersecurity databases and patterns as part of their generation scheme, the work of Byrans et al. [45] whose variable-based template libraries can be used with various system models, and the work of Pinchinat et al. [47] who delivers the only model-free generation method examined in this survey by relying on libraries and traces in place of a model.

New Directions

In order to examine the way in which attack trees can be generated, it would be informative to first consider how attack tree construction and analysis is performed by human experts. When introducing the attack tree, Schneider offered a fairly informal procedure for their construction [18]:

1. Identify possible attack goals.
2. Think of all attacks against each goal and add them to the tree.
3. Repeat this process down the tree until done.

This interpretation of attack tree construction as a top-down procedure is further corroborated by the ISO/SAE 21434 standard for automotive cybersecurity, which states that "attack paths are deduced (i.e. theorised, inferred, reasoned, conjectured) for the item or component based on historical knowledge of vulnerabilities in similar systems and components" [14]. Thus, the attack tree in practice can be seen as a hypothesis on the threats and vulnerabilities that may exist within a certain type of system.

As far as the analysis of attack trees goes (such as for risk assessments), Schneider [18] proposes augmenting the nodes with attributes, such as possibility and cost. In contrast to the construction process, this proceeds in a bottom-up fashion, starting with the assignment of values to the low-level leaf nodes and propagating them up the tree. This eventually results in the root node having a value that summarises the procedure as a whole (such as if an attack is possible or the cost of performing a successful attack). This upward propagation of information is significant, as it highlights the fact that certain information must be sourced from the lowest level and then fed to the high level more abstract nodes. Importantly, this suggests that information from a specific system-under-test is meant to be introduced in the analysis rather than construction stage, which contrasts with the way in which models of specific systems have been utilised in attack tree generation thus far, which take a bottom-up approach towards construction [19].

Therefore, a move away from strict formalisms and towards probabilistic AI driven approaches may be necessary to more faithfully capture the process behind creating and analysing attack trees by a human expert, specifically pertaining to the inherently hypothesis-driven nature of creating attack trees [18]. However, a natural objection to the proposition that attack trees should be generated in a probabilistic and/or model-free manner is that attack trees are an explicit representation of relationships between real entities, and subsequent attack tree analysis like risk assessments and testing will involve real systems, thus generating new attack trees with no explicit mapping to a real system seems counterproductive. Firstly, since there is a potentially limitless amount of possible attacks, the task of enumerating them all in a formal fashion is infeasible [8, 42] regardless of how expressive the system model is. Instead, the focus can be shifted towards vulnerabilities deemed likely exploitable by real attackers in order to strengthen the most vulnerable areas of a system, which is precisely the idea behind performing risk assessments and aligns with the way attack trees are used in practice within the automotive industry [14]. To that end, an attack tree generated in such a manner is not meant to represent a successful real-life attack for a specific system but to infer a potentially feasible attack on a given type of system. Thus, a generated attack tree can represent a hypothesis on where an attack might be possible, to enable an expert to judge how resilient a number of given systems are against this possible attack. Hence, the use of libraries of general patterns of attack is a useful way of integrating commonly used techniques against a wide variety of systems even if not all of these may be possible to perform on a given system. Concerning how this could be executed in practice, the use of vulnerability databases and further expert input should not be discounted as a possibility, and there are various ways in which such attack tree generators could find practical use, though naturally not without obstacles [24].

In regards to attack tree-based testing, it may seem difficult to imagine how an attack tree without explicit reference to the real system may possibly be used, but there are approaches which could be explored [24]. For instance, the risk assessment as outlined by Schneider [18] involving labelling leaf nodes can be performed as a penetration test where, as each attack action is performed on a system, it is marked with an appropriate annotation (such as "possible" or "impossible", as in Schneider's example [18]). As an example, when given the goal of performing a code injection on a remote server, a human expert would construct an attack tree for "code injection" with children such as: "SQL injection", "buffer overflow" and "XSS" with no reference to any specific server, rather encapsulating their collective experience with code injection into a single tree. During the subsequent analysis, an expert performing a penetration test will then attempt to compromise a specific system-under-test and judge if each of these actions is possible to perform. If there is no SQL capability on the server, then naturally "SQL Injection" is marked as impossible and so on. Thus, the attack tree can be constructed in a system-agnostic manner and an analysis can be performed in an empirical manner rather than as a pure risk assessment.

While there is some research into the use of attack trees in such a manner for the automotive domain [21–23], this direction is still not well attested enough to assert it as a practically viable methodology with automatically generated attack trees, particularly when even risk assessment cannot yet find practical use with automatically generated trees [19]. However, the potential is there for an increasingly automated procedure for automotive cybersecurity assurance [24], and further research into themes concerning model-free generation and a testing use case may prove fruitful. Of particular interest in this direction could be the work of Falco et al. [51], hitherto not studied in relation to other attack tree generation methodologies (see Fig. 7.1), which ties the attack tree generation process much closer to the types of tools it would require in order to perform a penetration test in the manner described by authors in attack tree-based testing [21–23]. In addition, allowing for the inclusion of certain degrees of information, such as in the work of Bryans et al. [45], who compared to Salfer et al. [35], require much less information from their model of the on-board network, allows for approaching testing from a grey-box rather than fully black-box manner, which could prove a more practically viable middle ground.

While the state of the art in attack tree generation developed, in parallel many breakthroughs have been made in the realm of machine learning graph generation [55]. In particular, the rise of graph neural networks [32] has opened the door for a wide variety of graph-based learning tasks, such as generation, hitherto unexplored due to the complexity of utilising traditional neural networks with graph-based structures.

Advances in fields such as molecule generation [56, 57] offer methods for learning semantically rich data from a given domain using typed nodes and edges. Further, methods using models such as recurrent neural networks [58], neural network driven sequential decision making processes [59] and Markov models [60] allow the learning of the procedure by which one generates a given graph or tree.

However, despite the promise of such approaches, one should not assume that transposing graph generation research into the field of attack tree generation is trivial, as there are many concerns and obstacles which would need to be surmounted before such a procedure could be developed for this domain. For instance, methods focusing on learning graph representations based on connectivity via adjacency matrices are not a good fit for attack trees given the relatively sparse amount of edges compared to a more general graph. Further, an attack tree may not offer learnable parameters based on structure alone but may have to be conditioned on cybersecurity domain information via natural language processing or other forms of data which would create a learnable space for a neural network to work in. Finally, there are concerns regarding the availability of training data, as many of the methods mentioned require a very large training set, which may be difficult to source for the cybersecurity field, particularly for the automotive field.

Despite these challenges, if executed correctly a machine learning-based method could offer opportunities that are not possible with traditional generation methods. It is feasible to imagine a hybrid method which may potentially combine the strengths of both while introducing the powerful ability of machine learning to generalise to new data, which for instance could be used to extend or classify elements within a library used to generate attack trees.

Conclusion and Future Research

In response to increasing calls for stricter cybersecurity assurance for modern vehicles [17], there is an increasing need for further automation and systematisation of methodologies, which render the task of assuring vehicle cybersecurity easier. To that end, this paper has explored attack tree generation as a potentially advantageous direction towards an automated cybersecurity assurance paradigm for vehicles.

A comprehensive survey of the literature on automatic attack tree generation was conducted, in which 15 different methods were collected and evaluated on their applicability to various aspects of assuring cybersecurity in automotive systems, such as aiding risk assessments and penetration tests. As such, this survey observes that there is a discordance between attack trees in cybersecurity practice and the directions taken by researchers in attack tree generation in general namely in the fact that the former consider attack trees system agnostic and utilise them as such, while the vast majority of literature on attack tree generation utilises models of a specific system to some extent. Not only does this violate the system-agnostic nature of attack trees, but it also renders them of little use for black-box scenarios, which are increasingly relevant in the automotive domain [13, 31, 37].

Therefore, more research in the direction of model-free generation methods and attack tree aided testing needs to be performed, alongside exploring alternative means of generating attack trees rather than only focusing on formal methods. For instance, several recent advances in machine learning generation of graphs have spawned methods for generating molecular structures [56, 57] or sequential graph generation methods [58–60], which could produce a truly model-free attack tree generation scheme. However, several hurdles would need to be overcome before such an approach becomes feasible, such as the problem of sourcing a rich and extensive data set of attack trees to train on.

References

1. Koscher, K., Czeskis, A., Roesner, F., Patel, S., Kohno, T., Checkoway, S., McCoy, D., Kantor, B., Anderson, D., Shacham, H., Savage, S.: Experimental security analysis of a modern automobile. In: 2010 IEEE Symposium on Security and Privacy. IEEE (2010)
2. Salfer, M., Eckert, C.: Attack graph-based assessment of exploitability risks in automotive on-board networks. In: Proceedings of the 13th International Conference on Availability, Reliability and Security—ARES 2018, pp. 1–10 (2018)
3. Möller, D.P.F., Haas, E., Akhilesh, K.B.: Automotive electronics, IT, and cybersecurity. In: IEEE International Conference on Electro Information Technology, pp. 575–580. IEEE Computer Society (2017)
4. Hee, K.K., Tae, S.K., Myoung, K.H.: A security risk assessment framework for smart car. In: Proceedings—2016 10th International Conference on Innovative Mobile and Internet Services in Ubiquitous Computing, IMIS 2016, pp. 102–108. Institute of Electrical and Electronics Engineers Inc. (2016)
5. Valasek, C., Miller, C.: Adventures in automotive networks and control units. Technical Report, IOActive (2014)
6. Checkoway, S., Mccoy, D., Kantor, B., Anderson, D., Shacham, H., Savage, S., Koscher, K., Czeskis, A., Roesner, F., Kohno, T.: Comprehensive experimental analyses of automotive attack surfaces. In: Proceedings of the 20th USENIX Conference on Security. USENIX Association, USENIX Association (2011)
7. Hoppe, T., Kiltz, S., Dittmann, J.: Security threats to automotive CAN networks Practical examples and selected short-term countermeasures. In: Reliability Engineering and System Safety, vol. 96, pp. 11–25. Elsevier (2011)
8. Cheah, M., Shaikh, S.A., Haas, O., Ruddle, A.: Towards asystematic security evaluation of the automotive Bluetooth interface. Veh. Commun. 9, 8–18 (2017)
9. Miller, C., Valasek, C.: Remote exploitation of an unaltered passenger vehicle. Technical Report, Black Hat USA, (2015)
10. Foster, I., Prudhomme, A., Koscher, K., Savage, S.: Fast and vulnerable: a story of telematic failures. Technical Report (2015)
11. Barisani, A., Bianco, D.: Injecting RDS-TMC traffic information signals. Technical Report, Inverse Path (2007)

12. Bozdal, M., Samie, M., Aslam, S., Jennions, I.: Evaluation of CAN bus security challenges. Sensors **20**(8), 2364 (2020)
13. Mahmood, S., Nguyen, H.N., Shaikh, S.A.: Automotive Cybersecurity Testing: Survey of Testbeds and Methods. Springer, Cham (2021)
14. ISO and SAE International. ISO/SAE 21434 Road Vehicles—Cybersecurity Engineering (2021)
15. VDA QMC Working Group 13 and Automotive SIG. Automotive SPICE Process Assessment: Reference Model (2015)
16. Fuchs, A., Sigrid, G.: D3.4.4: On-board architecture and protocols attack analysis. Technical Report December, Fraunhofer Institute for Secure Information Technology (2010)
17. World Forum for the Harmonization of Vehicle Regulations. Proposal for a new unregulation on uniform provisions concerning the approval of vehicles with regards to cyber security and cyber security management system (2020)
18. Schneider, B.: Attack trees. Dr Dobb's J. **24**, 12 (1999)
19. Gadyatskaya, O., Trujillo-Rasua, R.: New directions in attack tree research: catching up with industrial needs. LNCS, vol 10744, pp. 115–126. Springer (2018)
20. Widel, W., Audinot, M., Fila, B., Pinchinat, S.: Beyond 2014: formal methods for attack tree-based security modeling. ACM Comput. Surv. **52**, 8 (2019)
21. Cheah, M., Nguyen, H.N., Bryans, J., Shaikh, S.A.: Formalising systematic security evaluations using attack trees for automotive applications. In: Lecture Notes in Computer Science (including subseries Lecture Notes in Artificial Intelligence and Lecture Notes in Bioinformatics), LNCS, vol. 10741, pp. 113–129. Springer (2018)
22. Mahmood, S., Fouillade, A., Nguyen, H.N., Shaikh, S.A.: A model-based security testing approach for automotive over-the-air updates, vol. 10, pp. 6–13. Institute of Electrical and Electronics Engineers Inc. (2020)
23. Dürrwang, J., Braun, J., Rumez, M., Kriesten, R., Pretschner, A.: Enhancement of automotive penetration testing with threat analyses results. SAE Int. J. Transp. Cybersecur. Priv. **1**, 91–112, 11 (2018)
24. Sowka, K., Cobos, L.-P., Ruddle, A., Wooderson, P.: Requirements for the automated generation of attack trees to support automotive cybersecurity assurance. SAE Tech. Paper Series **1**, 3 (2022)
25. Kordy, B., Piètre-Cambacédès, L., Schweitzer, P.: DAG-based attack and defense modeling: don't miss the forest for the attack trees. Comput. Sci. Rev. **13–14**(C), 1–38 (2014)
26. Nagaraju, V., Fiondella, L., Wandji, T.: A survey of Fault and Attack Tree Modeling and Analysis for Cyber Risk Management, vol. 6. Institute of Electrical and Electronics Engineers Inc. (2017)
27. Gadyatskaya, O., Jhawar, R., Mauw, S., Trujillo-Rasua, R., Willemse, T.A.C.: Refinement-aware generation of attack trees. In: Security and Trust Management, vol. 10547, 1st edn. Springer International Publishing (2017)
28. Ivanova, M.G., Probst, C.W., Hansen, R.R., Kammüller, F.: Transforming graphical system models to graphical attack models. In: Graphical Models for Security. GraMSec 2015. Lecture Notes in Computer Science, vol. 9390, pp. 82–96. Springer, Cham (2016)
29. Vigo, R., Nielson, F., Nielson, H.R.: Automated generation of attack trees. In: 2014 IEEE 27th Computer Security Foundations Symposium, pp. 337–350 (2014)
30. Pinchinat, S., Acher, M., Vojtisek, D.: Towards synthesis of attack trees for supporting computer-aided risk analysis. In: Software Engineering and Formal Methods, vol. 8938, 1st edn. Springer International Publishing (2015)
31. Cheah, M., Shaikh, S.A., Bryans, J., Nguyen, H.N.: Combining third party components securely in automotive systems. In: Lecture Notes in Computer Science (including subseries Lecture Notes in Artificial Intelligence and Lecture Notes in Bioinformatics), LNCS, vol. 9895, pp. 262–269. Springer (2016)
32. Scarselli, F., Gori, M., Chung Tsoi, A., Hagenbuchner, M., Monfardini, G.: The graph neural network model. IEEE Trans. Neural Netw. 61–80 (2009)

33. Felderer, M., Büchler, M., Johns, M., Brucker, A.D., Breu, R., Pretschner, A.: Security Testing: A Survey, volume 101. Academic Press Inc., 1 2016.
34. Schieferdecker, I., Grossmann, J., Schneider, M.: Model-Based Security Testing, pp. 1–12 (2012)
35. Salfer, M., Schweppe, H., Eckert, C.: Efficient attack forest construction for automotive on-board networks. Lecture Notes in Computer Science (including subseries Lecture Notes in Artificial Intelligence and Lecture Notes in Bioinformatics), vol. 8783, pp. 442–453, 10 (2014)
36. Karray, K., Danger, J-L., Guilley, S., Abdelaziz Elaabid, M.: Attack tree construction and its application to the connected vehicle. In: Cyber Physical Systems Security. Springer International Publishing (2018)
37. Hoppe, T., Kiltz, S., Dittmann, J.: Automotive IT-security as a challenge: basic attacks from the black box perspective on the example of privacy threats. In: Lecture Notes in Computer Science (including subseries Lecture Notes in Artificial Intelligence and Lecture Notes in Bioinformatics), vol. 5775, pp. 145–158, LNCS. Springer, Berlin (2009)
38. Macher, G., Armengaud, E., Brenner, E., Kreiner, C.: A review of threat analysis and risk assessment methods in the automotive context, LNCS, vol. 9922, pp. 130–141. Springer (2016)
39. Fowler, D.S., Cheah, M., Shaikh, S.A., Bryans, J.: Towards a testbed for automotive cybersecurity. In: 2017 IEEE International Conference on Software Testing, Verification and Validation (ICST). IEEE (2017)
40. Hoffmann, J.: Simulated penetration testing: from "Dijkstra" to "Turing Test++". In: Proceedings International Conference on Automated Planning and Scheduling, ICAPS, pp. 364–372 (2015)
41. Kovacevic, I., Gros, S.: Red teams—pentesters, apts, or neither, vol. 9, pp. 1242–1249. Institute of Electrical and Electronics Engineers Inc. (2020)
42. Felderer, M., Zech, P., Breu, R., Büchler, M., Pretschner, A.: Model-based security testing: a taxonomy and systematic classification. Software Testing, Verification and Reliability (2015)
43. Sheyner, O., Haines, J., Jha, S., Lippmann, R., Wing, J.M.: Automated generation and analysis of attack graphs, vol. 2002, pp. 273–284. Institute of Electrical and Electronics Engineers Inc. (2002)
44. Nguyen, H.N., Bryans, J., Shaikh, S.A.: Attack defense trees with sequential conjunction. In: 2019 IEEE 19th International Symposium on High Assurance Systems Engineering (HASE), pp. 247–252 (2019)
45. Bryans, J., Liew, L.S., Nguyen, H.N., Sabaliauskaite, G., Shaikh, S., Zhou, F.: A template-based method for the generation of attack trees. In: Maryline, L., Giannetsos, T., (eds.) Information Security Theory and Practice, pp. 155–165. Springer International Publishing, Cham (2020)
46. Jhawar, R., Kordy, B., Mauw, S., Radomirovi´c, S., Trujillo-Rasua, R.: Attack trees with sequential conjunction. In: ICT Systems Security and Privacy Protection, vol. 455, 1st edn. Springer International Publishing, (2015)
47. Pinchinat, S., Schwarzentruber, F., Lê Cong, S.: Library based attack tree synthesis, LNCS, vol. 12419, pp. 24–44. Springer Science and Business Media Deutschland GmbH, 6 (2020)
48. Hong, J.B., Kim, D.S., Takaoka, T.: Scalable Attack Representation Model Using Logic Reduction Techniques, pp 404–411 (2013)
49. Paul, S.: Towards Automating the Construction & Maintenance of Attacktrees: A Feasibility Study, pp. 31–46 (2014)
50. Xu, J., Venkatasubramanian, K.K., Sfyrla, V.: A methodology for systematic attack trees generation for interoperable medical devices, vol. 6. Institute of Electrical and Electronics Engineers Inc. (2016)
51. Falco, G., Viswanathan, A., Caldera, C., Shrobe, H.: A masterattack methodology for an ai-based automated attack planner for smart cities. IEEE Access **6**, 48360–48373, 8 (2018)
52. Lemaire, L., Vossaert, J., De Decker, B., Naessens, V.: Security Evaluation of Cyber-Physical Systems Using Automatically Generated Attack Trees, LNCS, vol. 10707, pp. 225–228. Springer (2018)
53. Jhawar, R., Lounis, K., Mauw, S., Ramírez-Cruz, Y.: Semiautomatically Augmenting Attack Trees Using an Annotated Attack Tree Library, LNCS, vol. 11091, pp. 85–101. Springer, 9 (2018)

54. Kern, M., Liu, B., Betancourt, V.P., Becker, J.: Model based attack tree generation for cyber-security risk-assessments in automotive. In: 2021 IEEE International Symposium on Systems Engineering (ISSE), pp. 1–7, Vienna, Austria. IEEE (2021)
55. Faez, F., Ommi, Y., Baghshah, M.S., Rabiee, H.R.: Deep Graph Generators: A Survey (2020)
56. Ma, T., Chen, J., Xiao, C.: Constrained Generation of Semantically Valid Graphs via Regularizing Variational Autoencoders (2018)
57. de Cao, N., Kipf, T.: MolGAN: An Implicit Generative Model for Small Molecular Graphs (2018)
58. You, J., Ying, R., Ren, X., Hamilton, W.L., Leskovec, J.: GraphRNN: generating realistic graphs with deep auto-regressive models. In: 35th International Conference on Machine Learning, ICML 2018, vol. 13, pp. 9072–9081 (2018)
59. Li, Y., Vinyals, O., Dyer, C., Pascanu, R., Battaglia, P.: Learning Deep Generative Models of Graphs. arXiv, 3 (2018)
60. Bacciu, D., Micheli, A., Sperduti, A.: Compositional generative mapping for tree-structured data-part i: Bottom-up probabilistic modeling of trees. IEEE Trans. Neural Netw. Learn. Syst. **23**, 1987–2002 (2012)

Chapter 8
Malware Analysis Using Machine Learning Tools and Techniques in IT Industry

N. G. Bhuvaneswari Amma⊙ and R. Akshay Madhavaraj

Introduction

Malware continues to be one of the most potent hazards in cyberspace despite considerable advancements and the ongoing evolution of cybersecurity measures [23]. Malware analysis examines the malicious sample using techniques from other fields, such as network analysis and programme analysis, to better understand its behaviour and long-term evolution. Every development in security technology is typically promptly tracked by equivalent bypasses in the never-ending conflict between malware developers and malware analysts [15]. The features the new defences make use of have a role in how effective they are. For instance, obfuscation and more advanced techniques like polymorphism and metamorphism can both be used to easily avoid detection algorithms constructed on the message digest of known malware. For a thorough analysis of these methods, these techniques change the hash and binary of the malware, but not its behaviour. On the other hand, it is far more challenging to defeat detection criteria that capture the semantics of bad samples since malware creators must make more intricate alterations. The primary goal of malware is to discover new attributes that may be utilised to strengthen security controls and make evasion as challenging as feasible. A logical choice to assist such a knowledge extraction process is machine learning. In fact, a large body of literature points in this direction using a variety of methods, objectives, and conclusions [8].

In recent years, Android OS and smart city technologies have been widely used in a variety of informational industries, including smart governance, intelligent transportation, energy and resource management, etc. Because of the considerable ease

N. G. Bhuvaneswari Amma (✉) · R. Akshay Madhavaraj
Vellore Institute of Technology, Chennai, Tamil Nadu, India
e-mail: bhuvaneswariamma.ng@vit.ac.in

R. Akshay Madhavaraj
e-mail: akshay.madhavaraj2019@vitstudent.ac.in

© The Author(s), under exclusive license to Springer Nature Singapore Pte Ltd. 2023
V. Sarveshwaran et al. (eds.), *Artificial Intelligence and Cyber Security in Industry 4.0*, Advanced Technologies and Societal Change,
https://doi.org/10.1007/978-981-99-2115-7_8

that the Android platform offers, the malware that targets it also poses serious risks that result in monetary loss, data leaks, worries about national security, and terrorist attacks [24]. Due to Android's exceptional openness and flexibility, it has more than 80% of the smartphone market. On the other side, malware has significantly affected Android as well. Approximately 4 million new harmful APPs were created, and every eight seconds on average, hackers produce a malicious application [12]. Cyberse-curity is being threatened by the malware epidemic that is spreading quickly. To improve mobile security, effective detection techniques (or frameworks) for Android malware are urgently needed. To reduce the danger of harmful APPs, mobile security researchers have presented a variety of protection strategies, particularly the malware detection methods based on machine learning techniques that seem to be the answer to all problems [17]. Machine learning applications have drawn many security experts' interest to this magical technology. Modern malware detection approaches using machine learning techniques can be broadly categorised into three groups based on the underlying feature: static feature, dynamic feature, and mixed feature [7].

The majority of earlier research has shown that feature selection is an essen-tial step in the training of machine learning models [10]. Researchers use feature selection to minimise duplicate or irrelevant features and refine important features in order to enhance model performance. However, feature trimming procedure typically destroys the integrity of the underlying data, which is a time-consuming and difficult task on top of that [2]. TC-Droid, a robust Android malware detection framework feeding on reports of APK static analysis, eliminates the need for feature selection. AndroPyTool, which integrates popular Android APP analysis tools like DroidBox, FlowDroid, Strace, AndroGuard, or VirusTotal analysis, produces a series of words for each analysis report [4].

Over 90 eminent IT organisations, including Accenture, Acer, Bell, etc., were attacked with malware in 2021. While some businesses can recover from the harm done to them and the financial loss they suffered within a few days or weeks of cleaning, others will never be able to do so. An effective, reliable, and scalable malware recognition module is essential to every cybersecurity product. Malware identification modules determine whether an object poses a threat based on the information they have gathered about it. This information may be gathered at many stages:

1. Pre-execution phase data is anything you can learn about a file without actu-ally running it. This might include details gleaned from code emulation, text strings, statistics on binary data, explanations of executable file formats, code descriptions, and other information of the same type.
2. Post-execution phase data defines the behaviours or events that result from process activity in a system.

As malware threats were few and far between in the early days of the cyber age, simple pre-execution rules often sufficed to identify dangers. Because of the rapid development of the Internet and the resulting increase in malware, manually constructed detection rules are no longer practicable. This has necessitated the devel-opment of new, cutting-edge protection strategies. The use of machine learning in anti-malware organisations has been successfully applied to search, recognition, and decision-making in the past. Today, machine learning improves malware detection by utilising data from host, network, and cloud-based anti-malware components [6].

Malware Analysis Using Machine Learning Techniques

In this section, malware analysis using machine learning techniques is discussed. We have read a lot of articles to fully comprehend the implementation strategies that are now in use and to categorise the work that has been done, such as the machine learning algorithm utilised, the features employed. The sort of algorithm (machine learning) utilised for malware analysis, such as unsupervised learning, is examined in the first subsection. The second paragraph discusses the main goal of malware analysis, such as malware similarity analysis. The third and final subsections list the features taken into consideration, such as opcodes, as well as the feature extraction techniques used for the study, such as static analysis [8]. Figure 8.1 depicts the hierarchy of malware analysis using machine learning.

Algorithms for Malware Analysis

On the basis of how they learn, machine learning algorithms can be roughly divided into three categories: supervised, unsupervised, and semi-supervised.

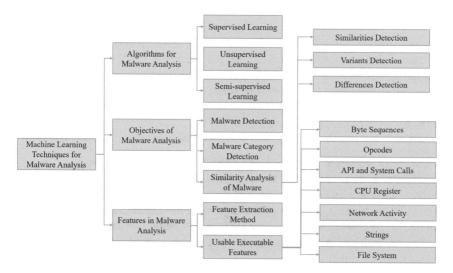

Fig. 8.1 Hierarchy of malware analysis

Supervised Learning An approach to building AI-called supervised learning uses input data that has been tagged for certain outputs to train computer algorithms [5]. Models are developed until they can identify underlying trends and connections between the input and output labels, allowing them to produce precise labelling results even confronted with unheard of amounts of data. Several supervised algorithms, including Bayesian network, Bayes classifier, Rule-based classifier, multiple kernel learning, decision tree, Naive Bayes, support vector machine, prototype-based classification, gradient boosting decision tree, random forest, logistic model tree, artificial neural network, multilayer perceptron neural network, and K-nearest neighbours, were used in recent research [9].

Unsupervised Learning The antithesis of supervised learning is unsupervised learning. In this method, algorithms are given unlabelled data and created to recognise patterns and similarities in a special way. The following were the unsupervised algorithms: expectation maximisation, self-organising maps, hierarchical clustering, prototype-based clustering, K-means clustering, clustering with locality-sensitive hashing, clustering with similarity, and distance metrics either Jaccard, Euclidean, Hamming distance, or cosine similarities), and K-medoids [21].

Semi-supervised Learning Between supervised and unsupervised learning algorithms, semi-supervised learning is a type of learning algorithm. During the training phase, this learning technique uses both labelled and unlabelled datasets.

Belief propagation and learning with local and global consistencies are the two algorithms that are employed [9].

Objectives of Malware Analysis

Strong detection capabilities are often needed for malware analysis to identify that match utilising information from studying prior samples. Finding these pairings has a different final objectives, though. As an illustration, malware analyst might be more interested in identifying whether a new suspicious sample is harmful, while another malware analyst looks into the category that the more recent malware finds are most likely to fall into. The three primary goals of the studies under review were malware identification, malware category detection, and malware similarity analysis [18].

Malware Detection Identifying if a specific sample is harmful is the most frequent objective of malware analysis. This objective is the most crucial because it allows us to stop a sample from becoming harmful if we are aware of its danger in advance. In reality, this is the main objective of the majority of the assessed works. The resulting output can be labelled with a confidence score depending on the machine learning technique employed. Analysts can use this to determine whether the sample deserves to be looked at further [9].

Malware Category Detection Based on its distinctive behaviour and intended victims, malware can be categorised. They might be motivated by the desire to take remote control of an infected machine, encrypt documents and demand a ransom, or spy on user activities to steal private data [20]. These categories offer a crude yet insightful way to categorise harmful samples. Although cybersecurity firms have not yet reached consensus on a common taxonomy for malware types, correctly classifying the samples can yield useful data for study [14].

Similarity Analysis of Malware Finding commonalities between various malware and comprehending how fresh samples differ from known samples are two other related objectives. It is discovered that similarity analysis is employed for four very distinct things: the detection of similarities, variations, differences, and families.

- *Similarities detection*: Analysts may find it interesting to note certain parallels and differences between the binary being examined and those that have already been examined. Similarity detection is the process of determining if a sample's constituent parts and distinguishing characteristics like those have previously been investigated. This enables us to concentrate on what is actually novel while excluding the rest as unworthy of further investigation.
- *Variants' detection*: By reusing existing code and resources to the greatest extent feasible, creating variations is one of the most efficient and affordable techniques for attackers to circumvent detection mechanisms. Such tactics are made more difficult by understanding how malware evolves over time as new varieties emerge and by realising that samples are actually variations of well-known malware. With multiple peer-reviewed studies focused on variation detection, this objective has also been thoroughly researched in the literature. Variant detection includes choosing patterns from the knowledge base that are variations of a dangerous pattern m. In order to lessen the strain of human analysts, it is crucial to identify recognised malware variants given the daily enormous influx of dangerous samples from top security firms.
- *Differences' detection*: In addition, it is important to notice the variations from others that have been seen in prior findings. In actuality, variations can result in the identification of fresh elements that demand in-depth investigation.

Features in Malware Analysis

This section describes the sample features considered in the analysis. The feature selection process is covered in the upcoming subsections.

Feature Extraction Method Static analysis, dynamic analysis, or a combination of both is used for the information extraction process, while machine learning techniques are used for testing and correlation. Static analysis-based methods look at the sample's contents without running it, whereas dynamic analysis runs the sample and looks how it behaves. Malware dynamic analysis can be done using a variety of ways. For analysis at the instruction level, a debugger is used. A simulator builds a simulation of the environment that malware anticipates existing in. Emulators, on the other hand, more closely mimic the behaviour of the system but demand more resources [13].

A sandbox is a virtualised operating system that offers a secure, private setting for the explosion of malware. When employing dynamics analysis, execution traces are frequently used to extract characteristics. Verified articles produce execution traces in a sandbox or emulator. Tools and methods for programme analysis are very helpful in the feature extraction process. For instance, it offers control and data flow diagrams as well as disassembled code. The ability to extract API and system calls depends on the ability to recover the correct byte sequences and opcode functions from accurately disassembled code [19].

Usable Executable Features An overview of the characteristics employed in the papers examined to accomplish the goals outlined in the aforementioned sections is provided.

- *Byte sequences*: When computing features are applied to byte-level content, binary content can be characterised. A common static approach is to analyse specific byte sequences within a PE. It is common to use specific-sized data chunks in some works, but n-grams are also used. An n-gram is a collection of n bytes and a set of attributes connected to different combinations of these n bytes. To put it another way, each feature displays the frequency with which a particular collection of n bytes appears in binary. The majority of the publications under evaluation demonstrate that the number of n-grams employed depends on three or fewer sequences. Be aware that as n increases, the number of features to be considered increases exponentially.

- *Opcodes*: The machine-level operations that the PE performs are identified by the opcode, which is obtained through static analysis by looking at the assembly code. One of the most often used functions is opcode frequency. It tracks the frequency with which each unique opcode appears in the assembly or is used by PE. Others use tools like memory access instructions and mathematical instructions to count opcode occurrences in a different method. The same manner that n-grams are used as features, so are sequences of opcodes.

- *API and system calls*: The examination of example behaviour is made possible through APIs and device calls, which are more like opcodes than not. To determine if they can be evoked statically or dynamically, the execution traces or the disassembled code can be analysed (for the listing of calls clearly invoked). While system call invocations offer insight into how the PE communicates with the working device, APIs allow for the pattern-based description of the activities executed. The ability to extract large volumes of data utilising device calls and APIs for seeing has led to numerous works undergoing further processing in order to lower function area using useful information systems. One of the most well-known data structures for illustrating PE behaviour and determining the programme's structure is the control flow graph. Compilers can provide an optimal representation of the linkages between the programme and model control flow thanks to this information shape [11]. In efforts to analyse samples, control flow graphs and their extensions are regularly paired with other feature classes.

- *CPU register*: Additionally, how the CPU registers are used, particularly the FLAGS register, and whether hidden registers are used can provide useful information.
- *Network activity*: By examining how the PE collaborates with the network, one can find a large variety of important records. Hidden information, such as details regarding communications with a command and control centre, can be discovered by analysing generated traffic and contacted addresses. Information on commonly used protocols, TCP/UDP ports, HTTP requests, and DNS-stage exchanges are all relevant components. Dynamic analysis is required to retrieve these types of records from the majority of the works assessed. Other articles extract inputs that are connected to networks by watching the network and looking at incoming and outgoing patterns. A related strategy is to examine users' download patterns on a network that is being watched. It no longer requires pattern execution and instead merely concentrates on community capabilities linked to pattern downloads.
- *Strings*: In order to specifically look for the strings that a PE contains—such as file names, resource details, code snippets, author signatures, it can be statically scanned.
- *File system*: The sample's file manipulations are essential for gathering proof of interaction with the environment and perhaps spotting attempts to become persistent. File types, directories, and files that are viewable on clean versus infected PCs are the key interesting elements. Modules that track file system interactions with file systems represented by counting the number of files generated, deleted, or modified by a PE are frequently included in sandboxes and storage analysis toolkits [16].

Challenges of Malware Analysis

It is clear how effective and beneficial machine learning can be for analysing malware. Saving time and valuable human resources are made possible through machine learning [13]. We must also take into account the reality that no machine or algorithm is perfect and does not have problems of its own. Some prominent issues or difficulties encountered in malware analysis are described in this section.

Analysis-Resistant Malware

A malware developer's objective is to prevent analysis of their samples as they are being developed; therefore, they design and enhance a number of anti-analysis approaches that are successful in preventing the reverse engineering of executables [1]. Many of the assessed publications assert that the answer they offer still holds true even after accounting for these anti-analysis tactics. Static evaluation is typically thwarted by obfuscating and encrypting sample binary and sources to render them

	Name	e_magic	e_cblp	e_cp	e_crlc	e_cparhdr	e_minalloc	e_maxalloc	e_ss	e_sp	...	SectionMaxChar	SectionMainChar
0	VirusShare_a878ba26000edaac5c98eff4432723b3	23117	144	3	0	4	0	65535	0	184	...	3758096608	0
1	VirusShare_ef9130570fddc174b312b2047f5f4cf0	23117	144	3	0	4	0	65535	0	184	...	3791650880	0
2	VirusShare_ef84cdeba22be72a69b198213dada81a	23117	144	3	0	4	0	65535	0	184	...	3221225536	0
3	VirusShare_6bf3608e60ebc16cbcff6ed5467d469e	23117	144	3	0	4	0	65535	0	184	...	3224371328	0
4	VirusShare_2cc94d952b2efb13c7d6bbe0dd59d3fb	23117	144	3	0	4	0	65535	0	184	...	3227516992	0

Fig. 8.2 Sample malware data

unreadable. To enable the accurate execution of the payload, code and other secret data must be both deobfuscated and decrypted during runtime. This indicates that any such anti-analysing tactics may be defeated by utilising dynamic evaluation to force the pattern to reveal secret data and move it into memory, from whence it could later be recovered using a dump. There are more effective anti-analysis approaches that hide a virus internals even when dynamic assessment is used [25].

The virus employs a technique called environmental awareness to determine if the sample is in debug mode or whether it is working in a controlled environment when an analyst is trying to analyse the malware, for example, by utilising a virtual machine. The malicious payload of the malware is no longer executed when any signs of a controlled environment are discovered. To study and improve the efficiency of malware detection and analysis, identifying and defeating those anti-analysis tactics are essential [22].

Accuracy

Question arises when the proposed methods and their results perform with the real-world samples. The accuracy may vary from a few percentages to a significant value. This kind of problems arises when the machines are not trained with proper and sufficient datasets. Most of the datasets used by the reviewed works are obtained from public repositories, and the balance between the malicious and benign samples varies according to their objectives. More than 78% of the works used dataset with both benign and malicious samples. Rest 28% used datasets only with malicious samples. Some works used datasets with less than 1500 samples. Only, a mere works of 29% used datasets with over 10,000 samples. Greater the number of sample, greater the accuracy remains (Fig. 8.2).

Case Study

This section discusses a case study pertaining to malware analysis using machine learning. The dataset used for case study is malware dataset from Kaggle [3]. The number of features in the dataset is 78. The features are listed as follows: *e_magic, e_cblp, e_cp, e_crlc, e_cparhdr, e_minalloc, e_maxalloc, e_ss, e_sp, e_csum, e_ip, e_cs, e_lfarlc, e_ovno, e_oemid, e_oeminfo, e_lfanew, machine,*

NumberOf Sections, TimeDateStamp, PointerToSymbolTable, NumberOf Symbols, SizeOf OptionalHeader, Characteristics, Magic, MajorLinkerV ersion, MinorLinkerV ersion, SizeOf Code, SizeOf InitializedData, SizeOf UninitializedData, AddressOf EntryPoint, BaseOf Code, ImageBase, SectionAlignment, FileAlignment, MajorOperatingSystemV ersion, MinorOperatingSystemV ersion, SizeOf Headers, CheckSum, SizeOf Image, Subsystem, DllCharacteristics, SizeOf StackReserve, SizeOf StackCommit, SizeOf HeapReserve, SizeOf HeapCommit, LoaderFlags, NumberOf RV aAndSizes, Malware, SuspiciousImportFunctions, SuspiciousNameSection, SectionsLength, SectionMinEntropy, SectionMaxEntropy, SectionMinRawsize, SectionMaxRawsize, SectionMinV irtualsize, SectionMaxV irtualsize, SectionMaxPhysical, SectionMinPhysical, SectionMaxV irtual, SectionMinV irtual, SectionMaxPointerData, SectionMinPointerData, SectionMaxChar, SectionMainChar, DirectoryEntryImport, DirectoryEntryImportSize, DirectoryEntryExport, ImageDirectoryEntryExport, ImageDirectoryEntryImport, ImageDirectoryEntryResource, ImageDirectoryEntryException, and *ImageDirectoryEntrySecurity.*

Figure 8.3 depicts the sample malware data. Figure 8.4 depicts the category of data.

The dataset consists of two classes, viz., benign and malware. The features such as *MajorLinkerV ersion, MajorSubsystemV ersion, NumberOf Symbols, SectionMaxChar, SizeOf Code,* and *SizeOf Headers* are visualised and depicted in Fig. 8.5. The importance of all the features are depicted in Fig. 8.6. It is observed that more than 50% of the features are important for malware detection.

The malware is classified into either benign or malware using random forest machine learning classifier. This classifier is a well-liked supervised learning approach. An ensemble of classifiers is known as a random forest, and it increases the amount of dependability over different classifiers. A tree-structured classification

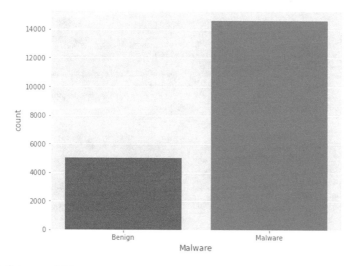

Fig. 8.3 Category of data

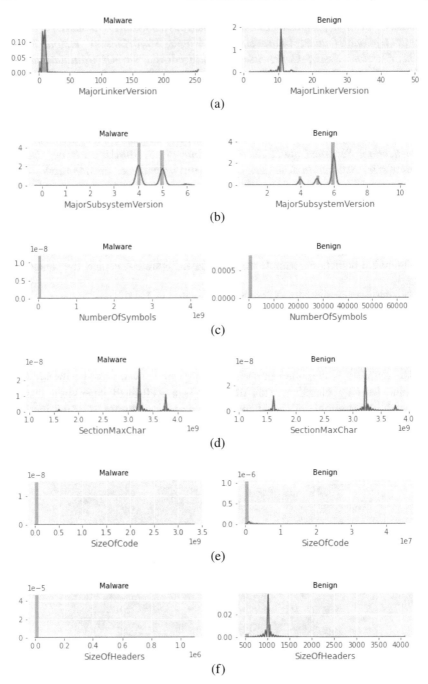

Fig. 8.4 Feature visualisation **a** major linker version, **b** major subsystem version, **c** number of symbols, **d** section maximum character, **e** section of code, **f** size of headers

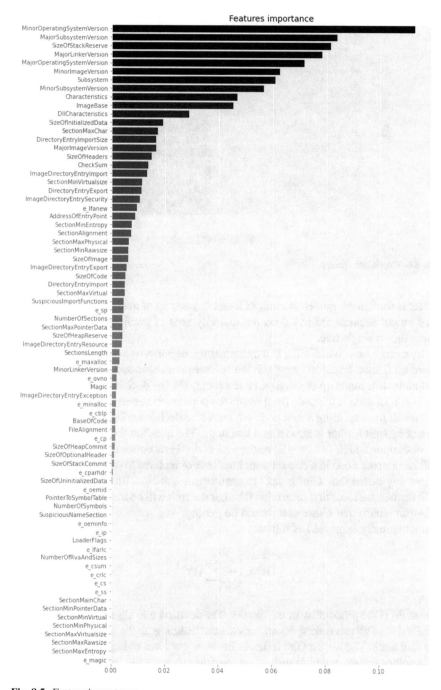

Fig. 8.5   Feature importance

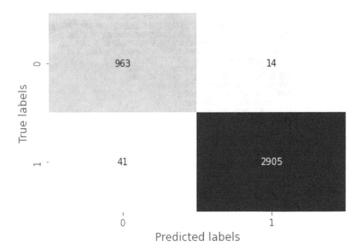

Fig. 8.6 Confusion matrix

model is used in the model. A random forest is made up of many decision trees. The final output is predicted based on the majority vote of predictions from each tree rather than a single tree.

By using a flowchart similar to a tree structure, decision trees illustrate predictions based on feature-based splitting's. Initially, there is a root node, and then, there is a leaf node. It is made up of three parts: the root node, the decision node, and the leaf node. A root node is the node from which the population begins to divide. The nodes that result from splitting a root node are known as decision nodes, and the node that cannot be split further is known as a leaf node. The question then becomes, how do we determine which feature will be the root node? How do we decide which feature will be our root node in a dataset with hundreds of features? We need to understand something called the "Gini Index" to answer this question. To choose a feature to split further, we must first determine whether the split will be impure or pure. A pure sub-split means that either you should be getting "yes" or "no". The Gini index is mathematically expressed as follows:

$$GI = 1 - \sum_{i=1}^{n} (\Pr_i)^2 \tag{8.1}$$

where $\Pr_i$ is the probability of the classes. The decision tree algorithm calculates the Gini index of all possible splits and selects the feature with the lowest Gini index as the root node. The lowest Gini index indicates a low level of impurity.

Another metric called "Entropy" is used to calculate the impurity of the split. Entropy can be calculated mathematically as follows:

$$\text{Etr}(S) = -\Pr_i \log \Pr_i \tag{8.2}$$

We typically use the Gini index because it is computationally efficient; it takes less time to execute because there is no logarithmic term as there is in entropy. It typically takes some time to perform logarithmic calculations. As a result, many boosting algorithms use the Gini index as a parameter. Another important point to remember is that if a node contains an equal number of both classes, the Gini Index will reach its maximum value, indicating that the node is highly impure. A random subspace concept is used to generate each tree. A training set with N examples should be used to construct the tree's resultant training set without replacing any cases. At each node, a number $K < M$ is assigned, wherein M is the variety of features. In order to find the best split among the K, a random subgroup of them is selected. An algorithm is made up of two elements: L and K, where L represents the size of the forest trees, and K represents the range of characteristics that should be considered when splitting the trees. The classifier would employ several decision trees on subsets of the dataset and average their results to boost the expected accuracy. Instead of relying solely on one decision tree, the random forest uses predictions from each decision tree and predicts the outcome based on the majority of projections. Overfitting and higher accuracy are prevented by the larger number of trees in the forest.

Figure 8.6 depicts the confusion matrix for the detection of malware using a random forest classifier. There are n columns and n rows in a confusion matrix, where each column represents a predicted classification and each row represents the true classification. In order to determine the model's accuracy, it is possible to examine the values along the diagonal—a good model will have a high diagonal value and low values of it. Furthermore, by examining the highest values that are not on the diagonal, one can determine where the model is having difficulty. These analyses are useful in identifying cases where the model's accuracy is high, but it consistently misclassifies the same data. The confusion matrix represents the performance measures such as true positive (T_P), false positive (F_P), true negative (T_N), and false negative (F_N). The performance metrics considered for evaluation include precision (Prec), recall (Rec), $F1$-score ($F1_{Sc}$), and accuracy (Acc).

$$\text{Prec} = \frac{T_P}{T_P + F_P} \times 100, \tag{8.3}$$

$$\text{Rec} = \frac{T_P}{T_P + F_N} \times 100, \tag{8.4}$$

$$F1_{Sc} = 2 \times \frac{\text{Precision} \times \text{Recall}}{\text{Precision} + \text{Recall}} \tag{8.5}$$

$$\text{Acc} = \frac{T_P + T_N}{T_P + F_P + T_N + F_N} \times 100 \tag{8.6}$$

Figure 8.7 depicts the classification report with the considered performance metrics. The precision, recall, F1-score, and support are computed for both the benign as well as malware classes. Furthermore, the macro average and weighted average are also computed to know the performance of the studied malware ensemble classifier.

	precision	recall	f1-score	support
Benign	0.99	0.96	0.97	1004
Malware	0.99	1.00	0.99	2919
accuracy			0.99	3923
macro avg	0.99	0.98	0.98	3923
weighted avg	0.99	0.99	0.99	3923

Fig. 8.7 Classification report

It is evident that the accuracy obtained is 99% using the random forest approach for the malware detection classifier.

Conclusion

In this chapter, we identified the importance of malware analysis in IT industry and how necessary machine learning is to the analysis. Machine learning is the ideal solution to keep pace with the malware developers. The techniques of machine learning was explained in three subsections, namely algorithms for malware analysis, objectives of malware analysis, and features in malware analysis. Then, the problems or challenges faced in malware analysis were discussed. The two main challenges identified are analysis-resistant malware and accuracy. Furthermore, a case study pertaining to malware analysis using machine learning was discussed. Thorough studies towards solving and discovering more challenges could further contribute to the advancements in malware analysis. More emphasis is to be laid on the objectives so that industries like IT can be safeguarded and work with confidence.

References

1. Aslan, Ö., Yilmaz, A.A.: A new malware classification framework based on deep learning algorithms. IEEE Access **9**, 87936–87951 (2021)
2. Batouche, A., Jahankhani, H.: A comprehensive approach to android malware detection using machine learning. Information Security Technologies for Controlling Pandemics, pp. 171–212 (2021)
3. Benign, Files, M.P.: Malware Dataset. https://www.kaggle.com (2022)
4. Damaševičius, R., Venčkauskas, A., Toldinas, J., Grigaliu¯nas, Š.: Ensemble-based classification using neural networks and machine learning models for windows pe malware detection. Electronics **10**(4), 485 (2021)
5. Damodaran, A., Troia, F.D., Visaggio, C.A., Austin, T.H., Stamp, M.: A comparison of static, dynamic, and hybrid analysis for malware detection. J. Comput. Virol. Hacking Tech. **13**(1), 1–12 (2017)

6. Feng, Z., Xiong, S., Cao, D., Deng, X., Wang, X., Yang, Y., Zhou, X., Huang, Y., Wu, G.: Hrs: A hybrid framework for malware detection. In: Proceedings of the 2015 ACM International Workshop on International Workshop on Security and Privacy Analytics, pp. 19–26 (2015)
7. Gibert, D., Mateu, C., Planes, J.: The rise of machine learning for detection and classification of malware: research developments, trends and challenges. J. Netw. Comput. Appl. **153**, 102526 (2020)
8. Huda, S., Abawajy, J., Alazab, M., Abdollalihian, M., Islam, R., Yearwood, J.: Hybrids of support vector machine wrapper and filter based framework for malware detection. Futur. Gener. Comput. Syst. **55**, 376–390 (2016)
9. Huda, S., Islam, R., Abawajy, J., Yearwood, J., Hassan, M.M., Fortino, G.: A hybrid-multi filter-wrapper framework to identify run-time behaviour for fast mal- ware detection. Futur. Gener. Comput. Syst. **83**, 193–207 (2018)
10. Kim, H.m., Lee, K.h.: Iiot malware detection using edge computing and deep learning for cybersecurity in smart factories. Appl. Sci. **12**(15), 7679 (2022)
11. Kumar, R., Zhang, X., Wang, W., Khan, R.U., Kumar, J., Sharif, A.: A multimodal malware detection technique for android iot devices using various features. IEEE Access **7**, 64411–64430 (2019)
12. Liu, K., Xu, S., Xu, G., Zhang, M., Sun, D., Liu, H.: A review of android malware detection approaches based on machine learning. IEEE Access **8**, 124579–124607 (2020)
13. Mahindru, A., Sangal, A.: Mldroid—framework for android malware detection using machine learning techniques. Neural Comput. Appl. **33**(10), 5183–5240 (2021)
14. Martín, A., Lara-Cabrera, R., Camacho, D.: Android malware detection through hybrid features fusion and ensemble classifiers: The andropytool framework and the omnidroid dataset. Information Fusion **52**, 128–142 (2019)
15. Naway, A., Li, Y.: A review on the use of deep learning in android malware detection. arXiv preprint arXiv:1812.10360 (2018)
16. O'Shaughnessy, S., Sheridan, S.: Image-based malware classification hybrid frame-work based on space-filling curves. Comput. Secur. **116**, 102660 (2022)
17. Ren, Z., Wu, H., Ning, Q., Hussain, I., Chen, B.: End-to-end malware detection for android iot devices using deep learning. Ad Hoc Netw. **101**, 102098 (2020)
18. Souri, A., Hosseini, R.: A state-of-the-art survey of malware detection approaches using data mining techniques. HCIS **8**(1), 1–22 (2018)
19. Surendran, R., Thomas, T., Emmanuel, S.: A tan based hybrid model for android malware detection. J. Inf. Secur. Appl. **54**, 102483 (2020)
20. Velliangiri, S., Manoharn, R., Ramachandran, S., Venkatesan, K., Rajasekar, V., Karthikeyan, P., Kumar, P., Kumar, A., Dhanabalan, S.S.: An efficient lightweight privacy-preserving mechanism for industry 4.0 based on elliptic curve cryptography. IEEE Trans. Ind. Inf. **18**(9), 6494–6502 (2021)
21. Venkatraman, S., Alazab, M., Vinayakumar, R.: A hybrid deep learning image- based analysis for effective malware detection. J. Inf. Secur. Appl. **47**, 377–389 (2019)
22. Verma, S., Muttoo, S.: An android malware detection framework-based on permissions and intents. Defence Sci. J. **66**(6) (2016)
23. Ye, Y., Li, T., Adjeroh, D., Iyengar, S.S.: A survey on malware detection using data mining techniques. ACM Comput. Surveys (CSUR) **50**(3), 1–40 (2017)
24. Zhang, N., Tan, Y.a., Yang, C., Li, Y.: Deep learning feature exploration for android malware detection. Appl. Soft Comput. **102**, 107069 (2021)
25. Zhu, H.J., Wang, L.M., Zhong, S., Li, Y., Sheng, V.S.: A hybrid deep network framework for android malware detection. IEEE Trans. Knowl. Data Eng. **34**(12), 5558–5570 (2021)

Chapter 9
Use of Machine Learning in Forensics and Computer Security

Nitish Ojha, Avinash Kumar, Neha Tyagi, Preetish Ranjan, and Abhishek Vaish

Introduction

Technology has changed the way of critical thinking and has evolved as magnanimous support to human basic to advanced needs. The archaic way of tackling daily life situations modifies and becomes more efficient when technology comes into the picture to tackle the social activities of a human being either as an individual or at a large scale.

The machine aims to become a synonym for the human working ability in sense of learning, understanding, processing, and executing tasks up to the most desirable level. This gave birth to the concept of 'machine learning.'

Machine Learning: Machine learning is a subdomain of artificial intelligence that can make the system learn automation and improve the result from previous learnings. Machine learning aims to develop a program that has the privilege to acquire the data and reuse it to learn itself.

N. Ojha (✉)
Department of Computer Science and Engineering, School of Information Technology, AI and Cyber Security, University of Stirling, Ras Al Khaimah, UAE
e-mail: nitishkumarojha@gmail.com

A. Kumar
Rashtriya Raksha University, Gandhinagar, Gujrat, India

N. Tyagi
IT Consultant, Cyber Security Division, Deloitte, Deloitte Consulting India, Gurugram, Haryana, India

P. Ranjan
Amity University, Patna, Bihar, India

A. Vaish
Indian Institute of Information Technology, Allahabad, India

© The Author(s), under exclusive license to Springer Nature Singapore Pte Ltd. 2023
V. Sarveshwaran et al. (eds.), *Artificial Intelligence and Cyber Security in Industry 4.0*, Advanced Technologies and Societal Change,
https://doi.org/10.1007/978-981-99-2115-7_9

The beginning of machine learning starts with recognizing the pattern via observations, and this helps the machine in making the decision more accurate, precise, and relevant according to the predefined objective of the system. The prime objective of machine learning is to make the system/computers learn and make the decision without human intervention.

History of Machine Learning

The foundation of machine learning roots deeply in the year 1943 when neurophysiologist Warren McCulloch and mathematician Walter Pitts introduced a paper on the working of neurons [1]. The working of a machine similar to a human was achieved by Alan Turing in the year 1950, through the creation of the Turing Test which helped to set a benchmark for a machine and finally concludes that a computer can work like a human being [2]. The concept of pattern recognition started with the contribution of Frank Rosenblatt by designing an artificial neural network (ANN) [3] which is one of the most desired steps of machine learning. Two scientists named Bernard Widrow and Marcian Hof of Stanford University created two powerful models named ADELINE and MADELINE, the first was capable of recognizing binary patterns, whereas the latter was able to remove echo over the phone [4]. The breakthrough came in the year 1997 when the chess-playing computer was developed by IBM [5].

Motivation

In this era of digital accessibility, cyber security is going to play a pivotal role in each segment of life, whereas in the same parallel world, artificial intelligence is getting involved with each element of technology whatever has its applicability in the real-time scenario. In the last 5 years, the IT industry is facing several challenges in the cyber security field; and after COVID-19, the pace of the involvement of IT in our daily life reached beyond its limit; and most of the revenues and manpower are getting wasted on handling those issues generated by cyber security measures while saving the assets of the organization. The cyber security report presented by *Sonicwall* in 2019 states that more than 10 billion direct attacks were there directly on core servers in most IT companies, and it was huge-size attacks based on data that makes any organization unable to perform its function as it is really big size data which is so big, which results from a nearly impossible situation to handle, to mitigate the risks and threats completely even if involving all manpower of that organization [6], that's why artificial intelligence is coming into the picture in most of the things as it becomes nearly impossible to have well-robust IT infrastructure which is full proof considering cyber security all elements. These days, each organization that is dealing with the newest technologies is in the danger zone of cyber-attacks irrespective of its size and turnover. One of the high-profile company 'Tesla' has also faced many issues even if well-packed with several cyber security measures that makes it highly immune to cyber-attacks [7]. When we talk about the versatility of cyber threats, it would be better if we can implement artificial intelligence-powered machine learning technologies in mainstream machines so that complex processes in large software packages have enough robustness to tackle the cyber threats lessening the dependency on human power.

Observing this much of the occurrence of inevitable formidable issues, it will not be wrong if it is said that artificial intelligence came as a boon. As a perfect example, in this case, we can take an example of an incident that happened where an attack was attempted on Microsoft corp. servers to install cryptocurrency mining software by penetrating the security layers of access control in 2018 (*termed as emoting attack*) by a team of unknown attackers driven by a malicious campaign; however, the after-effect was less because ML-trained AI packages where ML algorithms were working inside Microsoft defender software and that safeguarded against that attack [8].

When we talk about different applications, major intrusion detection systems (IDS) stand out which use machine learning algorithms at the primary stage, covering all its domains and types. The software then follows the established rules that allow a distinction to be made between good and bad files, without providing any guidance on what kind of similarities or data points to look for. A data point would be any unit of file-related data, including a file's internal structure, the converter used, text tools compiled into the file, and even more. The algorithm starts to be measured and refined as this ends up with an effective detection method (ideally) that should not identify any good programs as bad or bad programs [9]. As good as that develops the model by changing the weight or value of every data point. With each iteration, the model gets a little better at precisely identifying malicious and non-malicious data, and behind that, ML is working.

Anti-viruses are the most sought-after software that perfectly uses ML algorithms either for the detection or for recognition of several issues while relying on coding rules of programs. Dealing with several files and attacks and observing several data points generated from a variety of attacks, machine learning is going to help a lot while making categories of good and bad applications.

When we see the actual implementation and use of machine learning algorithms in nearly all major safeguarding tools of several domains of cyber security, then it can be classified into three segments—supervised learning algorithms, unsupervised algorithms, and reinforced learning. In the next chapters, we will see how these machine algorithms are playing their roles and what are others major segment-based applications in which ML is being used.

The Methods in Machine Learning

The algorithms used in machine learning are categorized into three types.

Supervised Machine Learning Algorithms

This concept involves predicting the future decision based on the previous learning of the dataset on labeled example [7]. The decision given by the learning algorithm is done by producing a relevant function that has been inferred from the training dataset. Moreover, this training makes the system target inputs that are new.

In supervised learning, a machine is designed to use 'label' information. The dataset is labeled when it consists of both input and yield parameters. In other words, the information is still labeled with the correct answer; therefore, the strategy mimics a classroom environment where understanding is within the grasp of an observer or teacher. On the other hand, untrained learning algorithms allow models to find data and learn from their claims.

Supervised machine learning is demonstrably helpful in keeping real-world computational issues. It predicts unpredictable information by learning from the information preparation label of the calculation. In this way, it takes highly skilled information researchers to construct and produce such models. Furthermore, information scientists use their special ability to recreate the model to preserve judgments of given experiences.

How Does It Work?

For the event, you will need to design a machine in anticipation of your commute time between your office and household. To begin with, you will set label information such as climate, time of day, and chosen course which will include your input information. And the output will be evaluated on the domestic duration of your trip on a particular day.

Once you set the readiness based on comparing variables, the machine looks at the connection between the information focus and uses it to find out how long it will take for you to drive back home. For illustration, a portable application can tell you that your travel time will be longer if there is more rainfall.

The machine can see other associations in your label information, just as you take time off work. If you come to the streets before the increased activity on the roads, you will be able to reach the domestic level. This phenomenon is investigated more to know how machine learning is done.

Currently, let us try to learn it with the help of another real-life example. Suppose you have found a natural product bushel, and you prepare the machine with all kinds of natural products. Creating information can include these scenarios:

1. If the protest is rigid in color, spherical in shape, and has a flatness on top of it, label it as 'Apple'
2. On the occasion that the green color includes yellow, and a barrel of the barrel becomes like, stamp it as 'Banana'

Suppose that, you distribute an unused question (test information) and query to the machine to isolate whether it could be a banana or an apple. It will learn from applying the information to formulate and classify the natural product for input colors and shapes.

Different Types of Supervised Learning

(a) *Regression*: In regression, single-yield respect is used to produce information. This respect can be a potential probability-based case paper, which is explored after considering the quality of the relationship between input factors. For illustration, a relay may aid estimate the cost of a house based on its area, estimate, etc.

   In logistic regression, the yield consists of discrete values based on a set of free factors. This strategy can be a flop when managing with nonlinear and varying option limits. Furthermore, it is not sufficiently suited to capture complex connections in datasets.

(b) *Classification*: It involves collecting information in classrooms. One of the chances is that you are considering extending credit to a person, you will be able to use the classification to decide whether a person will be an advance defaulter or not. When a supervised learning algorithm names input information into two special classes, it is called dual classification. Many classifications do the cruelty of classifying information into more than two classes.

(c) *Naïve Bayesian Model*: The Bayesian display of the classifier is used for a large dataset. This may be a strategy to re-enter the course name by employing a coordinated but non-cyclical chart. The chart has a parent hub and various child hubs. Each child-parent node is expected to be free and divided.

(d) *Decision Tree*: A decision tree can be a flowchart-like display with conditional control articulations, including options and their appreciable results. Output is deals with unexpected information among the tree representation, leaf nodes compare lesson names, and internal hubs to the symptom. An option tree can be used to illuminate issues with discrete properties as well as Boolean capabilities. Some excellent decision tree algorithms are ID3 and Cart.

(e) *Random Forest Model*: The arbitrary forest model is an organizational strategy. It works while constructing a large number of trees of choice, and it individually classifies the trees. Suppose that you need to predict which undergarments will perform well in the GMAT—a test taken for confirmation in undergraduate administration programs. An arbitrary case complements the misunderstanding, which gives the statistical and instructive variables of a set of perceivers already taking the test.

(f) *Neural Networks*: The calculation is planned to translate raw input, design recognition, or tactile information. Despite its various points of interest, the nervous system requires a remarkable computational property. It can be computed to fit a neural network when there are thousands of assumptions. Additionally, it is called 'black-box' calculations because translating the logic behind their forecasts can be a challenging issue.

(g) *Support Vector Machines*: The support vector machine (SVM) may be an administered learning calculation created within the year 1990. It draws from the factual learning hypothesis created by Vap Nick.

The SVM separates the hyperplane, making it a differential classifier. The yield is given within the shape of an ideal hyperplane that classifies unused images. SVM parts are closely connected with the system and used in various fields. Some illustrations incorporate bioinformatics, design acknowledgment, and interactive media data recovery.

Challenges in Supervised Machine Learning

Here are the challenges faced in administered machine learning:

1. Unimportant input involves designing displays in which information can produce inaccurate results.
2. Information arrangement and pre-processing are constant challenges.
3. The exact end occurs when the data given as input is non-redundant, impossible, and fragmented value.
4. If the corresponding master is not accessible, the other approach is "brute force" at that point. This means that you may wish to think that there are appropriate highlights (input factors) to prepare the machine. It can also be a wrong concept.

Pros and Cons of Supervised Learning

It allows for certain types of monitoring learning to receive and distribute information from past encounters. The administered learning has evolved as an effective tool within the AI field, from optimizing execution criteria to managing real-world issues. This is an additional reliable strategy compared to untested learning, which can be computationally complex and less accurate in some events.

As it may be, supervised learning is not achievable without its restrictions. Solid cases are required to prepare the classifier, and within the non-choice of the correct examples, the limits of choice can be extended. One may experience trouble classifying more information if that kind of situation exists where it is needed.

Unsupervised Machine Learning Algorithms

This concept of learning comes into play when the dataset given to training the system is not classified and not even labeled [10]. These kinds of algorithms aim to infer a more accurate function that could express the masked structure from the given unlabeled dataset.

Unsubstituted machine learning calculations without known or labeled results motivate us to find a new design from the dataset. Not like supervised machine learning, an unsupervised machine learning strategy can be directly linked to relapse or a classification issue because you haven't figured out what value can be good for yielding the information and preparing calculations. You will do this regularly to get it figured out. Unsupervised learning can be used to find the basic framework of information.

Unsupervised machine learning implies that fuzzy designs are already detected in the information, but most of the time these designs can meet fixed machine learning. Since you don't know what the results should be, so there is no way to decide how data is, making direct machine learning more suitable for real-world issues.

The most excellent time to use unsupervised machine learning is because you are not aware of the results you crave, such as setting a target market for a completely modern product that your company has never sold recently. As you are trying to get the farthest better, this can happen; much better; higher; strong; a better strong understanding of your existing customer base, supervised learning is the ideal process.

Some applications of unsupervised machine learning methods include:

(a) *Clustering*: This allows you to divide the user into bunches that agree with the concept. Routinely, in any case, cluster probes reduce the similarity between the clusters and do not focus on the information as the masses. For this reason, cluster exams can be a disjointed option for applications such as client division and focusing.

(b) *Anomaly detection*: This may result in abnormal information being focused on your dataset. This is often valuable in isolating exceptions due to inaccurate exchanges, finding flawed pieces of equipment, or a human disturbance between information blocks.

(c) *Association mining*: It identifies the set of things that regularly occur together in your dataset. Retailers often use it for the bushel exam, as it allows investigators to routinely find merchandise received at the same time and to promote more viable performance and methodology.

(d) *Latent variable model*: It is commonly used for pre-processing information, such as decreasing the number of highlights in the dataset (decreasing the dimension) or breaking the dataset into separate components. With unsupervised machine learning strategies, the designs you reveal may still be convenient when supervised machine learning strategies are later actualized. For illustration, you can use an unusable process to do a cluster exam on information, at that point use that cluster, with each push having a place within the supervised learning to show an additional highlight.

Advantages of Unsupervised Learning

1. Unsupervised learning is used for more complex epochs, as compared to supervised learning, in unsupervised learning; we do not have input data.

2. Unsupervised teaching is best because it is easier to motivate unlisted information than labeled information.

Disadvantages of Unsupervised Learning

1. Unsupervised learning is inherently more troublesome than supervised learning because it does not compare the output.
2. The result of unsupervised learning calculations may be less accurate because the input information is not labeled, and the calculation does not reveal the exact yield in development.

Reinforcement Machine Learning Algorithms

This concept deals with interaction with the environment and consequently produces relevant actions and tries to find rewards or errors. The characteristics are sketches, and this concept involves trial and search for errors as well as rewards [11]. This trial and error help the system or the software to learn, and consequently the system or software becomes more capable of determining the ideal nature for a specific domain that helps to maximize the performance.

Reinforcement learning is a limitation of machine learning. This is an appropriate activity to maximize remuneration in a specific situation. It is used by various computer programs and machines to discover the best conceivable behavior one can carry in a particular situation. Monitoring reinforcement learning is one way to prepare for supervised learning paradoxes. Information monitoring has an answer key with it, so the show is prepared with the correct answer, while in reinforcement learning, there is no answer, but the fortification operator chooses what to do to perform the given mistake. Within the non-proliferation of the preparing dataset, it is bound to be remembered by its participation.

Main Points in Reinforcement Learning

1. Input: The input ought to be a beginning state from which the show will start.
2. Yield: There's much conceivable yield as there is an assortment of arrangements for a specific problem.
3. Preparing: The preparation is based upon the input; the show will return to a state, and the client will choose to remunerate or rebuff the demonstration based on its output.
4. The demonstration keeps proceeding to learn. The best arrangement is chosen based on the most extreme reward.
5. Contrast between support learning and supervised learning.

Types of Reinforcement

There are two types of reinforcement:

(a) *Positive*: Positive reinforcement is portrayed as an opportunity, caused by specific behavior, increasing the quality, and repetition of the behavior. In other words, it has a positive effect on behavior.

- The advantages of reinforcement learning are:
- Maximizes performance
- Maintains alter for a long time
- Disadvantages of reinforcement learning:
- Applying too many iterations of reinforcement learning can lead to an over-burden of states which can decrease the output.

(b) *Negative:* Negative reinforcement is known as behavioral fortification because a negative state is not eliminated or a strategic distance from it is maintained.

- Advantages of reinforcement learning:
- Increments behavior
- Gives insubordination to the least standard of performance
- Disadvantages of reinforcement learning:
- It releases sufficiently to meet up the least behavior at the endpoint.

Various Viable Applications of Reinforcement Learning

1. RL can be used in mechanical technology for mechanical automation.
2. RL can be used in machine learning and information processing.
3. RL can be used to design outlines that agree with custom instructions and prerequisite understanding of materials.

Cyber Threat Intelligence

Cyber insights are the data that grants you to maintain a strategic distance from or direct cyber-attacks by considering the hazardous data and donating information to adversaries. It makes a distinction to recognize, get prepared, and maintain a strategic distance from attacks by giving information on aggressors, their method of reasoning, and their capabilities. Danger bits of knowledge plan organizations to be proactive with prescient capabilities instead of responsive to future cyber-attacks. Without understanding security vulnerabilities, peril pointers, and how perils are carried out, it is unfathomable to combat cyber-attacks effectively. Utilizing cyber insights, security specialists can expect and contain attacks speedier, possibly sparing the taken toll inside the event of cyber-attacks. Threat experiences can raise wander security at each level, checking orchestrate and cloud security.

What Does Threat Intelligence Do?

Risk insights make contrast organizations with imperative data around these threats, develop compelling protection components, and direct the threats that might cause budgetary and reputational hurt. Peril Bits of knowledge are the prescient capability to ensure the longer-term ambushes that the organization is revealed to, so they can proactively tailor their resistances and pre-empt future attacks.

Who Is a Cyber Threat Intelligence Analyst?

A cyber insights examiner may be security capable who screens and investigating exterior cyber hazard data to supply critical experiences. These masters triage data of security scenes collected from unmistakable hazard experiences sources and consider the plan of ambushes, their methodology, basis, earnestness, and hazard scene. This data is at that point examined and filtered to provide threat experience feeds and reports that offer to help organizations (security officers) in making choices concerning organizational security. Routinely, these individuals are Certified Threat Bits of Knowledge Examiners who come with the data and aptitudes required for the work portion.

What Does Threat Intelligence Do?

Danger insights make a difference in organizations with profitable information, almost these dangers, construction of successful resistance components, and relief of the dangers that might cause money-related and reputational harm. Risk insights are the prescient capability to guard the long-run assaults that the organization is uncovered to, so they can proactively tailor their resistances and pre-empt future assaults.

Who Is a Cyber Threat Intelligence Analyst?

A cyber insights examiner could be a security proficient who monitors and investigates outside cyber danger information to supply significant insights. These pros triage data of security scenes collected from differing peril experiences sources and consider the plan of ambushes, their technique, thought preparation, earnestness, and threat scene. This information is at that point examined and sifted to create dangerous insight nourishes and reports that offer assistance to administration (security officer) in making choices concerning organizational security. As often as possible, these individuals are Certified Risk Insights Examiners who come with both the data and aptitudes required for the work portion.

Cyber Security

The term cyber security deals with standard practices that are implemented to defend computers, mobile digital devices, servers, communicating channels, data, and all other digital devices from malicious cyber-attacks that could be executed either remotely or locally.

The first trace of cyber-attack was seen in the year 1988 when a student at Cornell University named Robert Tappan Morris programmed a worm whose nature is that of the present Trojan Horse, which affected [12] nearly 6000 computers.

Cyber-attacks affect human life directly or indirectly from time to time. The Stuxnet, which is believed to be developed jointly by the USA and Israel to disrupt the functioning of the Nuclear Project of Iran, was one of the deadliest malware that attacked the Supervisory Control and Data Acquisition (SCADA) [13]. The cyber-attacks affected network devices when BrickerBot hit nearly 60,000 modems and routers of Bharat Sanchar Nigam Limited (BSNL) and Mahanagar Telephone Nigam Limited (MTNL), and it lasted for six days in a stretch [14]. This affected communication, and hence the common people suffered as a result of this attack. The famous WannaCry attack that shook entire Europe is an aver to the devastation caused by cyber-attacks and their afterward effects.

The Domains of Cyber Security and Its Role in Society

Cyber security branches in various subdomains, these branches are distinct from one another in their nature of defending the system. The few more dominant ones are:

1. Penetration Testing: This branch deals with finding a vulnerability in the system. It helps the company in fighting cyber-attacks by removing the vulnerability found using penetration testing.
2. Digital Forensics: This deals with resolving cybercrimes by finding digital evidence and producing documentation in a court of law.
3. Cryptography: It deals with securing the information or data when it is in static and transit mode.
4. Information Security Governance and Risk Management: This deals with the technical as well as non-technical aspects that could pose a threat to the system.

The above domains and history of attacks show that the system is exploited by an attacker based on a vulnerability existing in the system. This has given birth to the idea that a machine could detect a vulnerability and could try to stop the cyber-attacks accordingly. Various research works have been proposed where the concept of machine learning is brought as one of the solutions that could stop cyber-attacks.

The concept of reinforcement learning has been introduced, the machine learned the behavior of the ransomware, and it was able to defend the system based on three parameters [15]. Another aspect of machine learning that was used to counter cyber-attacks has been researched using the concept of neural networks where the behavior of the malware was considered [16]. The deep learning concept of machine learning has been used to detect the crypto-jacking that is executed by an attacker for mining cryptocurrency on target systems [17]. This concept is based on profiling the context, and thus, it shows that the machine is learning the nature of the malware.

This shows that machine learning is one of the potential defensive concepts that could counter cyber-attacks, and thus, it could save e-commerce, education, defense, healthcare, and other domains where humans are involved. AI plays could be a major role player in tackling cyber-attacks in the coming years [18]. The power of learning and making the decision would result in a mechanism that would be more resilient to cyber-attacks.

Principles of Cyber Security

The three fundamental properties must exist in any domain, i.e., accessibility, secrecy, and astuteness, a few simple but compelling concepts can be taken after.

1. *Centers on Earlier Systems*: Stabilizing the degree of accessibility, security, and adroitness of sources arises underneath the most noteworthy challenges; and as a result, it is completed with the assistance of centering on the fundamental structures and appearing fine security guarantee to it, whereas differing techniques are associated for the security of less-prior frameworks.
2. *Differing Clients, Particular level of Openness*: What information is open through whom on a very basic level based on what kind of person he is, and no single individual gets right of the area to all the bits of knowledge and information. This potentially slightest benefits particular commitments. In this way, the trade-in obligation is directly relative to the exchange of privileges.
3. *The course of action of Free Assurance (Traditions)*: Many affirmation traditions for a single work may be a long superior thought than a single tradition. It exceedingly diminishes the chance of viable cyber-attacks, and the essential rule is creating the work of the attacker as he has had to perform extraordinary errands to wreck through diverse security layers.
4. *Fortifications*: Dissatisfaction can happen, but arranging the results after disillusionment can lessen the beat harmed to the framework, community, or individual. This could be a shockingly prominent strategy and is sharpened in a couple of fields.

Keeping records of all breaches: Cybersecurity staff keeps up records of all the breaches, these must ceaselessly think almost, and security measures have to be made. This method got to be a convenient one due to the truth software engineers are not holding up; they are creating their capabilities as wonderfully as overhauling cyber-attack gadgets.

The Role of ML in Cybersecurity

In the domain of cybersecurity and the science of fake experiences, machine learning is the preeminent common approach and term utilized to depict its application in cybersecurity. Indeed, even though several significant learning techniques are being utilized under the umbrella of ML as well, various would say DL is getting to be out of date in cybersecurity applications. Machine learning shows up a magnificent ensure in cybersecurity, indeed even though it does have many downsides. The work can help to examine the advancement that could be achieved in cyber security to get more adequate results and practical implementation for the future.

Why Has Machine Learning Become So Basic for Cyberspace?

Cybercity with machine learning can analyze framework designs and help them anticipate comparative attacks and respond to changing behavior. It can help cyber security groups to become more proactive in avoiding threats and reacting to dynamic attacks in real time. This can reduce the amount of time to use their assets more deliberately through schedule schedules and empowering organizations. In short, machine learning can make cyber security less difficult, more proactive, less expensive, and remotely viable. But it can be the same as doing things on a closed chance that basic knowledge of machine learning gives a total picture of the environment. As they say, rub inside, rub outside.

Why Focuses on Basic Information to Win Machine Learning in Cyberspace?

Machine learning is almost creating designs and controlling those designs with calculations. To create a design, you wish for everyone to have a parcel of rich data because the information must speak as conceivable as the many possible outcomes from many possible scenarios.

It is not about the amount of information; it is almost about quality. There should be a total, significant, and rich setting collected from each potential source for information—whether it is at the endpoint, on the network, or within the cloud. In addition, you need to be central to clearing information so that you can make sense of the information you capture can mark the results.

Importance of Threat Intelligence in Cybersecurity

Cyber danger insights make a distinction in an organization by giving them encounters into the rebellious and recommendations of threats, allowing them to develop defense methods and frameworks, and reduce their attack surface with the conclusion destinations of soothing harmed and guaranteeing theirs arrange. The elemental objective of cyber threat bits of knowledge is to supply organizations with a more significant understanding of what's happening exterior their organization, giving them better penetrability of the cyber dangers that bring the first chance to their framework. You require hazard experiences for a reasonable defense. It's besides prioritizing: emptying unfaithful positives that continuously hit SOCs and recognizing the advanced threats and abuses the organization is most defenseless to, so bunches can take action against them. With cyber risk insights, you will be able to choose on the off chance that your security defense system can truly handle those perils and advance it as vital.

Here are other major benefits to extraordinary cyber risk insights in your organization, as well.

Security Team Efficiency

When an irregularity in your organization is hailed and your security gathering is cautioned, they need to be known if it's a genuine peril or as it were an off-base positive. Joining risk insights will give your bunches more understanding into what ought to be tended to, advance their response rate, and allow them to center on what truly things. This will not be because it was making strides in its efficiency in managing security cautions and minimizing its workload, but to cut down the requirement for more staff.

Collaborative Knowledge

Information is because it was the thing that creates once it's shared. The same rings are veritable with perilous insights. The same outline shows that 66% of cybersecurity choice makers in organizations with threat experiences programs said their commerce looks to the government for information or data on cyber threats. To keep up with wafers, and the strategies they utilize that are getting more progressed each day, organizations share their data on the techniques and vulnerabilities they see inside the wild making a difference for others to secure themselves against them as well.

Top Machine Learning Use Cases for Security

Here, we break down machine learning beat use cases in security.

Using Machine Learning to Detect Malicious Activity and Stop Attacks

Machine learning calculations will assist businesses to stop the fast pace of movement and stop attacks for some time starting recently. David Palmer should know. As an executive of innovation at UK-based start-up Darktrace—a firm that has seen a parcel of victory around its machine learning-based Undertaking Safe Arrangement since the firm's inception in 2013—he has impacted on such innovations have seen.

Palmer says Darktrace made changes to a casino in North America after its calculations recognized an information exclusion attack that "used a connected angle tank as an entryway into the network." The firm claims that the WannaCry ransomware emergency is feared a comparative attack between last summer.

"Our systems observed the attack within seconds at an NHS agency event and addressed the threat without causing any harm to that organization," he said of ransomware, which has seen more than 200,000 people in more than 150 countries covering many level of casualties. "None of our customers were hurt by the WannaCry attack, which was not set against it."

Using Machine Learning to Analyze Mobile Endpoints

Machine learning is now running as a standard on versatile gadgets, but so the furthest of this action has been to drive further voice-based encounters at the likes of Google currently, Apple's Siri, and Amazon's Alexa. However, there is also a security application. As stated, Google is using machine learning to analyze threats against portable endpoints, while the undertaking is seeing an opportunity to ensure a developing number of its versatile gadgets.

In October, MobileIron and Zimperium reported a collaboration to assist enterprises to adopt a versatile anti-malware system involved in machine learning. MobileIron said it would coordinate Zimperium's machine learning-based threat discovery with MobileIron's security and compliance motor and offer a joint mechanism that would address challenges such as identifying gadgets, system, and application threats, and the company will immediately take up robotic activities to ensure that.

Other merchants want to support their portable system, as well. Zimperium, along with Post, Skycure (which has been purchased by Symantec), and Wandera, is considered pioneers within portable disk space and defense advertising. Each employment consists of machine learning calculations to identify potential hazards. In this case,

Wandera late discharged its Risk Discovery Motor MI: RIAM, which is believed to have identified more than 400 strains of SLacker ransomware focusing on the business's portable armadas.

Using Machine Learning to Enhance Human Analysis

At the heart of machine learning in security is the belief that it makes a difference in human investigators with all approaches to work, analyzing how to identify dangerous attacks, arrange counting, endpoint assurance, and powerlessness assessment. Despite being around risk insight is the most exciting.

For example, in 2016, MIT's Computer Science and Fake Insights Lab (CSAIL) created a framework called AI2, a versatile machine learning security stage that has made a difference in examinees searching for needles within the ack haystack We do. Looking at millions of logins each day, the framework was able to channel the information and passed it to the human investigator, which was reduced to around 100 per day. Investigations done by CSAIL and start-up PatternX revealed that the attack rate increased to 85%, with a fivefold decrease in incorrect positivity.

Using Machine Learning to Automate Repetitive Security Tasks

The real advantage of machine learning is that it seems to computerize tedious tasks, empowering employees to center on more important work. Palmer says that machine learning should ultimately "remove the need for people to remove insights" from a low-risk decision-making movement, such as at-risk insights. "Machines must handle deserted work and strategic firefighting such as ransomware to allow people time to bargain with core issues—such as modernizing Windows XP—instead."

Booz Allen Hamilton has turned this course down, believed to use AI tools to distribute human security assets more efficiently, putting threats at risk so that experts focus on basic rights Do it.

Using Machine Learning to Close Zero-Day Vulnerabilities

Some acknowledge that machine learning can provide support near vulnerabilities, particularly zero-day threats, and others typically target unsafe IoT gadgets. Active work has been done in this range: a group at Arizona State College used machine learning to screen activity on the dim web to identify information related to zero-day abuse for Forbes. Equipped with such an understanding, organizations may possess vulnerabilities and prevent misuse at some time because of recent information violations.

Hype and Misunderstanding Muddy the Landscape

In any case, machine learning is no silver bullet, not an industry still testing these advances in the validation of concepts. There are various disadvantages. In some cases, the machine learning framework reports untrue positives (from the unpublished learning framework where computation collects categories based on information), while some testers have spoken out explicitly about how machine learning in security can speak to the 'black-box' system, where CISOs originate is not entirely beyond any doubt 'under the hood.' They are thus forced to put their faith and duty on the shoulders of businessmen and machines.

Applications of Machine Learning in Cybersecurity

Cyber security threats that can protect against machine learning.

Spear Phishing

Traditional phishing location methods rely less on speed and accuracy, searching for all-male joints that threaten customers away. The mechanism of the issue lies within the presenter URL classification model, which is based on recent machine learning calculations, which can search for designs that display a mail sender's mail. Those models are organized to recognize small-scale behavior such as mail headers, body data, and plans. These systematic models can be used to identify whether an email is real or not.

Watering Hole

Software engineers arrange to track the districts that customers regularly visit and are external to the user's private organization. Machine learning calculations can ensure the security standard of web application organizations by analyzing the traversals of the location. It can identify whether malevolent websites are facilitated when customers search purposefully. The method of machine learning can be used to calculate the trajectory revelation to identify these malevolent places. Machine learning can occur in addition to screens for exceptional or extraordinary business plans near and near a site.

The Cyber Defense Benefit supplier Palladian arranged for an elite RisqVU to water the pierced donkeys. It can be a combination of simulated insights (AI) and colossal data analysis. A burning crack attack requires a synchronous examination of data from a middleman, mail action, and stash. RisqVU can be a call for data analysis arranging to apply exams from various sources.

Webshell

This can be a piece of code that is stacked pending in a web location to allow the attacker to make changes to the server's webroot catalog. This completely gets to the system's database that infers have picked up. If it is an e-commerce location, aggression may well occur on a journey introduction into the database to accumulate credit card information of the customer base.

The targets of attackers using web shells regularly support eCommerce steps. The major threat of eCommerce phases is related to online installments which are expected to be safe and mysterious.

Ransomware

Ransomware can be a combination of ransom + software. This allows for any type of computer program, which calls for any type of delivery in exchange for the encryption key of the seized records of the user. The encryption key is a key to open the client's shot record. Catapulted records can be mixed media records, office records, or system records that depend on the user's computer.

There are two types of ransomware:

1. A record coder that scrambles the records (a change in the information in a puzzle code)
2. A jolt screen locks a computer and prevents the client from using it until the content is paid.

Remote Exploitation

He has the final list of applications in machine learning in cyberspace. What is additionally considered an insurmountable attacker may be an agitated movement that targets one or the target of the computer. Through the defenseless centers of the machine or organization, the attacker arrives at the pick system. The goal of a remote attack is to tamper with the system and inflict focused injury on a computer orchestra by displaying an illegal computer program. This can happen in several ways:

Discontinuance of Benefit Attack

A strategy to shut down the server to block the server from flooding customers with unfit customer requests. This creates a huge usage spike that sets the server up and serves them with a high number of pending requests to proceed.

DNS Harming

DNS servers are systems that translate human-memorable space names such as face-book.com to compare numeric IP addresses. DNS systems are used to identify and favor resources on the net. Hurting DNS servers in a general sense leads them to realistically detect malformed data starts and redirect DNS servers that hurt those clients to areas that inadvertently download dangerous programs or diseases into the system.

Harbor Checking

The computer port is used to send and receive data. Harbor scanners can be used to identify data vulnerabilities and choose to control the computer by removing those vulnerabilities.

Machine learning calculations can be used to analyze system behavior and to identify unusual events that occur at interfaces with standard events. Calculations can be organized for specific datasets so that they can track an abuse payload in progress.

(a) *Malware Analysis*—If we go through the Symantec Annual Security Report 2018, malicious attacks on website-based resources were highest, and in comprise of data's integrity malware attacks stand first ahead of all and we compare simply it is 493% more than the year 2012. Each year malware attacks are on top, and it is increasing gradually, so there is a need for categorization of this huge incoming data of malware attacks as we know, the first and fore-most characteristic of the unsupervised learning algorithm is finding all types of unknown patterns in data and real-time scenario, unsupervised algorithms help in making the category of data based on the features of input data. The clustering algorithm is one of the highly used algorithms when we try to find the pattern or structure in the collection of uncategorized data. Clustering is a highly used algo-rithm in statistical analysis-based algorithms where it is used for determining the class of malware, i.e., signature-based virus attacks or behavioral-based Trojan horses [1]. Clustering has been also used for hybrid malware execution and multi-path-based malware analysis. In his report, Lindsay et al. pointed out the clustering-based analysis for Stuxnet malware which was very instrumental some time ago [2]. Some other highly used techniques in malware analysis are dynamic and static-based approaches. In the static approach model, i.e., in Fraley et al., the researcher discussed its impact and increasing volume of malware-based attacks [3], in their dynamic approach model conveyed in that paper.

(b) *Unknown Rootkit Analysis*—Depending on the outcome, a new approach has been developed in recent years that focuses on the unknown outcome when there's no correct set of labels and the errors are not meaningful. One of the highly famous works in this approach is discussed the use of Kohenon's Self Organizing Map (KSOM) [19]. A convolutional neural network-based hybrid approach is also being favored by many researchers in the last decade, Raff et al. discussed a similar kind of work and suggested using a multi-level-based hybrid model for early detection of rootkits where a machine learning approach is required [20], whereas Kracl et al. proposed a new model for the early prediction of rootkits along with the system scanning process [21].

(c) *PE Analysis*—Seeing the advanced use of AI-based models to detect and run suspicious software, some new approaches were also proposed such as automation-based malware analysis with the use of Potential Executables files where we focus to run that suspicious software in sandbox software [22]. Multi-tasking and multi-operational operating systems were also used in some continuing research work using the same line of study [23]. Shabtai et al. discussed a dynamic approach for their PE analysis while focusing on machine learning techniques, and a specially featured taxonomy model was prioritized in this work [24]. Source et al. discussed a new approach in which his team added a multi-level taxonomy-based dynamic model [25]. Razak et al. proposed a new model having the combination of static and dynamic approaches with a balanced volume of analysis, but the multi-level multi-model hybrid concept was a missing issue in his work [26]. Some other heuristic-based methods were also proposed by several researchers in this direction of research, i.e., Bazfarshan et al. discussed the same concept [27].

(d) *Requirement-Based Attack Analysis*—Some new work and their findings have been reported in the last 5 years where researchers came out with new approaches depending on the type of attack being received. Yegulalp [28] presented a new approach where methodological testing based on an in-line concept was used to detect and defend the raised security concern, maybe in terms of rootkit, etc.

(e) *File Masquerading*—In this type of serious security concern, machine learning is again playing a major role. Several case studies based on new findings are being reported, such as deep learning-based machine intelligence has been proposed by many authors where anomaly-based masquerade detection using three deep learning models has been quite helpful in the prediction of real-time masquerading attacks [29]. Deep learning-based techniques have been also reported in the Unix environment too [30]. A new model was suggested by

Kim et al. in their research paper citing memory-based methods for detection and prediction [31].

(f)   *Kernel Object Manipulation*—Machine learning algorithms have also been used in kernel object manipulation too. OS security features may be compromised with help of rootkit-based kernel object manipulation, so in this type of security attack, unsupervised approaches play a major role in early detection as suggested by Ries et al. [32].

(g)   *Hooking Challenges*—Hooking-based rootkits are also being reported these days in a high volume where security is being compromised especially confidentiality and integrity are mostly at stake. Again, in this type of security concern, several tools are available in the market which has been designed with the approach of machine learning in itself intrinsically. Tools like Rootkit Unhooker and GMERK Ver-1 [33, 34].

(h)   *Post-Execution Behavior Application Analysis*—Some new machine learning-based approaches have been suggested to analyze the application's behavior after its execution in a multi-platform-based environment. Hashing-based similarity analysis and signature-based analysis and prediction have been reported in many studies focusing on "Two-stage pre-execution detection on users' computers" [35].

(i)   *Outlier Analysis in Network Usage*—Agglomerative hierarchical clustering where in this clustering technique, every data is a cluster and divisive hierarchical clustering-based research findings have been reported very worthy in detection and prediction purpose with the help of several tools in case of network usage. In some research papers, two-factor authentication and its negative impact on efficiency in the resulting occurrence of exploits have been reported, and again here, clustering algorithms have been highly used [36].

(j)   *Multiple Fraud Detection*—Many findings and tools-based studies have been reported using machine learning approaches [37]. DataVisor is one of them which has been very much instrumental in finding the signatures of online frauds generated for credit scams. UML along with a mixed approach of GMM is also being reported in many studies [38–40].

How Does Machine Learning Benefit Cybersecurity?

For a long time, a wide assortment of occupations has been molded by the presence of artificial intelligence (AI) in new forms, which already paid attention to labor and expansion. Computerization of these forms through AI methods such as machine learning can be a huge victory for cyber security, where they can offer help capture and avoid all forms of cyber-attack [41–46].

While machine learning is not the same as AI as it is regularly caught, both systems allow computer forms to be memorized as a human computes a complex scale that no human achieves. Perhaps, there is no place more important within the growing world of cyberspace [47–50].

Conclusion

Because of the developing centrality of cyber security and informatics, and machine learning innovation, in this paper, we have talked about how cyber security informatics applies to data-driven clever selection in savvy cybersecurity and governance. We have investigated how this can affect security information, in terms of understanding security events and the datasets themselves. We indicated working on cyber security informatics by talking about the state of craftsmanship in security incidents information and comparison with security administrations. We also talked about how machine learning processes can affect the space of cyberspace and looked at security challenges [51–55]. In the context of the current investigation, the machine security process-based security framework has a lot of focus on traditional security systems, with less accessible work. For each common method, we have talked about important security checks. The reason for this article is to consider conceptualization, understanding, modeling, and almost cyber information informatics [56–60].

We have recognized and talked about various key issues in the security examination to understand the signposts of future investigation bearings within the space of cyber security informatics. Based on the information, we must deliver a non-multi-level system of cybersecurity informatics performance based on machine learning processes, where information is accumulating from mixed sources, and analytics to deliver the most recent data-driven design is complimentary [61–66]. Security Administration: The system includes some fundamental steps—security information gathering, information systems, machine learning-based security modeling, and incremental learning and mobility for incremental framework and administration. We are particularly focused on experiences gained from security information, in which an investigation plan is established with specific consideration for data-driven smart security systems [67–70].

Future Scope

By and large, this paper pointed out that it was not meant to talk about cybersecurity informatics and critical strategies, but rather about the suitability of a cybercity framework from a machine point of view and the selection of data-driven brilliance in governance. To talk about, our examination and talk may offer some suggestions for both security analysts and experts [71–76]. For analysts, we have highlighted some issues and headlines for future inquiries. Another range for potential investigation includes experimental evaluation of recommended data-driven shows and comparative investigation with other security frameworks. For professionals, multi-layered machine learning-based shows can be used as a reference in the planning of clever cybersecurity frameworks for organizations. We acknowledge that our thinking about cybersecurity informatics opens up a promising path and used it as a reference direct to both the scholarly world and industry for future investigations and applications within the field of cyber investigation can go [77–80].

References

1. Fielding, A.H.: An introduction to machine learning methods. In: Fielding, A.H. (eds.) Machine Learning Methods for Ecological Applications. Springer, Boston, MA. (1999). https://doi.org/10.1007/978-1-4615-5289-5_1
2. Woollacott, P.: Cybercrime Comes of Age. IEEE
3. *The New York Times* June 30, 2014. https://www.nytimes.com/2014/07/01/technology/energy-sector-faces-attacks-from-hackers-in-russia.html
4. Malware affects thousands of BSNL broadband modems. https://www.thehindu.com/news/national/karnataka/malware-affects-thousands-of-bsnl-broadband-modems/article19381410.ece, The Hindu Newspaper
5. Adamov, A., Carlsson, A.: Reinforcement Learning for Anti-Ransomware Testing. IEE
6. A Behavior-Based Ransomware Detection Using Neural Network Models. IEEE
7. Singh, A., Thakur, N., Sharma, A.: A review of supervised machine learning algorithms. In: 2016 3rd International Conference on Computing for Sustainable Global Development (INDIACom), New Delhi, pp. 1310–1315 (2016)
8. Zoppi, T., Ceccarelli, A., Bondavalli, A.: Into the unknown: unsupervised machine learning algorithms for anomaly-based intrusion detection. In: 2020 50th Annual IEEE-IFIP International Conference on Dependable Systems and Networks-Supplemental Volume (DSN-S), Valencia, Spain, pp. 81–81 (2020). https://doi.org/10.1109/DSN-S50200.2020.00044
9. Elgabli, A., Khan, H., Krouka, M., Bennis, M.: Reinforcement learning based scheduling algorithm for optimizing age of information in ultra reliable low latency networks. In: 2019 IEEE Symposium on Computers and Communications (ISCC), Barcelona, Spain, pp. 1–6 (2019).https://doi.org/10.1109/ISCC47284.2019.8969641
10. Vähäkainu, P., Lehto, M.: Artificial intelligence in the cyber security environment. In: The 14th International Conference on Cyber Warfare and Security ICCWS2019, Stellenbosch, South Africa (2019)
11. Cyber Threat Report by Sonicwall. Available at https://www.sonicwall.com/2020-cyber-threat-report/#threat-report-form. Accessed on 15 Nov 2020
12. Cyber-attacks on Tesla Amazon Cloud. Available on https://www.wired.com/story/cryptojacking-tesla-amazon-cloud/accessed on 15 Nov 2020
13. Attack on Microsoft Corp and its mitigation. Available on https://www.microsoft.com/security/blog/2018/02/14/how-artificial-intelligence-stopped-an-emotet-outbreak/. Accessed on 15 Nov 2020
14. Aycock, J.: Computer Viruses and Malware Report. Springer (2006)
15. Lindsay, J.R.: Stuxnet and the limits of cyber warfare. Secur. Stud. **22**:3, 365–404 (2013) https://doi.org/10.1080/09636412.2013.816122
16. Fraley, J.B., Cannady, J.: The promise of machine learning in cybersecurity. Southeast Con. **2017**, 1–6 (2017)
17. A Comparative Analysis of Rootkit Detection Techniques By Thomas Martin Arnold, The University Of Houston-Clear Lake, May 2011
18. Raff, E., Barker, J., Sylvester, J., Brandon, R., Catanzaro, B., Nicholas, C.K.: Malware detection by eating a whole EXE. The Workshops of the Thirty-Second AAAI Conference on Artificial Intelligence, New Orleans, Louisiana, USA, 2–7 Feb 2018, pp. 268–276

19. Krl, M., vec, O., Blek, M., Jaek, O.: Deep Convolutional Malware Classifiers Can Learn from Raw Executables and Labels Only (2018)
20. Ligh, M., Adair, S., Hartstein, B., Richard, M.: Malware Analyst's Cookbook and DVD: Tools and Techniques for Fighting Malicious Code. Wiley Publishing (2010)
21. Monnappa, Learning Malware Analysis: Explore the Concepts, Tools, and Techniques to Analyze and Investigate Windows Malware. Packt Publishing (2018)
22. Shabtai, A., Moskovitch, R., Elovici, Y., Glezer, C.: Detection of malicious code by applying machine learning classifiers on static features: a state-of-the-art survey. Information Security Technical Report 14
23. Souri, A., Hosseini, R.: A state-of-the-art survey of malware detection approaches using data mining techniques. HCIS 8(1), 3 (2018). https://doi.org/10.1186/s13673-_018-_0125-_x
24. Razak, M.F.A., Anuar, N.B., Salleh, R., Firdaus, A.: The rise of malware: a bibliometric analysis of malware study. J. Netw. Comput. Appl. 75, 58–76 (2016)
25. Bazrafshan, Z., Hashemi, H., Fard, S.M.H., Hamzeh, A.: A survey on heuristic malware detection techniques. In: The 5th Conference on Information and Knowledge Technology (2013), pp. 113–120
26. Review, Y.S.: Six rootkit detectors protect your system. In: Information Week 2015
27. Bertacchini, M., Fierens, P.: A survey on masquerader detection approaches. In: Proceedings of V Congreso Iberoamericano de Seguridad Informática, Universidad de la República de Uruguay (2008)
28. Erbacher, R.F., Prakash, S., Claar, C.L., Couraud, J.: Intrusion detection: detecting masquerade attacks using UNIX command lines. In: Proceedings of the 6th Annual Security Conference, Las Vegas, NV, USA, April 2007
29. Kim, J., Thu, H.L.T., Kim, H.: Long short term memory recurrent neural network classifier for intrusion detection. In: Proceedings of the 3rd International Conference on Platform Technology and Service, PlatCon 2016, Republic of Korea, February 2016
30. Ries, C.: Inside Windows Rootkits. Vigilant Minds, Inc., 22 May 2006, Retrieved from http://madchat.fr/vxdevl/library/Inside%20Windows%20Rootkits.pdf
31. Rootkit Unhooker (3.8.388.480 SR2) [Software]. http://www.antirootkit.com/software/RootKit-Unhooker.htm
32. Gmerek, P.: GMER (1.0.15.15281) [Software] (2010). Locked Down: Information Security for Attorneys (American Bar Association 2012, ISBN: 978-1-61438-364-2)
33. Kaspersky Lab White Paper. Available From https://media.kaspersky.com/en/enterprise-security/Kaspersky-Lab-Whitepaper-Machine-Learning.pdf
34. Martinelli, F., Saracino, A., Sheikhalishahi, M.: Modeling privacy-aware information sharing systems: a formal and general approach. In: 15th IEEE International Conference on Trust, Security and Privacy in Computing and Communications (2016)
35. Zhang, F., et al.: GMM-based undersampling and its application for credit card fraud detection. In: 2019 International Joint Conference on Neural Networks (IJCNN), pp. 1–8 (2019)
36. Abomhara, M., Koien, G.M.: Cyber Security and the Internet of Things: Vulnerabilities, Threats, Intruders and Attacks (2015)
37. Anderson, B., Storlie, C., Lane, T.: Multiple kernel learning clustering with a malware application. In: IEEE 12th International Conference on Data Mining, pp. 804–809 (2012)
38. Andreev, S., Koucheryavy, Y.: Internet of Things, Smart Spaces, and Next-Generation Networking. Springer International Publishing (2012)
39. Atzori, L., Iera, A., Morabito, G.: The internet of things: a survey. Comput. Netw. 54(15), 2787–2805 (2010)
40. Bagirov, A.M., Ugon, J.: Supervised data classification via max-min separability. In: Jeyakumar, V., Rubinov, A., (eds.) Continuous Optimization: Current Trends and Modern Applications, pp. 175–207. Springer, US (2005)
41. Bagirov, A., Ugon, M.J., Webb, D.: An efficient algorithm for the incremental construction of a piecewise linear classifier. Inf. Syst. 36(4), 782–790 (2011)
42. Benazzouz, Y., Munilla, C., Gnalp, O., Gallissot, M., Grgen, L.: Sharing user iot devices in the cloud. In: 2014 IEEE World Forum on Internet of Things (WF-IoT), pp. 373–374 (2014)

43. Buczak, A.L., Guven, E.: A survey of data mining and machine learning methods for cyber security intrusion detection. IEEE Commun. Surv. Tutorials **18**(2), 1153–1176 (2016)
44. Cheng, L., Liu, F., Yao, D.: Enterprise data breach: causes, challenges, prevention, and future directions. Wiley Interdisc. Rev. Data Min. Knowl. Discov. **7**(5)
45. Cheng, Y., Naslund, M., Selander, G., Fogelstrm, E.: Privacy in machine-to-machine communications a state-of-the-art survey. In: 2012 IEEE International Conference on Communication Systems (ICCS), pp. 75–79 (2012)
46. Choras, M., Kozik, R.: Machine learning techniques applied to detect cyber-attacks on web applications. Logic J. IGPL **23**(1), 45–56 (2015)
47. Delaney, P.R., Fersi, G.: A distributed and flexible architecture for the internet of things. Proc. Comput. Sci. **73**, 130–137 (2015)
48. Fraley, J.B., Cannady, J.: The promise of machine learning in cybersecurity. In: SoutheastCon 2017, pp. 1–6 (2017)
49. Gluhak, A., Krco, S., Nati, M., Pfisterer, D., Mitton, N., Raza find ralambo, T.: A survey on facilities for experimental internet of things research. IEEE Commun. Mag. **49**(11), 58–67 (2011)
50. Gubbi, J., Buyya, R., Marusic, S., Palaniswami, M.: Internet of things (iot): a vision, architectural elements, and future directions. Future Gener. Comput. Syst. **29**(7), 1645–1660 (2013)
51. Kent, A.D.: Cybersecurity data sources for dynamic network research. In: Dynamic Networks in Cybersecurity. Imperial College Press (2015)
52. Kokila, R.T., Selvi, S.T., Govindarajan, K.: Ddos detection and analysis in sdn-based environment using support vector machine classifier. In: Sixth International Conference on Advanced Computing (ICoAC), pp. 205–210 (2014)
53. Kolias, C., Kambourakis, G., Stavrou, A., Voas, J.: Ddos in the iot: Mirai and other botnets. Computer **50**(7), 80–84 (2017)
54. Kotpalliwar, M.V., Wajgi, R.: Classification of attacks using support vector machine (svm) on kddcup'99 ids database. In: Fifth International Conference on Communication Systems and Network Technologies, pp. 987–990 (2015)
55. Karimipour, H., Dinavahi, V.: Robust massively parallel dynamic state estimation of power systems against cyber-attack. IEEE Access **6**, 2984–2995 (2018)
56. Karimipour, H., Dinavahi, V.: Parallel domain-decomposition-based distributed state estimation for large-scale power systems. IEEE Trans. Ind. Appl. **52**(2), 1265–1269 (2016)
57. Karimipour, H., Dinavahi, V.: On false data injection attack against dynamic state estimation on smart power grids. In: 2017 IEEE International Conference on Smart Energy Grid Engineering (SEGE), pp. 388–393 (2017)
58. Chen, P., Yang, S., McCann, J.A., Lin, J., Yang, X.: Detection of false data injection attacks in smart-grid systems. IEEE Commun. Mag. **53**(2), 206–213 (2015)
59. Esmalifalak, M., Zheng, R., Han, Z.: Detecting stealthy false data injection using machine learning in smart grid. In: 2013 IEEE Global Communications Conference (GLOBECOM), pp. 808–813 (2013)
60. Mohammadi, S., Desai, V., Karimipour, H.: Multivariate mutual information-based feature selection for cyber intrusion detection, vol. 10, pp. 1–6 (2018)
61. Ozay, M., Esnaola, I., Yarman Vural, F.T., Kulkarni, S.R., Poor, H.V.: Machine learning methods for attack detection in the smart grid. IEEE Trans. Neural Netw. Learn. Syst. **27**(8), 1773–1786 (2016)
62. Yan, J., Tang, B., He, H.: Detection of false data attacks in smart grid with supervised learning. In: 2016 International Joint Conference on Neural Networks (IJCNN), July 2016, pp. 1395–1402
63. Ozay, M., Esnaola, I., Yarman-Vural, F.T., Kulkarni, S.R., Poor, H.V.: Sparse attack construction and state estimation in the smart grid: centralized and distributed models. IEEE J. Sel. Areas Commun. **31**, 1306–1318 (2013)
64. Karimipour, H., Dinavahi, V.: Extended kalman filter-based parallel dynamic state estimation. IEEE Trans. Smart Grid **6**(3), 1539–1549 (2015)

65. Karimipour, H., Dinavahi, V.: Parallel relaxation-based joint dynamic state estimation of large-scale power systems. IET Gener. Transm. Distrib. **10**(2)
66. Kramer, F.D., Starr, S.H., Wentz, L.: Cyberpower and national security. Potomac Books Inc. (2009)
67. United States General Accounting Office, Cybersecurity for Critical Infrastructure protection. Accessed on 26 Feb 2014 on http://www.gao.gov/new.items/d04321.pdf
68. Brown, G., Carlyle, M., Salmeron, J., Wood, K.: Defending critical infrastructure. Informs J. (2006)
69. Jonas, J., Hassel, H.: Impact of functional models in a decision context of critical infrastructure vulnerability. In: Second International Conference on Vulnerability and Risk Analysis and Management (2014)
70. Takahashi, T., Kadobayashi, Y., Fujiwara, H.: Ontological Approach toward Cybersecurity in Cloud Computing (2010)
71. Simmons, M., Chi, H.: Designing and implementing cloud-based digital forensics. In: Proceedings of the 2012 Information Security Curriculum Development Conference, pp. 69–74 (2012)
72. Ouyang, M.: Review on modeling and simulation of independent critical infrastructure systems. In: Reliability Engineering and System Safety, pp. 43–60 (2014)
73. Volatility tool, https://code.google.com/p/volatility/
74. Linux Unified Key Setup. https://code.google.com/p/cryptsetup/
75. Hastie, T., Tibshirani, R., Friedman, J.: The Elements of Statistical Learning-Data Mining, Inference and Prediction, 2nd edn., vol. II, pp. 465–576. Springer, Stanford, (2008)
76. Datta, R., Joshi, D., Li, J., Wang, J.Z.: Image retrieval: ideas, influences, and trends of the new age. ACM Comput. Surv. **40**(2), Article 5 (2008)
77. Huang, Y., Liu, Q., Lv, F., Gong, Y., Metaxas, D.N.: Unsupervised image categorization by hypergraph partition. IEEE Trans. Pattern Anal. Mach. Intell. **33**(6) (2011)
78. Wang, H.H., Mohamad, D., Ismail, N.: Semantic gap in CBIR: automatic objects spatial relationships semantic extraction and representation. Int. J. Image Process. (IJIP) **4**(3) (2010)
79. Lowe, D.G.: Distinctive image features from scale-invariant keypoints. Int. J. Comput. Vis. (2004)
80. Bay, H., Tuytelaars, T., Gool, L.V.: SURF: speeded up robust features. ETH Zurich, Zurich (2005)

Chapter 10
Control of Feed Drives in CNC Machine Tools Using Artificial Immune Adaptive Strategy

A. Lavanya, S. Revathi, N. Sivakumaran, and K. Rajkumar

Introduction

Industry 4.0 paves the way to develop and distribute the products of the manufacturing industry. Industry 4.0 is also referred to as smart factories. The Industrial transformation uses the combined technologies like machine learning, cloud computing, artificial intelligence, IOT, and analytics to support production. To provide better decision-making sensors, embedded software, and robotics are used in smart factories to collect the data and analyze it. A completely new level of visibility and insight is produced by fusing operational data from production operations with operational data from supply chain, customer service, and another business system. These digital technologies have enabled process automation, predictive maintenance, self-optimization of process improvements, and most importantly, a new level of efficiency and customer response. As more smart industries emerge as a result of the development of smart factories, the manufacturing sector has a great opportunity to enter the fourth industrial revolution [1]. Based on the data collected from the sensor used in the factory floor helps to ensure that manufacturing assets are working properly and it can provide a tool for performing predictive maintenance in order to minimize equipment downtime. Recent developments in the field of power electronic switches and permanent magnets improved the application of BLDC motors, particularly in CNC machines. In CNC machine, feed drive is a crucial part because the performance and characteristics of the driving system have a significant impact

A. Lavanya · S. Revathi (✉)
Department of Electrical and Electronics Engineering, National Institute of Technology
Puducherry, Karaikal 609609, India
e-mail: revathi.s@nitpy.ac.in

N. Sivakumaran · K. Rajkumar
Department of Instrumentation and Control Engineering, National Institute of Technology,
Tiruchirappalli, India

on the machine precision and repeatability. CNC machine tool employs various types of drive for spindle and feed motion. Brushless DC motors are typically utilized in servo, actuation, positioning, and variable speed applications in industrial settings. In this chapter, the performance of BLDC motor is analyzed for the suitability for feed drives in CNC machine. Due to its benefits, including dependability, high power factor, compact size, and low maintenance requirements, the BLDC motor has been used in numerous sectors [2]. Torque ripples are produced by cogging torque and non-sinusoidal back electromotive force (EMF) at the BLDC motor, and the torque follows the amplitude of the stator current. Maintenance of constant torque and its smoothness determines the cutting performance of the spindle.

CNC Machine

The numerical control, which served as the forerunner to the contemporary CNC machine, was invented by John T. Parsons (1913–2007), of Parsons Corporation in Traverse City, Michigan. John Parsons has been called as the "father of the second industrial revolution" for his contributions [3]. Around the world, CNC machines produce parts for practically every sector. Although everyone now refers to it as CNC, the term "CNC" stands for computer numerical control. Applications for CNC can be found in both machine tool and non-machine tool fields. CNC is used in various process like for drill presses, lathes, milling machines, lasers, grinding sector, sheet-metal press working machines, tube bending machines, etc. in the machine tool category. Machine tools that change the cutting tools automatically while being controlled by CNC, such as turning centers and machining centers, have been produced. The advantages of CNC include high manufacturing accuracy, quick output, enhanced manufacturing flexibility, simpler fixturing, contour machining (2–5 axis machining), and reduced human error.

Principles of CNC

Every machine that uses a computer is capable of precisely and repeatedly controlling motion in a variety of directions. Each of these motion directions is referred to as an axis. There are typically two and five axes, depending on the type of equipment. There are two kinds of axis: they are linear axis and rotary axis.

When using a CNC machine, the first thing to do is write and input the part programme, which consists of G-code and M-code. This will be carried out in the CNC machine control unit, which enables all data processing to occur. It will then be fed to the driving system, which aids in regulating the machine tool motion and speed. A feedback system exists that records the machine tool's position and velocity measurements and transmits a feedback signal to the MCU [4]. The MCU then checks for faults by comparing the feedback signals with the reference signal,

Fig. 10.1 CNC feed drive

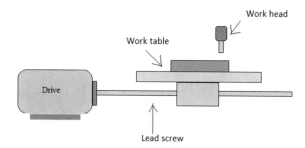

and if any are found, it corrects them before sending a fresh signal to the machine tool. In addition, display unit is provided on CNC machine to view commands, programming. Figure 10.1 shows the CNC feed drive.

Principle and Operation of Brushless DC Motor

A permanent synchronous motor with rotor position feedback is termed as brushless DC motor. Typically, a three-phase power semiconductor bridge is used to operate brushless motors. BLDC motor has two parts: electrical and mechanical part. The electrical part determines current and torque. The BLDC motor has three phases: incoming phase, outgoing phase, and floating phase. It is common to have two phases of the BLDC motor running in series during bipolar operation. A partial amount of Dc voltage is given to the two-phase so that a high starting torque is achieved. However, the speed of the motor is inversely proportional to its torque, a full DC bus voltage is given to each phase. To energize the motor and to give supply to the correct commutation sequence on the device in the inverter bridge, it is necessary to provide rotor position sensor. It is an electronic motor because electronic commutation is utilized to switch the armature current instead of brushes. Since electronic commutation is implemented, BLDC motor is more durable than a DC motor because of the absence of brushes and commutator arrangement so that it eliminate sparking and wear out. The BLDC motor has three phases: incoming phase, outgoing phase, and floating phase. It is common to have two phases of the BLDC motor running in series during bipolar operation. A partial amount of Dc voltage is given to the two-phase so that a high starting torque is achieved. However, the speed of the motor is inversely proportional to its torque, a full DC bus voltage is given to each phase.

Figure 10.2 shows permanent magnet synchronous machine (PMSM), sensors, control algorithm, and power converter are the four main parts of the BLDC motor. The PMSM transfers electrical energy into mechanical energy by transferring power from the source to the power converter. The brushless DC motor's rotor position sensors are one of its standout features.

The control algorithm choose the gate signals for each semiconductor in power electronic converter based on the command signal, rotor position, torque, speed, and

Fig. 10.2 Block diagram of
Brushless DC motor

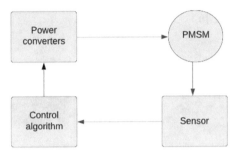

voltage of the motor. There are two basic kinds of brushless DC motors voltage source-based drives and current source-based drives which are determined by the structure of the control algorithms. Permanent magnet synchronous machines are driven by both voltage source and current source-based drives, with either sinusoidal or non-sinusoidal back emf waveforms. The sinusoidal back emf is controlled to attain a constant torque. To attain constant torque, a machine sinusoidal back EMF must be controlled. However, a machine with a non-sinusoidal back emf reduces inverter sizes and losses for the same power level.

Mathematical Model of BLDC Motor

Assumptions made in the circuit

The following presumptions form the foundation of the modeling.

- The resistance of the stator for all windings is equal.
- The self- and mutual-inductance of all windings are equal.
- Rotor reluctance and electrical angle are both equal.
- The inverter should have optimum power semiconductor components.
- The motor is not saturated.
- Losses of iron are minimal.

Figure 10.3 shows the equivalent circuit of BLDC motor. A BLDC motor has a trapezoidal flux distribution. It is not appropriate to use the d-q rotor reference frame for permanent magnet synchronous motor. Given the non-sinusoidal flux distribution, a model of the BLDC motor should be derived in phase variables. The model development is predicated on the suppositions that iron and stray losses, as well as the induced current in the rotor caused by harmonic fields in the stator, are to be ignored. The BLDC motor often lacks damping windings because damping is provided by the inverter control. The derivation technique can be used for any number of stages; however, the motor is assumed to have three phases. The coupled stator winding circuit equations in terms of the motor electrical constant are

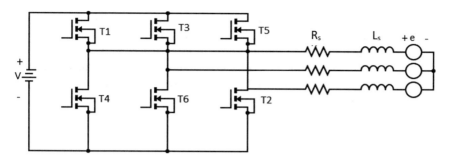

Fig. 10.3 Equivalent circuit of Brushless DC motor

$$
\begin{bmatrix} V_a \\ V_b \\ V_c \end{bmatrix} = \begin{bmatrix} R_a & 0 & 0 \\ 0 & R_a & 0 \\ 0 & 0 & R_a \end{bmatrix} \begin{bmatrix} i_a \\ i_b \\ i_c \end{bmatrix} + P \begin{bmatrix} L_{aa} & L_{ba} & L_{ca} \\ L_{ba} & L_{bb} & L_{bc} \\ L_{ca} & L_{cb} & L_{cc} \end{bmatrix} \begin{bmatrix} i_a \\ i_b \\ i_c \end{bmatrix} \begin{bmatrix} e_a \\ e_b \\ e_c \end{bmatrix} \tag{10.1}
$$

where R_s is the resistance of stator per phase, V_{as}, V_{bs}, V_{cs} are the phase voltage of stator i_a, i_b, i_c are phase current. L_{aa}, L_{bb}, L_{cc} and L_{ab}, L_{bc}, L_{cc} are self and mutual inductance of phase a, b, c respectively.

$$L_{aa} = L_{bb} = L_{cc} = L$$

$$L_{aa} = L_{ba} = L_{ac} = L_{bc} = L_{cb} = M$$

$$
\begin{bmatrix} V_a \\ V_b \\ V_c \end{bmatrix} = \begin{bmatrix} R_a & 0 & 0 \\ 0 & R_a & 0 \\ 0 & 0 & R_a \end{bmatrix} \begin{bmatrix} i_a \\ i_b \\ i_c \end{bmatrix} + d/dt \begin{bmatrix} L & M & M \\ M & L & M \\ M & M & L \end{bmatrix} \begin{bmatrix} i_a \\ i_b \\ i_c \end{bmatrix} \begin{bmatrix} e_a \\ e_b \\ e_c \end{bmatrix}
$$

$$V_{as} = V_{oa} - V_{no}$$

$$V_{bs} = V_{ob} - V_{no} \tag{10.2}$$

These voltages were in relation to the DC link midpoint zero reference potential. The stator phase currents must be balanced, which means that

$$i_a + i_b + i_c = 0 \tag{10.3}$$

this results in the models' inductances matrix being simplified as the

$$M i_c + M i_b = -M i_a \tag{10.4}$$

here is the state space form:

$$
\begin{bmatrix} V_a \\ V_b \\ V_c \end{bmatrix} = \begin{bmatrix} R_a & 0 & 0 \\ 0 & R_a & 0 \\ 0 & 0 & R_a \end{bmatrix} \begin{bmatrix} i_a \\ i_b \\ i_c \end{bmatrix} + d/dt \begin{bmatrix} L-M & 0 & 0 \\ 0 & L-M & 0 \\ 0 & 0 & L-M \end{bmatrix} \begin{bmatrix} i_a \\ i_b \\ i_c \end{bmatrix} \begin{bmatrix} e_a \\ e_b \\ e_c \end{bmatrix}
$$

$$\tag{10.5}$$

where r is the rotor position in radians, and m is the flux linkage. The functions (f_{as}, f_{bs}, and f_{cs}) have similar shape as a, b, and c, e with 1 as the maximum magnitude. Due to their trapezoidal shape, the induced emfs lack sharp corners. The electromagnetic fields, which are a continuous function, are produced by the flux connections derivatives. The flux density function is also smoothed out by fringes, resulting in no sharp edges. The Newton's definition of the electromagnetic torque is

$$T_e = [e_{as}i_{as} + e_{bs}i_{bs} + e_{cs}i_{cs}]\frac{1}{\omega_m}(\text{Nm}) \tag{10.6}$$

it is crucial to note that the armature-voltage equation of a DC machine and the phase voltage equation are the same. The moment of inertia is described as

$$J = J_m + J_i \tag{10.7}$$

when inertia J, friction coefficient B, and load torque T_i are included in the simple motion system equation is given by

$$J\frac{d\omega_m}{dt} + B\omega_m = (T_e - T_l) \tag{10.8}$$

the position and speed of rotor are correlated by

$$d\frac{\theta_r}{dt} = \frac{P}{2}\omega_m \tag{10.9}$$

although damping coefficient B is small and negligible, it affects the system performance. The potential of the neutral point with regard to the zero potential must be taken into consideration in order to avoid applied voltage imbalance and accurately replicate the drive performance.

Thus

$$V_{no} = (V_{ao} + V_{bo} + V_{co}) - \frac{(e_a + e_b + e_c)}{3} \tag{10.10}$$

when all pertinent equations are combined, the system in state-space form is

$$x = AX + BU + Ce$$

$$A = \begin{bmatrix} \frac{-R}{L-M} & 0 & 0 & \frac{-\lambda m}{f}fas(\theta r) & 0 \\ 0 & \frac{-R}{L-M} & 0 & \frac{-\lambda m}{j}fbs(\theta r) & 0 \\ 0 & 0 & \frac{-R}{L-M} & \frac{-\lambda m}{j}fcs(\theta r) & 0 \\ \frac{\lambda}{j}fas(\theta r) & \frac{\lambda m}{j}fbs(\theta r) & \frac{\lambda m}{j}fcs(\theta r) & \frac{-B}{j} & 0 \\ 0 & 0 & 0 & \frac{P}{2} & 0 \end{bmatrix} \tag{10.11}$$

$$B = \begin{bmatrix} \frac{1}{L-M} & 0 & 0 & 0 \\ 0 & \frac{1}{L-M} & 0 & 0 \\ 0 & 0 & \frac{1}{L-M} & 0 \\ 0 & 0 & 0 & \frac{1}{L-M} \end{bmatrix} \qquad (10.12)$$

$$C = \begin{bmatrix} \frac{-1}{L-M} & 0 & 0 \\ 0 & \frac{-1}{L-M} & 0 \\ 0 & 0 & \frac{-1}{L-M} \end{bmatrix} \qquad (10.13)$$

$$U = [V_{as} \, V_{bs} \, V_{cs} \, T_l]^t \qquad (10.14)$$

$$e = [e_a e_b e_c]^t \qquad (10.15)$$

Commutation Torque Ripple

For a three phase PMBCM drive, a rectangular 120° wide current waveform is desired. The leakage inductance L distorts the ideal waveform into a trapezoidal shape, causing the stator currents to rise and fall in a finite length of time. The phase current wave and rotor flux linkage of the BLDC is depicted in Figs. 10.4 and 10.5. For every 360° electrical. The torque factor is significantly impacted by the current ripple. The waveform of the power supply to BLDC motors is a trapezoidal waveform [5].

Permanent magnets cause a voltage to be induced in the winding that deviates from the supply voltage's waveform, which results in a sizable ripple component in the supply current.

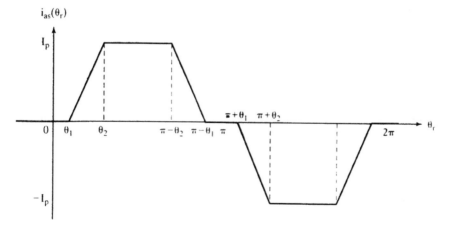

Fig. 10.4 Phase current of BLDC motor

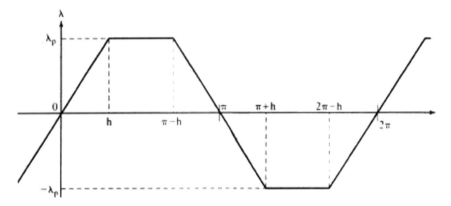

Fig. 10.5 Rotor flux linkage of BLDC motor

$$T_{el} = \frac{16 I_p \lambda_p}{\pi^2(\theta_2 - \theta_1)}[\sinh(\sin\theta_2 - \sin\theta_1)\sin^{2\theta r} + \sinh(\sin\theta_2 \sin\theta_1)\sin^{2(\theta r - \frac{2\pi}{3})}$$
$$+ \sinh(\sin\theta_2 - \sin\theta_1)\sin^{(\theta r + \frac{2\pi}{3})} \qquad (10.16)$$

and the basic torque normalized in p.u. as a function of h is equal to 30° for a current of 1 p.u. Note that increases: thus further reducing the base torque of the motor drive.

Control of Feed Drives

The feed drives of CNC machines are controlled by different controllers like PI controller, fuzzy logic controller, and fuzzy immune PI controller. The actual and reference parameter from the motor is given to the controller. Control signal is produced from the controller then it is fed to the motor.

(a) PI Controller

One of the most crucial controllers used in industrial applications is the PI controller. PI, which stands for proportional integral. It regulates multiple process variables, such as pressure, speed, temperature, and flow in industrial settings. Its name is proportional integral. Using a control loop feedback method, this controller controls all of the process variables. The torque ripples in BLDC motors are managed with a PI controller. Lowering the error signal between the desired and measured signals is the controller's objective. The Ziegler-Nichols tuning method is used to adjust the PI controller's parameters. The PI loop's best disturbance rejection performance is provided by the Ziegler-Nichols tuning rule. Figure 10.6 represents the block diagram of PI controller.

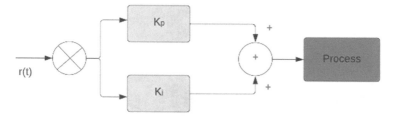

Fig. 10.6 PI controller

(b) **Fuzzy logic controller**

Fuzzy set theory, first proposed by Lofti Zadeh in 1965, is the foundation of fuzzy logic control. Fuzzy logic is a paradigm for a different approach to design that can be used to create both linear and nonlinear systems. Fuzzy control algorithms were created as a result of the realization that combining human intelligence into automated control systems would be a more effective solution. Fuzzy control, which has its origins in Professor Zadeh's fuzzy set development proposal, enables the use of experience and information from earlier systems for the control of operations. In particular, fuzzy logic controllers (FLC) are useful for compensating nonlinearity. Fuzzy logic's systematic characteristic can transform language control rules based on professional judgment into automatic control schemes. Fuzzy logic controller is a type of nonlinear intelligent controller based on fuzzy set theory, fuzzy logic control, and fuzzy language is a technique for employing human intelligence to regulate the system. The following features of the fuzzy controller are present: excellent robustness, easy control, and strong adaptability. The fuzzy controller can execute effective control whether the controlled item is linear or nonlinear. The fuzzy process, the knowledge base (database and rule base), fuzzy reasoning, and defuzzification are the four components that make up the fuzzy control system. It is not necessary to understand the precise mathematical model of the controlled object for the intelligent fuzzy control system function. Whether the controlled object is linear or nonlinear, the fuzzy controller can be efficiently controlled and has high resilience and adaptability [6]. Figure 10.7 represents the block diagram of fuzzy logic controller.

(i) **Fuzzifier**

Input values are converted to domain values in the appropriate proportions. Colloquial variables are used to describe the process of measuring physical quantities. The corresponding degree of membership is derived from the corresponding linguistic value. This colloquial variable is called a fuzzy subset.

(ii) **Rule base**

A rule-base system is one that uses rules created by humans to store, sort, and manage data.

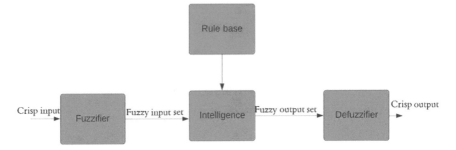

Fig. 10.7 Fuzzy logic controller

Table 10.1 Rule base for torque controlled signal

CE	E						
	NL	NM	NS	ZE	PS	PM	PL
NL	PL	PB	PM	PM	PS	PS	ZE
NM	PL	PM	PM	PS	PS	ZE	NS
NS	PM	PM	PS	PS	ZE	NS	NS
ZE	PM	PS	PS	ZE	NS	NS	NM
PS	PS	PS	ZE	NS	NS	NM	NM
PM	PS	ZE	NS	NS	NM	NM	NL
PB	ZE	NS	Ns	NM	NM	NB	NL

(iii) **Defuzzifier**

The process used to obtain crisp output is called defuzzification. This process is also called anti-fuzzification. It is used to convert the results of a fuzzy inference into a clear output. In other words, a decision-making algorithm that chooses the optimal crisp value based on a fuzzy set achieves defuzzification. Based on the fuzzy set, the decision-making algorithm chooses the optimal crisp value [7].

The rule base for torque-controlled signal is shown in Table 10.1. Typically, seven terms are chosen for every input and output variable. The E and CE fuzzy subsets are mapped to the universe and defined as NL, NM, NS, ZE, PS, PM, and PL. Each input is represented by one of seven membership functions: Negative Large (NL), Negative Medium (NM), Negative Small (NS), Zero (ZE), Positive Large (PL), Positive Small (PS), and Positive Medium (PM) [8].

Artificial Immune System

AIS is focused with the application of the immune system structure and function to computational structures as well as the investigation of how these techniques are used to address analytical issues in the fields of mathematics, engineering, and information science. Artificial immune systems (AIS) are sophisticated algorithm-based ideas drawn from the functioning of human immune system. It should be emphasized that the biological immune system functions effectively and powerfully to maintain the body's strength and health in its whole. The primary type of immune cell involved in the immune response and possessing the qualities of specificity, diversity, memory, and adaptability is the lymphocyte. Other immune cells that are referred to as phagocytic cells, such as neutrophils, eosinophils, basophils, and monocytes, provide as support for the elimination of antigens. T cells and B cells make up the majority of the two different types of lymphocytes. Sites where lymphocytes develop and become antigenically committed can be found in the primary lymphoid organs. In particular, T cells develop in the bone marrow and travel to the thymus to mature, whereas B lymphocytes develop and mature within the bone marrow. In order to trigger an immune response, lymphocytes need somewhere to interact with antigen, which the secondary lymphoid organs provide to do. However, the functions of these two types of lymphocytes in the immune response differ.

Only a few of these immune cells are able to identify the peptides of invaders when antigens enter the body. The cells that create matching clones are proliferated and differentiated as a result of this recognition. This procedure, known as clonal expansion, produces a sizable population of antigen-specific antibody-producing cells. Immune cells that have grown clonally kill or neutralize the antigen [9]. Additionally, some of these cells are kept in immunological memory so that any future exposure to an identical antigen triggers an immediate immune response. Figure 10.8 represents the mechanism of immune feedback system.

We illustrate how T cells control their behavior using the straightforward immunological feedback law.

$$u(K) = K\{(1 - \eta f[\Delta u(k)]\} \, e(k) \tag{10.17}$$

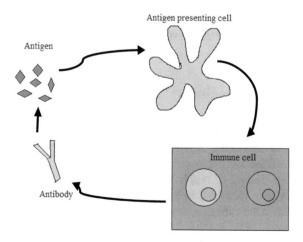

Fig. 10.8 Mechanism of immune feedback system

Antigen presenting cell

Antigen

Immune cell

Antibody

Table 10.2 Fuzzy rules for fuzzy immune frame work

$f(u(k), u(k))$	P	ZE	N
P	NS	NB	PB
ZE	NS	ZE	PB
N	NS	PS	PS

where $k = k_1$

$$u(k) = k_p e(k) + k_i \sum e(i) \tag{10.18}$$

where k_p, k_i are the coefficient of proportional, integral. Combining with conventional PI control algorithm gives the following PI control algorithm.

$$\Delta u(k) = k_{p1}(e(k) - e(k-1)) \tag{10.19}$$

where

$$K_{p1} = K(1 - \eta f(u(k), \Delta u(k))) \tag{10.20}$$

The speed of a system is controlled by the parameter K and stabilization effect is controlled by the parameter η. An immune PI controller is a nonlinear controller that is realized mainly by $f(.)$. The ability of this type of controller depends on the choice of parameter K, η and function $f(.)$ in a large extend. Fuzzy controllers, a kind of nonlinear controllers, have been shown to be effective for general functional approximation. Fuzzy controllers may actualize any linear or nonlinear control law when the design parameters are properly chosen. On the other hand, fuzzy control is a useful technique for creating a nonlinear functional approximation of control and modeling applications. Two input variables are fuzzified by two fuzzy sets; called "positive" (P) "negative" (N) and Zero (Z). The output variable is composed by three fuzzy sets: called "positive" (P) "negative" (N) and Zero (ZE). Fuzzy rules for fuzzy immune frame work are tabulated in Table 10.2.

Simulation Study

The value of the rated voltage (V), rated power, rated speed (RPM), and pole pair are specified in the simulation of BLDC are tabulated in Table 10.3. The speed can be increased or decreased by increasing or decreasing the supply voltage as the speed is a function of supply voltage. Torque is directly proportional to current. Therefore, in open loop control the output depends on the supply voltage and load condition. Since the current and torque are inversely proportional to the speed. The torque of a motor remains same for a speed range up to the rated speed. The BLDC motor can run up to maximum speed and the torque starts reducing [10].

Parameters	Values
Rated voltage (V)	24
Rated power (W)	300
Rated speed (RPM)	3000
Pole pair	2

Table 10.3 System parameters

(a) *Performance of BLDC using PI controller*

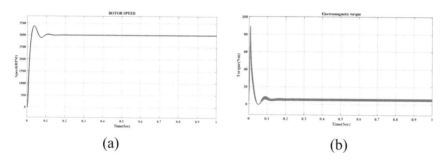

(a)   (b)

The speed and torque of BLDC motor is shown in Fig (a) and (b) with PI controller. The speed of the motor is 2994 RPM with overshoot. The torque obtained by the motor is shown in Fig (b). Torque of the motor is 5 Nm with initial overshoot and ripples.

(b) *Performance of BLDC with fuzzy logic controller*

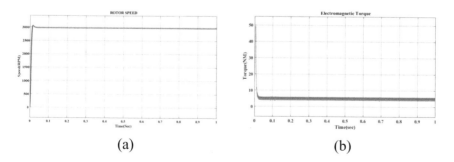

(a)   (b)

The speed and torque of BLDC motor is shown in Fig (a) and (b) with fuzzy logic controller. The speed of the motor is 3000 RPM with slight overshoot. The torque obtained by the motor is shown in Fig (b). Torque of the motor is 5 Nm with initial overshoot and ripple.

(c) *Performance of BLDC with fuzzy immune controller*

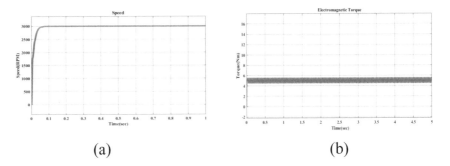

(a)   (b)

The speed and torque of BLDC motor is shown in Fig (a) and (b) with fuzzy immune controller. The speed of the motor is 3000 RPM without any overshoot and oscillations. The torque obtained by the motor is shown in Fig (b). Torque of the motor is 5 Nm with initial overshoot and ripples.

Comparative Study

Torque ripples produced by the motor are calculated by using the method described in [11] and the performance of the controllers are compared and the same is listed in Table 10.4

$$T_{error} = \frac{T_+ - T_-}{T_a} * 100$$

T_+ = maximum point.
T_- = minimum point.

Table shows that BLDC motor with PI controller runs at 3000 rpm with 5.5 Nm torque which has initial oscillations in the speed and torque ripples of about 62%. The motor with fuzzy controller runs at 2994 rpm with 5 Nm with slight initial oscillations and 48% of ripples present in it. The motor with fuzzy immune PI controller achieves a speed of 3000 rpm as a smooth curve, and it also achieves torque of 5 Nm with less torque ripples as compared with other to controller.

Table 10.4 Performance of controllers

Controller	Speed (RPM)	Torque (Nm)	Torque ripples (%)
PI	3000	5.5	62
Fuzzy	2994	5	48
Fuzzy immune PI	3000	5	20

Conclusion

The performance of feed drive used in CNC machine is tested with different controllers. Since the smoothness of torque produced by the motor used as a feed drives in CNC machine decides the cutting accuracy of the work piece. The performance of motor is analyzed with different types of controller like PI controller, fuzzy controller, and fuzzy immune controller and the ripples in torque are minimized. Hence, the BLDC motor satisfies the requirement of CNC feed drives and by reducing the torque ripple the machine overcomes the friction. As a result, the fuzzy immune PI controller achieves better torque and speed as compared with other controllers.

References

1. Lasi, H., Fettke, P., Kemper, H.G., Feld, T., Hoffmann, M.: Industry 4.0. Bus. Inf. Syst. Eng. **6**(4), 239–242 (2014)
2. Krishnan, R.: Permanent magnet synchronous and brushless DC motor drives. CRC Press (2017)
3. Keller, A.Z., Kamath, A.R.R., Perera, U.D.: Reliability analysis of CNC machine tools. Reliab. Eng. **3**(6), 449–473 (1982)
4. Xu, X.W., Newman, S.T.: Making CNC machine tools more open, interoperable and intelligent—a review of the technologies. Comput. Ind. **57**(2), 141–152 (2006)
5. Fang, J., Li, H., Han, B.: Torque ripple reduction in BLDC torque motor with nonideal back EMF. IEEE Trans. Power Electron. **27**(11), 4630–4637 (2011)
6. Ramesh, M.V., Amarnath, J., Kamakshaiah, S., Rao, G.S.: Speed control of brushless DC motor by using fuzzy logic PI controller. ARPN J. Eng. Appl. Sci. **6**(9), 55–62 (2011)
7. Baharudin, N.N., Ayob, S.M.: Brushless DC motor drive control using single input fuzzy PI controller (SIFPIC). In: 2015 IEEE Conference on Energy Conversion (CENCON), pp. 13–18, IEEE (2015)
8. Lee, C.C.: Fuzzy logic in control systems: fuzzy logic controller. I. IEEE Trans. Syst. Man Cybern. **20**(2), 404–418 (1990)
9. Gupta, R.A., Kumar, R., Bansal, A.K.: Artificial intelligence applications in permanent magnet brushless DC motor drives. Artif. Intell. Rev. **33**(3) (2010)
10. Song, J.H., Choy, I.: Commutation torque ripple reduction in brushless DC motor drives using a single DC current sensor. IEEE Trans. Power Electron. **19**(2), 312–319 (2004)
11. Xia, K., Ye, Y., Ni, J., Wang, Y., Xu, P.: Model predictive control method of torque ripple reduction for BLDC motor. IEEE Trans. Magn. **56**(1), 1–6 (2019)

Chapter 11
Efficient Anomaly Detection for Empowering Cyber Security by Using Adaptive Deep Learning Model

Balasubramanian Prabhu Kavin, Jeeva Selvaraj, K. Shantha Kumari, Rashel Sarkar, S. Rudresha, and Hong-Seng Gan

Introduction

A variety of supervised procedures have been used to notice cyber-attacks [1, 2], and this emerging field of study uses machine learning techniques [3]. The difficulty in actually collecting a large number of labelled samples of attacks is a key drawback of these methods. Additionally, in future, there will be an increase in the prevalence of assaults of hitherto unseen varieties. Consequently, anomaly detection [4–8] and other unsupervised methods are gaining traction in the detection of these varied and novel forms of assault. Anomaly detection methods often use a broad "statistical measure," which regrettably might mistakenly label harmless behaviour as suspicious even when it has nothing to do with potential security issues. Thus, they generate a lot of false positives, which necessitate a security professional to physically check each one through an expensive and time-consuming inquiry. Analysts are often left in the dark when confronted with a batch of anomalies since the finding approach is a black box that does not expose the attributes that lead to each sample's assortment. Some fresh efforts [9–12] have tried to address this problem by offering rationales

B. P. Kavin (✉) · J. Selvaraj · K. S. Kumari
Department of Data Science and Business Systems, SRM Institute of Science and Technology, Kattankulathur, India
e-mail: ceaserkavin@gmail.com

R. Sarkar
Department of Computer Science, University of Science and Technology (USTM), Baridua, Meghalaya, India

S. Rudresha
Department of Mechanical Engineering, JSS Academy of Technical Education, Noida, India

H.-S. Gan
School of AI and Advanced Computing, XJTLU Entrepreneur College (Taicang), Xi'an Jiaotong - Liverpool University, Suzhou 215400, Jiangsu, P.R., China

© The Author(s), under exclusive license to Springer Nature Singapore Pte Ltd. 2023   253
V. Sarveshwaran et al. (eds.), *Artificial Intelligence and Cyber Security in Industry 4.0*, Advanced Technologies and Societal Change,
https://doi.org/10.1007/978-981-99-2115-7_11

for anomaly identification. After reading the description, the analyst would know exactly which aspects to look at in order to except time: those that match the given scenario.

The analyst would typically classify each instance as conducting their inquiry. A case is marked as "unknown" if there is insufficient data to determine whether it is a TP or FP within the time restrictions of the inquiry. Incorporating this kind of user input has been suggested as a way to enhance the presentation of anomaly detection in a number of recent publications [13–17]. As a consequence of encouraging outcomes in image processing, CV, NLP, and other areas, deep learning (DL) algorithms have lately acquired favour as potent algorithms [18]. For academics, DL stands out owing to its ability to understand temporal pattern relationships across time and its use of hierarchical feature representations. As a result, DL approaches have lately been proposed to boost the intelligence of ID techniques, despite the lack of investigate to assess such ML approaches using publicly available datasets. In a word, DL's high-quality learning for complicated data processing is made possible by its nonlinear structural architecture. Increased hardware support for DL techniques has been made possible by the rapid advancement of parallel computing technologies.

Keeping up with the ever-increasing high-speed network traffic in today's diverse environments necessitates fixing three fundamental problems: (1) The current generation of ML-based models has a high FPR for a wider variety of hostile incursions [19]. Third, we need state-of-the-art solutions since (2) present ML-based models are not generalisable, and (3) existing ID systems overlook fresh threats because of stale ID datasets.

- To address these issues, we build a hybrid CNN-based ID system with an emphasis on testing the performance of ID area by utilising a real-world dataset. As was previously noted, ID techniques have certain drawbacks. To get around this, we suggest a new classical procedure that syndicates the assistances of two approaches, each of which offers enhanced performance over more conventional methods. To boost the ID scheme's ability to learn and perform, we present a new IDS that combines state-of-the-art DL techniques like convolutional neural networks (CNNs) with traditional ML techniques like recurrent neural networks (RNNs). Summarised below are the major findings from our study.
- To cut down on analytical costs while increasing returns, we came up with the CRNN-POA, a hybrid model that combines deep and shallow learning. Since attacks may be categorised into the consistent intrusion class, the proposed CRNN-POA emphases on determining whether network behaviour is usual or hostile. The suggested strategy is compared to well-liked ML procedures. Using 10-cross-validation, the empirical results demonstrate that the CRNN-POA is highly suitable for attack finding and can properly classify the misuses in 97.75% of occurrences.

Experiments on the widely used and cutting-edge CSE-CIC-IDS2018 real-world dataset show that the CRNN-POA produces better results than conventional classification algorithms, leading to a more accurate ID and a new avenue for ID research.

This is how the article progresses from here on out. Section "Related Works" provides a summary of the NID's history. Section "Proposed Methodology" explains the ID data in detail and gives an overview of the planned CRNN-POA structure. In Sect. "Results and Discussion", we detail the CRNN-POA simulation. In Sect. "Conclusion", we graphically present the findings of our study.

Related Works

Sharma et al. [20] suggested a fog layer-based anomaly detection architecture for IoT networks. For simplicity, the method only uses a single variable, the CRPS measure. As a result, we have provided a method for utilising numerous variables and demonstrated why doing so is necessary in a network with as much diversity as the IoT. For our purposes, we have used DARPA 99 for detection (testing data) because it includes a single TCP SYN attack that lasts for 6 min and fifty-one seconds, and for ICMP, we have used DARPA 99 because it includes two attacks that last for a total of one second. The programme can accurately recognise various types of assaults.

For distributed anomaly discovery in WSNs, Yang et al. [21] suggest using a blockchain-based discovery (BCEAD), which uses the blockchain to hold the perfect of a conventional anomaly technique. A better global perfect for detection may be updated repeatedly by developing an appropriate block structure and consensus method. Since the blockchain ensures the network's trustworthy environment, the detection method is protected against attacks from within the system. Finally, the findings show that BCEAD outperforms competing schemes in terms of performance, cost, etc.

Anomaly-based intrusion detection is the main emphasis of Saif et al. [22], which aims to identify security assaults on cloud systems. The proposed HIIDS's performance has been measured against that of the widely used NSL-kDD dataset, which consists of 125,973 samples. Supervised learning algorithms like known nearest neighbour (kNN) features, while metaheuristic algorithms like PSO and genetic algorithm (GA) are used to reduce computation cost through best feature selection. In addition, a hybrid strategy for feature selection and categorisation has been introduced. Six variations of the proposed hybrid procedures integrating GA, PSO, DE with kNN, DT are implemented in MATLAB 2019b after the dataset has been pre-processed in Python. A thorough analysis of the system's efficiency has been completed, including measurements of its memory footprint, CPU load, and the speed with which it completes tasks. When compared to GA-kNN, DE-DT, the GA-DT variation achieves the maximum accuracy of 99.88% for the DoS class, U2R class, normal class, respectively, using just 8–10 characteristics.

To improve upon previous attempts at network anomaly identification, Xu et al. [23] offer a novel 5-layer autoencoder (AE)-based perfect. To back up our suggestion, we conducted a thorough and exhaustive analysis of many key performance metrics in an AE model. To mitigate model bias caused by data, we employ a novel data pre-processing approach in our proposed model, which involves transforming and removing the greatest impacted outliers from the input tasters. In order to determine whether or not a given network traffic taster is typical or out of the ordinary, our suggested model makes use of the most efficient reconstruction error function available. Our model achieves higher detection accuracy and $F1$-score thanks to a combination of these novel methods with the best-possible model design, which improves its capacity for reduction. On the NSL-KDD dataset, we found that our suggested model had the greatest accuracy (90.61%) and $F1$-score (92.26%) of all the models we tested.

Pre-processing, classification is only some of the procedures in Sathya et al. [24] that help pinpoint an attack's origin. An unbalanced dataset is handled after being pre-processed using a random oversampler, which handles NaN values, and unbalanced datasets. The dataset is passed to KS test technique for feature extraction after it has been pre-processed. Following is a detailed explanation of classification algorithm used to sort the data into attack and non-attack groups. The experimental evaluation shows that the suggested technique is greater to the approaches in terms of spotting intrusions and attacks. The suggested approach detects attacks with a success rate of 98.1% in 77 s.

An anomaly detection system based on a combination of traditional methods and deep learning was proposed by Osamor and Wellman [25] to progress the detection precision and proficiency of anomaly detection schemes. To reduce the dimensionality of the system call traces, the raw sequence of traces is first fed into a CNN network. In order to learn the sequences of the system calls and provide the final detection result, the LSTM network is given this reduced trace vector. The hybrid model was trained with the help of TensorFlow-GPU and tested with the ADFA-LD dataset. The suggested strategy improved the rate at which anomalies could be detected during training, and the experimental findings revealed that it required less time to implement. Consequently, the number of unfounded alarms is reduced using this strategy.

On the basis of deep belief networks (DBN) and LSTM networks, Chen et al. [26] presented an effective network behaviour anomaly detection (NBAD) technique. The first step is to apply a DBN-based nonlinear feature extraction algorithm to automatically extract features and shrink the original dataset in dimension without sacrificing precision. The categorisation outcomes are then obtained by employing a light-structure LSTM network. Multiple studies demonstrate that the suggested method achieves great performance in feature learning and accuracy, while also being fast to acquire results and simple to update.

This approach [27] suggests a deep learning-based intrusion detection methodology as a solution to the problems of low accuracy and feature extraction. This strategy employs a recurrent neural network after three stages of pre-processing—numerical conversion of data, normalisation of data, and balancing of data. It improves anomaly detection and can accurately depict network traffic flow. The proposed model is evaluated using a publicly available benchmark dataset, and the results demonstrate that it outperforms the alternatives. The examination of the results using the proposed method reveals an average accuracy of 99.56%, a TPR of 99.55%, and a false negative rate of 99.32%.

The massive increase [28] of data being transmitted via IoT devices to end-user devices has increased the importance of developing intrusion detection systems. There is no smart home, smart city, smart farm, or smart business without an intrusion detection system. The intruders use the IoT sensor gadget as a conduit for their attacks, which begin in the intruders' crate. In order to detect intrusion's activities in the IoT environment, many deep learning models have been developed and deployed. This literature review looks at the IoT environment's intrusions detection system, which makes use of deep supervised learning, deep unsupervised learning, and a specific dataset. In the end, the unsolved issues in intrusion detection systems for the Internet of Things are provided for further study.

Proposed Methodology

Datasets

Selecting the right ID data for use in evaluating the ID system is crucial, thus we did so before running simulations of the suggested method.

Explanation of the ID Data

Numerous ID databases may be downloaded for free; however, some of them include inaccuracies that are out of date, rigid, poorly validated, and impossible to replicate. In order to address these issues and provide up-to-date traffic patterns, the Amazon Web Facilities (AWS) stage industrialised the widely used CSE-CIC-DS2018 [29] dataset. It includes several types of datasets for testing anomaly-based methods. The dataset shows network activity in real time and includes many incursion conditions. In addition, the whole network's distribution encloses all of the inner network suggestions used to calculate data packet payloads. Due to its unique properties, the CSE-CIC-DS2018 dataset was chosen to test our proposed intrusion detection method. As a result of the projected CRNN-based ID system, the field of network security is hoped to go in a more logical and beneficial path.

Several intrusion profiles applicable to various network protocols and topologies are included in this dataset for use in the security field. These improvements to the dataset were made using the IDS2017 standards. Currently, there are two profiles and seven intrusion techniques accessible in the publically available IDS2018 dataset. Multiple data sets were collected, and the raw data was continuously revised. Since volume and sum of bytes were computed in both onward and opposite mode, IDS2018 contains 80 statistical features. In the end, the dataset containing about 5 million entries was uploaded to the internet and made available to academics worldwide. You may download the CSE dataset in either PCAP or CSV format. CSV is the most common format used in AI, while PCAP is used for feature extraction [28, 29].

Table 11.1 Impression of the extracted features of dataset

Explanation	Features
Complete data packets in a back technique	Tot-bw-pk
The bottommost volume of the packet in a further way	Fw-pkt-l-min
Maximum period among two packets advanced in a back way	Bw-iat-max
Least time among two packets transported in a forward way	Bw-iat-min
Collective data packets in a onward way	Tot-fw-pk
Complete size of the packet in an up way	Tot-l-fw-pkt
Flow interval	Fl-dur
Extreme time between two flows	Fl-iat-max
Overall time among two packets transported in a back way	Bw-iat-tot
Mean period among two packets carried in a back way	Bw-iat-avg
Regular period among two packets advanced in a back way	Bw-iat-std
The regular quantity of data in the packet in an up way	Fw-pkt-l-avg
Smallest period among two packets brought in an onward way	Fw-iat-min
Lowest time among two packets forwarded in a reverse way	Bw-iat-min

Fifty computers make up the infrastructure used in attacks on datasets, while thirty servers and four hundred and twenty workstations make up the attacking businesses. Data collected by CSE-CIC IDS2018 represents AWS network traffic using CICFlowMeter-V3. The CSE-CIC IDS2018 dataset is over 400 GB in size, making it significantly larger than the CIC-IDS 2017 dataset. A few characteristics of the CSE-CIC2018 dataset have been retrieved and shown in Table 11.1.

When compared to the CIC-IDS 2017 ID dataset, CSE-CIC-IDS2018 features a much larger sample size, especially when it comes to Botnet and Penetration attacks (where it grows by 143 and 4497, correspondingly). Though, in CSE-CIC-IDS2018, there are only 929 Web Attacks accessible.

Data Pre-processing

In the stage, the network traffic was first sorted and pre-processed. To ensure that HCRNNIDS and IDS-friendly data formats could be used and benefited from, all required conversions were performed during pre-processing. Certain characteristics, such as IP addresses and timestamps, play a little role in determining whether base CSE-CIC IDS2018. Since timestamp characteristics are mainly useful for keeping track of when malicious traffic occurred and offer minimal help when training the algorithm, we omitted them during the pre-processing stage. We also eliminated IP address characteristics since were used to carry out the necessary pre-processing operations on the data.

After the data was cleaned and prepared, it was divided into three sets: one for testing, one for training, and one for validation. Each set represented the ultimate evaluation of the model. Furthermore, we found that there were excessively many examples of typical network traffic in the dataset, which might easily skew the model's preference for categorisation.

Classification

It was proposed that a CRNN-POA-based deep learning technique be used to build an IDS. Compared to other methods, our suggested CRNN-POA requires less computations, works well with full-feature datasets, and improves accuracy while reducing the likelihood of FAR. A huge data processing architecture is used by CRNN-POA-based deep learning to address actual ID challenges. A scarcity of resources, including time and space, makes this a challenging challenge to solve. Even if big data is growing rapidly, it requires a lot of processing power, specialised resources, and a computing expedient to aid in the learning procedure in order to be effectively managed.

Using an RNN in conjunction with a CNN DL model, CRNN-POA-based deep learning reduces these difficulties. The experiment's origin may be traced back to a central component of the CRNN-POA found at this particular location. The CRNN-POA summary reveals that there are two primary parts to a CNN: the feature extractor and the classifier. We use convolution and pooling layers to separate features for extraction. The feature map, the extracted output, is sent into the second part of the classification process. In this way, CNN is able to get a deep understanding of the distinctive characteristics of the area in which it is operating. However, it has the flaw of failing to account for the time dependence of critical traits. Because of this, we added recurrent layers following the CNN layers to better capture temporal as well as spatial data.

By doing so, we were able to properly deal with the vanishing and expanding gradient difficulties, enhancing our capacity to effectively capture spatial and temporal correlations and quickly learn from orders of varying extents. To simulate both spatial and temporal information, the CRNN network first processes the input using a CNN and then uses the CNN's output to feed into recurrent layers, which create sequences at each timestep. The arrangement vector is then sent into a fully connected layer, after which a softmax layer is used to calculate over the classes.

In order to summarise temporal patterns in 2D, CRNN employs a 2-layer RNN with gated recurrent units (GRU) on top of a 4-layer CNN. This approach is predicated on the idea that RNNs are more suited for aggregating the temporal pattern than CNNs, while still utilising CNNs on the input side for extraction. Instead of averaging the results from smaller segments subsampling like other CNNs, RNNs are utilised to collective the temporal patterns in CRNN. The CNN component uses max-pooling layers and convolutional layers measuring 3×3 and 2×2, respectively, followed by max-pooling layers measuring 4×4 and 4×4, respectively. As a consequence of this subsampling, the total size of the feature map is N115 (feature map size frequency time). Next, they are input into a 2-layer RNN, the output of which is coupled to the final concealed state. This study describes how to use POA to pick the best momentum, epochs, or learning rate for CRNN models.

Mathematical Model of the Proposed POA

An example of a population-based algorithm, the proposed POA uses a flock of pelicans as its building blocks. Every individual in a population represents a potential answer in algorithms that are based on populations. According on their relative location in the search space, members of the population provide suggestions for the optimisation problem variables. The problem's lower and upper bounds are used in Eq. 11.1 to generate a random sample of population members (11.1).

$$x_{i,j} = l_j + \text{rand} \cdot (u_j - l_j), \quad i = 1, 2, \dots, N, \quad j = 1, 2, \dots, m, \qquad (11.1)$$

where N is the total number of individuals in the population, m is the total sum of variables in the problem, as stated by the ith candidate explanation, where l_j is the jth lower certain and u_j is the jth upper bound.

A matrix called the population matrix is used in equation to determine which pelican birds belong to the proposed POA (11.2). These values for the issue variables are shown in the columns of the matrix, and each row of the matrix indicates a possible solution.

$$X = \begin{bmatrix} X_1 \\ \vdots \\ X_i \\ \vdots \\ X_N \end{bmatrix}_{N \times m} = \begin{bmatrix} x_{1,1} & \cdots & x_{1,j} & \cdots & x_{1,m} \\ \vdots & \ddots & \vdots & & \vdots \\ x_{i,1} & \cdots & x_{i,j} & \cdots & x_{i,m} \\ \vdots & & \vdots & \ddots & \vdots \\ x_{N,1} & \cdots & x_{N,j} & \cdots & x_{N,m} \end{bmatrix}_{N \times m} \qquad (11.2)$$

If X is a pelican population matrix, X_i is the index of the ith bird.

Each member of the population is represented by a pelican in the planned POA. Each potential answer may then be assessed in terms of how well it meets the problem's objective function. In equation, a vector denoted as the objective function vector is used to calculate the resulting values of the objective function (11.3).

$$F = \begin{bmatrix} F_1 \\ \vdots \\ F_i \\ \vdots \\ F_N \end{bmatrix}_{N \times 1} = \begin{bmatrix} F(X_1) \\ \vdots \\ F(X_i) \\ \vdots \\ F(X_N) \end{bmatrix}_{N \times 1} \tag{11.3}$$

where F is the vector of objective functions and F_i is the value of the objective function for the ith possible solution.

In order to improve potential answers, the suggested POA models the actions of pelicans during attacks and hunts. There are two phases to this virtual hunting method:

(i)   Touching towards prey.
(ii) Winging on the water surface.

Phase 1: Moving Towards Prey

The first step is for the pelicans to locate where the food is, and then fly there. Exploring new regions of search space is made possible by the robust exploratory capabilities of the proposed POA, which are revealed via modelling the pelican's behaviour. The key concept behind POA is that the prey's position in the search space is produced at random. This improves POA's ability to precisely search the space of possible solutions to a problem. Mathematical modelling of the aforementioned ideas and the pelican's approach to the location of its prey may be found in Eq. (11.4).

$$x_{i,j}^{P_1} = \begin{cases} x_{i,j} + \text{rand.}(p_j - I.x_{i,j}), & F_p < F_i \\ x_{i,j} + \text{rand.}(x_{i,j} - p_j), & \text{else} \end{cases} \tag{11.4}$$

where I is a random integer which is the position of prey in the jth dimension, F_p is the value of its impartial function based on phase 1, and $x_{(i,j)}^{(P1)}$ is the novel status of the ith pelican in the jth dimension. Parameter I is a random integer between 1 and 2. Each member's value for this parameter is chosen at random throughout each cycle. When this parameter's value is set to two, additional space is displaced for a given member, potentially expanding the search space in which that member may operate. As a result, the ability of the POA to correctly explore the search space is affected by the value of parameter I.

The suggested POA allows a pelican to move to a new location if the value of the objective function increases there. Effective updating is an approach where the algorithm is not allowed to wander into suboptimal regions. Equation is used to model this process (11.5).

$$X_i = \begin{cases} X_i^{P_1}, & F_i^{P_q} < F_i \\ X_i, & \text{else} \end{cases} \tag{11.5}$$

where $X_i^{P_1}$ is the new status of the ith pelican and $F_i^{P_q}$ is its objective function value based on phase 1.

Phase 2: Winging on the Water Surface (Exploitation Phase)

When the pelicans reach the surface, the second step begins: They extend their wings on the water's surface to propel the fish upward, at which point they scoop the fish up into their neck pouch. As a result of this tactic, the pelicans are able to capture a larger sum of fish in the assaulted region. The suggested POA improves upon its initial hunting location as a result of simulating this pelican behaviour. This procedure enhances POA's capacity for localised searching and resource utilisation. For the algorithm to converge to a better answer, it must first investigate the locations close to the pelican's location, from a mathematical point of view. Using equation, we may mathematically model the way in which pelicans behave when hunting (11.6).

$$x_{i,j}^{P_2} = x_{i,j} + R.\left(1 - \frac{t}{T}\right).(2 \text{ rand} - 1).x_{i,j} \tag{11.6}$$

where $x_{i,j}^{P_2}$ is the novel status of the ith pelican in the jth dimension based on phase 2, R is a constant, which is equal to 0.2, $R \cdot (1 - t/T)$ is the neighbourhoods radius of $x_{i,j}$, , while (t) counts up to T, and t is the iteration counter. Specifically, the radius "$R(1 - t/T)$" denotes the neighbourhood of the population members to search locally near each member to converge to a better solution. By increasing the POA's exploitation power, this coefficient helps us approach the globally optimum solution. For the first few iterations, when this coefficient is given a big value, a greater region surrounding each member is taken into account. The "$R(1 - t/T)$" coefficient falls as the number of copies of the method rises, leading to lower neighbourhood radii for each individual. To get the POA to converge to explanations closer to the global ideal based on the utilisation notion, we can scan the region surrounding steps.

At this juncture, the new pelican posture, represented in equation, has been accepted or rejected via effective updating (11.7).

$$X_i = \begin{cases} x_i^{P_2}, & F_i^{P_2} < F_i \\ X_i, & \text{else} \end{cases} \tag{11.7}$$

where $x_i^{P_2}$ is the new status of the ith pelican and $F_i^{P_2}$ is its objective function value based on phase 2.

After the first and second phases have been applied to all members of the population, the best candidate solution so far will be updated based on the current state of the population and the values of the objective function. The algorithm then starts a new iteration, at which point the various operations of the projected POA based on Eqs. (11.4)–(11.7) are recurrent. To conclude, a near-optimal solution to the problem of identifying the best CRNN model solution is offered, which is the best applicant solution produced during the procedure iterations.

Results and Discussion

Evaluation Metrics

A variety of criteria, including accuracy, precision, recall, and $F1$-score, were used to assess the efficiency of the suggested approach. False positive (FP), false negative (FN), and true positive (TP) rates are used in the analysis (FN). Equations (11.8)–(11.10) may be used to quantify all of these performance indicators (11.11).

$$\text{Accuracy} = \frac{TP + TN}{TP + TN + FP + FN} \tag{11.8}$$

$$\text{Precision} = \frac{TP}{TP + FP} \tag{11.9}$$

$$\text{Recall} = \frac{TP}{TP + FN} \tag{11.10}$$

$$F1\text{-Measure} = \frac{2 * (\text{Pression} * \text{Recall})}{\text{Pression} + \text{Recall}} \tag{11.11}$$

Tables 11.2 and 11.3 present the comparative analysis of projected model in terms of binary class and multi-class attacks.

Table 11.2 represents that the analysis of projected model for binary classification; in this analysis, we have evaluated the results parameter of recall, f-measure, precision, and accuracy measurement and also appraised the methods of RF, LSTM, CNN, RNN, CRNN with CRNN-POA; in this comparisons analysis of these methods, the CRNN-POA reached the better results, respectively.

Figure 11.1 represents that the analysis of projected model for binary classification of precision comparison; in this analysis, we have evaluated the results parameter

Table 11.2 Analysis of projected model for binary classification

Algorithms	Precision	Recall	$F1$-score	Accuracy
RF	0.8928	0.8924	0.8925	0.8925
LSTM	0.9007	0.8750	0.9127	0.9104
CNN	0.9293	0.9125	0.9305	0.9217
RNN	0.9425	0.9249	0.9537	0.9316
CRNN	0.9619	0.9474	0.9646	0.9564
CRNN-POA	0.9851	0.9660	0.9852	0.9951

Table 11.3 Multi-class attack classification

Algorithms	Precision	Recall	F1-score	Accuracy
RF	0.7903	0.7681	0.7790	0.7192
LSTM	0.9003	0.9099	0.8745	0.9181
CNN	0.9279	0.9266	0.9167	0.9354
RNN	0.9394	0.9497	0.9395	0.9697
CRNN	0.9540	0.9563	0.9351	0.9752
CRNN-POA	0.9797	0.9698	0.9597	0.9898

of precision measurement and also evaluated the methods of RF, LSTM, CNN, RNN, CRNN with CRNN-POA; in this comparisons analysis of these methods, the CRNN-POA reached the better precision comparison results, respectively.

Figure 11.2 represents that the analysis of projected model for binary classification of recall analysis; in this analysis, we have evaluated the results parameter of precision measurement and also evaluated the methods of RF, LSTM, CNN, RNN, CRNN with CRNN-POA; in this comparisons analysis of these methods, the CRNN-POA reached the better recall analysis results, respectively.

Figure 11.3 represents that the analysis of projected model for binary classification of F1-score analysis; in this analysis, we have evaluated the results parameter of precision measurement and also evaluated the methods of RF, LSTM, CNN, RNN, CRNN with CRNN-POA; in this comparisons analysis of these methods, the CRNN-POA reached the better F1-score analysis results, respectively.

Figure 11.4 represents that the analysis of projected model for binary classification of comparison of accuracy analysis; in this analysis, we have evaluated the results parameter of comparison of accuracy measurement and also evaluated the methods

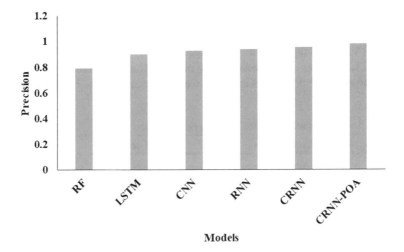

Fig. 11.1 Precision comparison

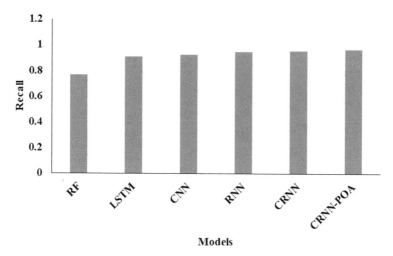

Fig. 11.2 Recall analysis

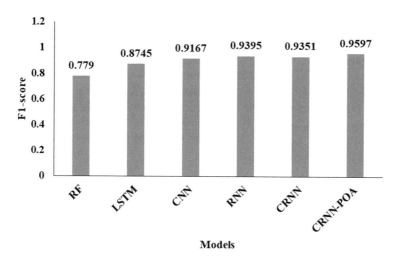

Fig. 11.3 F1-score analysis

of RF, LSTM, CNN, RNN, CRNN with CRNN-POA; in this comparisons analysis of these methods, the CRNN-POA reached the better comparison of accuracy results, respectively.

Table 11.3 represents that the analysis of proposed model for multi-class attack classification; in this analysis, we have evaluated the results parameter of recall, f-measure, precision, and accuracy measurement and also evaluated the methods of RF, LSTM, CNN, RNN, CRNN with CRNN-POA; in this comparisons analysis of these methods, the CRNN-POA reached the better results, respectively.

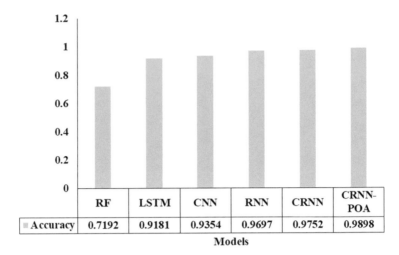

Fig. 11.4 Comparison of accuracy

Figure 11.5 represents that the analysis of proposed model for multi-class attack classification; in this analysis, we have evaluated the results parameter of precision comparison measurement and also evaluated the methods of RF, LSTM, CNN, RNN, CRNN with CRNN-POA; in this comparisons analysis of these methods, the CRNN-POA reached the better precision comparison results, respectively.

Figure 11.6 represents that the analysis of proposed model for multi-class attack classification; in this analysis, we have evaluated the results parameter of recall analysis comparison measurement and also evaluated the methods of RF, LSTM,

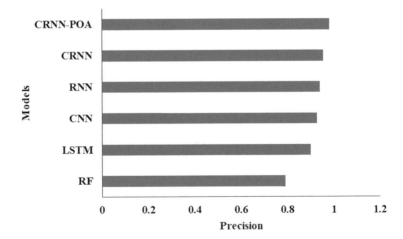

Fig. 11.5 Precision comparison

CNN, RNN, CRNN with CRNN-POA; in this comparisons analysis of these methods, the CRNN-POA reached the better recall analysis comparison results, respectively.

Figure 11.7 represents that the analysis of proposed model for multi-class attack classification; in this analysis, we have evaluated the results parameter of $F1$-score analysis comparison measurement and also evaluated the methods of RF, LSTM, CNN, RNN, CRNN with CRNN-POA; in this comparisons analysis of these methods, the CRNN-POA reached the better $F1$-score analysis comparison results, respectively.

Figure 11.8 represents that the analysis of proposed model for multi-class attack classification; in this analysis, we have evaluated the results parameter of comparison

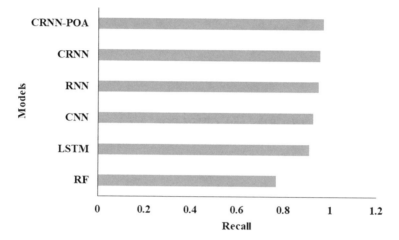

Fig. 11.6 Recall analysis

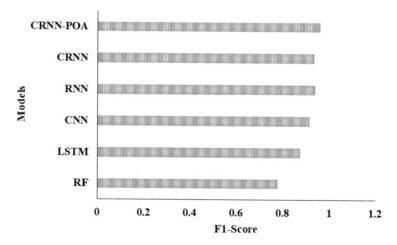

Fig. 11.7 $F1$-score analysis

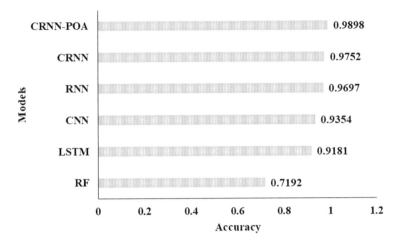

Fig. 11.8 Comparison of accuracy

of accuracy analysis comparison measurement and also evaluated the methods of RF, LSTM, CNN, RNN, CRNN with CRNN-POA; in this comparisons analysis of these methods, the CRNN-POA reached the better comparison of accuracy results, respectively.

Conclusion

The CRNN used to create the NIDS in this article has proven useful in the realm of cyber security. The ID system's framework was trained using the CSE-CIC-DS2018 dataset. After the CNN layers, we added recurrent layers to improve the robustness with which we collect spatial and temporal data. Our goal was to efficiently learn from sequences of varying extents while simultaneously capturing spatial and temporal connections, which necessitated a solution to the vanishing and expanding gradient issues. To capitalise on strengths of both anomaly-based (AB) and signature-based (SB) approaches, a DL-based identification system is presented. By including an optimum solution (POA) into the suggested model, the projected ID system helps to minimise computing complexity while simultaneously improving accuracy and DR for intrusion detection. Famous classification measures were used to evaluate both classical ML and deep learning approaches (accuracy, precision, recall, and $F1$-score). The results of the simulations prove that the proposed CRNN is capable of achieving the calcification of attack events that are malevolent. Conferring to CSE-CIC-IDS2018 statistics, the average accuracy of both standard and non-standard assaults is close to 98%. The simulation results show that the CRNN deep learning model may be used to successfully protect against harmful assaults. Nonetheless, we have only tried the CRNN out on one ID dataset; thus, this may be a weakness of the suggested method. Since the signature of the connected traffic frequently changes, it will be crucial to evaluate it on a more current dataset. In the event of additional ID challenges in modern, realistic datasets, we will focus on researching different deep learning approaches in conjunction with a feature extraction strategy to develop informed data drawings.

References

1. Buczak, A.L., Guven, E.: A survey of data mining and machine learning methods for cyber security intrusion detection. IEEE Commun. Surv. Tutorials **18**(2), 1153–1176 (2016)
2. Dua, S., Du, X.: Data Mining and Machine Learning in Cybersecurity. Auerbach Publications (2016)
3. Murugesan, M., Thilagamani, S.: Efficient anomaly detection in surveillance videos based on multi layer perception recurrent neural network. Microprocess. Microsyst. **79**, 103303 (2020)
4. Mishra, S., Sagban, R., Yakoob, A., Gandhi, N.: Swarm intelligence in anomaly detection systems: an overview. Int. J. Comput. Appl. **43**(2), 109–118 (2021)
5. Grill, M., Pevný, T.: Learning combination of anomaly detectors for security domain. Comput. Netw. **107**, 55–63 (2016)
6. Rahman, M., Halder, S., Uddin, M., Acharjee, U.K.: An efficient hybrid system for anomaly detection in social networks. Cybersecurity **4**(1), 1–11 (2021)
7. Injadat, M., Salo, F., Nassif, A.B., Essex, A., Shami, A.: December. Bayesian optimization with machine learning algorithms towards anomaly detection. In: 2018 IEEE Global Communications Conference (GLOBECOM), pp. 1–6. IEEE (2018)
8. Kim, S., Jo, W., Shon, T.: APAD: autoencoder-based payload anomaly detection for industrial IoE. Appl. Soft Comput. **88**, 106017 (2020)
9. Siddiqui, M.A., Fern, A., Dietterich, T.G., Wong, W.-K.: Sequential feature explanations for anomaly detection (2015). arXiv:1503.00038
10. Nawir, M., Amir, A., Yaakob, N., Lynn, O.B.: Effective and efficient network anomaly detection system using machine learning algorithm. Bulletin EEI **8**(1), 46–51 (2019)
11. Vinh, N.X., Chan, J., Bailey, J., Leckie, C., Ramamohanarao, K., Pei, J.: Scalable outlying-inlying aspects discovery via feature ranking. In: Pacific-Asia Conference on Knowledge Discovery and Data Mining, pp. 422–434. Springer (2015)
12. Duan, L., Tang, G., Pei, J., Bailey, J., Campbell, A., Tang, C.: Mining outlying aspects on numeric data. Data Min. Knowl. Disc. **29**(5), 1116–1151 (2015)
13. Siddiqui, M.A., Fern, A., Dietterich, T.G., Wright, R., Theriault, A., Archer, D.W.: Feedback-guided anomaly discovery via online optimization. In: Proceedings of the 24th ACM SIGKDD International Conference on Knowledge Discovery & Data Mining. ACM (2018)
14. Das, S., Wong, W.-K., Dietterich, T., Fern, A., Emmott, A.: Incorporating expert feedback into active anomaly discovery. In: IEEE 16th International Conference on Data Mining (ICDM), pp. 853–858. IEEE (2016)
15. Das, S., Wong, W.-K., Fern, A., Dietterich, T.G., Siddiqui, M.A.: Incorporating feedback into tree-based anomaly detection (2017). arXiv:1708.09441
16. Hwang, R.H., Peng, M.C., Huang, C.W., Lin, P.C., Nguyen, V.L.: An unsupervised deep learning model for early network traffic anomaly detection. IEEE Access **8**, 30387–30399 (2020)
17. Veeramachaneni, K., Arnaldo, I., Korrapati, V., Bassias, C., Li, K.: Ai^2: training a big data machine to defend. In: 2016 IEEE 2nd International Conference on Big Data Security on Cloud (BigDataSecurity), IEEE International Conference on High Performance and Smart Computing (HPSC), and IEEE International Conference on Intelligent Data and Security (IDS), pp. 49–54. IEEE (2016)

18. Avci, O., Abdeljaber, O., Kiranyaz, S., Hussein, M., Gabbouj, M., Inman, D.J.: A review of vibration-based damage detection in civil structures: from traditional methods to machine learning and deep learning applications. Mech. Syst. Signal Process. **147**, 107077 (2021)
19. Kumar, K.P.M., Saravanan, M., Thenmozhi, M., Vijayakumar, K.: Intrusion detection system based on GA-fuzzy classifier for etecting malicious attacks. Concurr. Comput. Pr. Exp. **33**, 5242 (2021)
20. Sharma, D.K., Dhankhar, T., Agrawal, G., Singh, S.K., Gupta, D., Nebhen, J., Razzak, I.: Anomaly detection framework to prevent DDoS attack in fog empowered IoT networks. Ad Hoc Netw. **121**, 102603 (2021)
21. Yang, X., Chen, Y., Qian, X., Li, T., Lv, X.: BCEAD: a blockchain-empowered ensemble anomaly detection for wireless sensor network via isolation forest. Secur. Commun. Netw. (2021)
22. Saif, S., Das, P., Biswas, S., Khari, M., Shanmuganathan, V.: HIIDS: hybrid intelligent intrusion detection system empowered with machine learning and metaheuristic algorithms for application in IoT based healthcare. Microprocess. Microsyst. 104622 (2022)
23. Xu, W., Jang-Jaccard, J., Singh, A., Wei, Y., Sabrina, F.: Improving performance of autoencoder-based network anomaly detection on nsl-kdd dataset. IEEE Access **9**, 140136–140146 (2021)
24. Sathya, M., Jeyaselvi, M., Krishnasamy, L., Hazzazi, M.M., Shukla, P.K., Shukla, P.K., Nuagah, S.J.: A novel, efficient, and secure anomaly detection technique using DWU-ODBN for IoT-enabled multimedia communication systems. Wireless Commun. Mobile Comput. (2021)
25. Osamor, F., Wellman, B.: Deep learning-based hybrid model for efficient anomaly detection. Int. J. Adv. Comput. Sci. Appl. **13**(4) (2022)
26. Chen, A., Fu, Y., Zheng, X., Lu, G.: An efficient network behavior anomaly detection using a hybrid DBN-LSTM network. Comput. Secur. **114**, 102600 (2022)
27. Rajasekar, V., Sarika, S, Velliangiri, S., Kalaivani, K.S.: An efficient intrusion detection model based on recurrent neural network. In: IEEE International Conference on Distributed Computing and Electrical Circuits and Electronics (ICDCECE), pp. 1–6. IEEE (2022)
28. Balaji, R., Deepajothi, S., Prabaharan, G., Daniya, T., Karthikeyan, P., Velliangiri, S.: Survey on intrusions detection system using deep learning in IoT environment. In: International Conference on Sustainable Computing and Data Communication Systems (ICSCDS), pp 195–199. IEEE (2022)
29. A Collaborative Project between the Communications Security Establishment (CSE) and The Canadian Institute for Cybersecurity (CIC). Available online: https://www.unb.ca/cic/datasets/ids-2018.html. Accessed on 31 Mar 2021

Chapter 12
Intrusion Detection in IoT-Based Healthcare Using ML and DL Approaches: A Case Study

Priya Das and Sohail Saif

Introduction

The advent and widespread use of Internet of Things (IoT) and wireless sensor network (WSN) have enabled the virtual connection of various devices [1–3]. The use of these cutting-edge technologies to monitor a person's health has become quite popular because of its accessibility, widespread support, and real-time performance. Small sensor nodes gather data on the patient's electrocardiogram (ECG), body movement, blood pressure temperature, blood glucose, heart rate, etc., and upload it to a medical server that the doctor may view remotely [4]. Both implanted and wearable varieties of tiny sensors are available for use in data collection-based three-tier health monitoring framework is shown in Fig. 12.1.

However, the danger of data theft and cyber-security threats rises when we adopt such a paradigm change in our everyday life [5]. Cyberattacks tailored to IoT devices and security countermeasures to protect this bridge between the digital and physical realms have received a lot of attention because of the IoT's ability to link the two in such a direct manner. One of the dangerous cyberattacks occurred at a German steel factory and was triggered by an infected IoT device. In 2015, the German Federal Office for Information Security confirmed in an investigation that intruders had broken into a popular German steel company, compromising several systems and parts of the production network. As a consequence, milling industry employees missed an important deadline and therefore caused significant system damage [6]. Unfortunately, it's not the first instance of a cyberattack on an Internet of Things

P. Das
Department of Computer Science, Chakdaha College, Chakdaha, West Bengal, India

S. Saif (✉)
Department of Computer Applications, Maulana Abul Kalam Azad University of Technology, Haringhata, West Bengal, India
e-mail: sohailsaif7@gmail.com

© The Author(s), under exclusive license to Springer Nature Singapore Pte Ltd. 2023
V. Sarveshwaran et al. (eds.), *Artificial Intelligence and Cyber Security in Industry 4.0*, Advanced Technologies and Societal Change,
https://doi.org/10.1007/978-981-99-2115-7_12

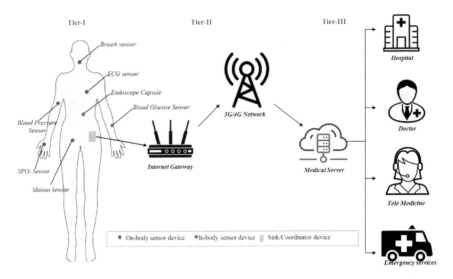

Fig. 12.1 Three-tier architecture of IoT-enabled health monitoring system

system or devices in recent memory. Take, for example, the recent cyberattack on the Ukrainian power system [7], the shutdown of a floating oil rig [8], the hacking of kitchen appliances in the UK [9], and the conversion of smart refrigerators into spambots [10]. In 2017, the Mirai virus infected over 380,000 IoT systems to establish massive hostile botnets, making it one of the most notable breaches in the history of the Internet of Things. These are "networks of bots," or infected devices directed by a central server to launch a DDoS attack. According to a study by the United States Computer Emergency Readiness Team (US-CERT) [11], this virus is still regarded the greatest DDoS attack ever documented, illustrating the severity of the attack.

Security frameworks for sensor-based health monitoring systems have been the subject of a number of studies in recent years [12–15], with a focus on cryptographic methods. Unfortunately, these measures are not adequate to thwart Insider threats. A node compromise is an insider attack that may cause the sensor network to behave erratically. An Intrusion Detection System (IDS) is essential in this setup for spotting and eliminating any suspicious data packets sent or received by the sensor nodes. In order to identify any potential cyber threats, an IDS constantly analyzes data coming in from many sources. This book chapter offers a critical exploration of the deep learning and machine learning techniques used in the development of IoT-IDS. We review the various approaches used in each method, along with the detection strategies, validation strategies, and deployment strategies.

The rest of the paper is structured as outlined below. Section "Attacks on IoT-Based Healthcare Ecosystem" presents various security attacks in IoT-based healthcare. Section "Intrusion Detection in IoT-Based Healthcare" describes intrusion detection system for IoT-enabled healthcare. Anomaly-based intrusion detection systems are explored in Sect. "Anomaly Based Intrusion Detection System (AIDS)".

Various deployment techniques of IDS have been discussed in Sect. "AIDS Deployment Techniques in IoT". Validation techniques are presented in Sect. "AIDS Validation Techniques". Finally, this book chapter is concluded in Sect. "Conclusion".

Attacks on IoT-Based Healthcare Ecosystem

Connected medical devices and systems are a vital component of the Internet of Things (IoT) in healthcare. Because of the interconnected nature of the many gadgets, the safety of the sent data is heightened to the point of becoming a potential disaster. Safeguarding data sent across networks by means of Internet of Things (IoT) devices is what "IoT security" is all about. Due to the interconnected nature of these devices through the Internet, hackers are able to exploit security flaws and gain access to sensitive health information. Without proper safeguards, sensitive information poses serious risks to businesses and people alike, with the potential for irreparable harm should it be lost [16]. Figure 12.2 shows different types of cyberattack on IoT system.

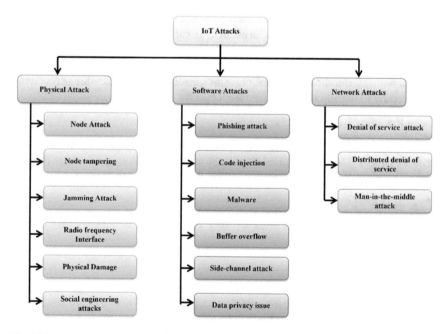

Fig. 12.2 Taxonomy of attack in IoT

Physical Layer

Attacks rely on undiscovered vulnerabilities in electronic hardware. These attacks may compromise the device's security by altering its physical components. When an attacker is in close proximity to a target network or Internet of Things device, a physical attack is carried out. Major dangers at the physical/perception level include:

Node Attack

It is possible for the cybercriminal to gain complete control of the sensor nodes. Tags are vulnerable to physical attacks since IoT devices are spread out in various areas. If a hacker wanted to exploit an RFID system, all they would have to do is take these tags, which are considered legitimate tags, and manufacture copies of them.

Node Tampering

Node tampering describes the act of breaking into a system in order to discover the keys needed to decipher the encrypted data.

Jamming Attack

As a variant of denial-of-service attack, "jamming" occurs when an attacker sends out a signal from a great distance in an effort to disrupt a signal lower on the spectrum. Jamming attacks include a malicious node in the sensor network sending out a signal that overlaps with the sensors' signals in terms of frequency. By disrupting the IoT network with interference, this attack prevents sensor nodes from sending or receiving data, thereby rendering these services inaccessible.

Radio Frequency (RF) Interface

In the Internet of Things (IoT), wireless data transmissions occur through radio frequency (RF). There are a number of exploits that might easily damage the IoT devices because of the wireless nature of this data transmission method.

Physical Damage

With the intent to alter data or steal sensitive information, the attacker takes active involvement in the attack.

Social Engineering Attacks

An attacker gains unauthorized access to a system using social engineering and discreetly install malware. Internet of Things (IoT) devices, especially wearables, collect massive amounts of personally identifiable information PII to tailor the user experience to each individual user. Such Internet of Things devices also make use of consumers' individual data to provide user-friendly services, such voice-controlled online goods buying. However, attackers may breach PII systems to acquire access to users' passwords, purchase histories, and other sensitive information.

Software/Application Layer

Applications in IoT technology are built with application programming interfaces (APIs) and are web-based programs that need special software to be installed before they can be used. Phishing, trojans, ransomware, worms, viruses, and other forms of malicious software and content, such as spyware and adware, are all examples of software intrusions.

Phishing Attack

Attacker sets up a trap using an IoT edge node. The objective is to get sensitive data such as login credentials.

Code Injection

An unprotected sensor node may have code injected into it to alter its execution. For instance, the inaudible attack, also known as the Dolphin Attack, uses the ultrasonic channel to insert inaudible speech instructions into voice-controllable devices. Voice squatting is another kind of attack where the attacker takes control of the VPA service and steals the user's information by inserting malicious code into the conversation.

Malware

Malware refers to any kind of malicious software that is designed to cause harm or damage to an IoT infrastructure. There is a wide variety of malware, from viruses and worms to Trojan horses and ransomware as well as spyware and adware.

Buffer Overflow

When data are written to a buffer on a sensor node, an overflow may occur if the bounds of the buffer aren't properly validated, causing data values at memory addresses near the destination buffer to become corrupted.

Side-Channel Attack

In order to decipher a cryptosystem, a side-channel attack exploits data leaked by the system itself.

Data Privacy Issue

There is a risk that the attackers can use RFID tags on a wide variety of common home products. Users' privacy may be at risk if their IoT devices are tracked via RFID tags, since this data may be used to compile a detailed profile of their whereabouts and activities.

Network Layer

The network layer is where data transmission takes place, making it vulnerable to attackers. Instances of eavesdropping, man-in-the-middle attacks, denial of service attacks, storage attacks, exploit attacks, spoofing, and so on may be included. There are several varieties of infosec attacks that may be launched against the IoT devices, each of which could be tailored to a particular component, network, or data source. Physical security attacks might be carried out against the devices that are part of IoT networks. The bulk of IoT threats are either network-based or are carried out to harm certain information attributes. In most cases, they are malicious attempts to compromise the uptime of the IoT application or the security of the associated data. Important network layer risks include.

Denial of Service (DoS) Attack

When a system is affected by a denial-of-service attack, it stops providing the services it was previously capable of. The system's legitimate users are unfairly denied of its benefits. Distributed denial of service describes an attempt that is undertaken by many malicious nodes (DDoS). The victim of a denial-of-service attack may incur costs in addition to any data loss that may occur. Network and processing power may be impacted by a denial-of-service attack [17].

Distributed Denial of Service (DDoS)

A distributed denial of service (DDoS) attack occurs when an adversary temporarily compromises a network of Internet of Things (IoT) devices to form an arrangement known as a botnet, which then forwards coordinated requests to a server or network of servers for a specific service, flooding the server and forcing it to prioritizes the DDoS attack requests over those from actual users. This occurs often when all devices are hijacked and messages are overloaded by Internet of Things devices in order to cause a traffic congestion in the devices.

Man-in-the-Middle (MITM) Attack

Wireless sensor connections are susceptible to man-in-the-middle attacks, which may compromise the security of the information sent between devices in the Internet of Things (IoT) [18]. Attacks on wireless networks may take several forms, such as decryption tools, malicious wireless equipment, eavesdropping, MAC spoofing, packet sniffing, and so on. As part of a man-in-the-middle (MITM) attack, an adversary listens in on a conversation between two parties who believe they are speaking over a secret service and then makes unauthorized changes to the conversation an attacker may essentially spy on a conversation and put themselves into it.

Intrusion Detection in IoT-Based Healthcare

Any type of unauthorized action is known as network intrusion or any attack that affects any of four basic principles (Authentication, confidentiality, integrity, and availability) of security is called intrusion. When this intrusion harms the IoT system that refers to the intrusion in IoT and intrusion detection system (IDS) is a type of system that detects any type of illegal activity after monitoring the network traffic. IDS is a type of software or hardware system that continuously scan the network traffic and alert after detecting an anomaly and maintain security of the entire system. IDS is divided into two main categories: signature-based intrusion detection system and anomaly-based intrusion detection system. Figure 12.3 describes different placement and detection technique of IoT.

Signature-Based Intrusion Detection System (SIDS)

SIDS [19] uses pattern recognition method to detect an intrusion. It finds a specific pattern like number of zero or one present in the network and try to match the pattern to the known sequences. Already detected pattern is known as signature of IDS. This method is well suited for known pattern that is already present in the system, but this is quite difficult for unknown attack because their pattern is new for the system.

Figure 12.4 demonstrates the working principle of signature-based intrusion detection system. SIDS maintains a database of intrusion signature and compares this existing signature with upcoming new malware pattern, and it alert the system if any match is found.

Some tools are employed in SIDS such as Snort [20] and NetSTAT. Snort is a type of network-based intrusion detection system which is used to scan each network packet. Snort database is also updated time to time which is helpful to store new signature into the database. The major disadvantage of SIDS is to identify attack where multiple packets are present. Malware is increasing day by day. So, extraction of signature from new malware is also increasing. There are many techniques for

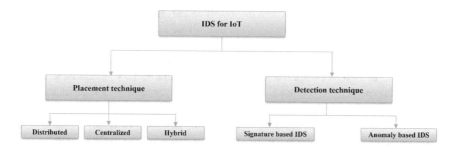

Fig. 12.3 Classification of IDS for IoT

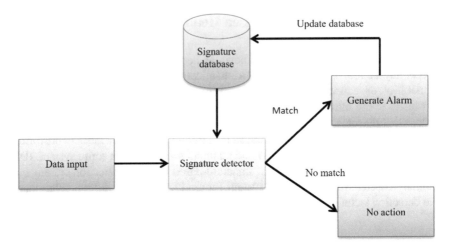

Fig. 12.4 Signature-based intrusion detection system

creation of signature for SIDS which includes finite state machine and hashing technique. [21] SIDS technique has become less efficient because it is unable to detect unknown malware attack and variant of known attack. Solution of this problem is anomaly-based intrusion detection system (AIDS) where an efficient technique is used.

Anomaly-Based Intrusion Detection System (AIDS)

Anomaly-based IDS perform better than SIDS because it is based on machine learning algorithms which can be developed to detect unknown modified attack. Machine learning algorithm is more trustful as compared to signature-based IDS because those algorithms are trained according to the datasets. Entire process of AIDS [22] is divided into two steps. First step is training phase, and second phase is testing phase. In the first or training phase, a dataset is trained according to the algorithm and in the testing phase, a new unknown dataset is used to check the accuracy of the algorithm. SIDS only detects that attack whose signature is already present in the database. Unlike SIDS, anomaly-based IDS can detect unknown malware attacks which is the main advantage. But there are some disadvantages of AIDS such as false positive rate is high as compared to SIDS. Anomaly-based IDS generates an alarm after any deviation from normal activity. Means after generation of false alarm, it is the task of security administrator to check why this false alarm is generated, which is a time-consuming task. Training is also a difficult task in AIDS because it is time-consuming process and it requires more hardware and software resource.

Classification of AIDS

Anomaly-based IDS is categorized into four types. Supervised learning, unsupervised learning, deep learning, reinforcement learning, and ensemble learning as shown in Fig. 12.5.

Machine learning is a field where machines learn from a data, build a model, and give output of a new data from that predicate model. Machine learning technique is applied in different area of anomaly-based IDS. Many algorithms of classification, clustering, and neural network are used for this purpose. Machine learning (ML) is used in AIDS because through these ML algorithms a machine is trained and gives output of new data. The accuracy also gets improved through this machine learning algorithms.

Supervised learning is one of the types of machine learning technique, where machine is trained using some labeled data. Labeled means input is already trigger into correct output. The task of the machine is to predict the result of another dataset whose output is not labeled. Unlike supervised learning, there is no guidance in unsupervised learning. It tries to identity the output from existing system. Reinforcement learning method uses an agent to learn from environment by trial-and-error method. The task of the agent to maximize the learning outcome. Deep learning is completely based on artificial neural network.

Supervised Learning in AIDS

In this section, we have discussed about supervised learning and details about different supervised learning algorithms. IDS with supervised learning technique uses only labeled data for intrusion detection. This approach has two stages namely

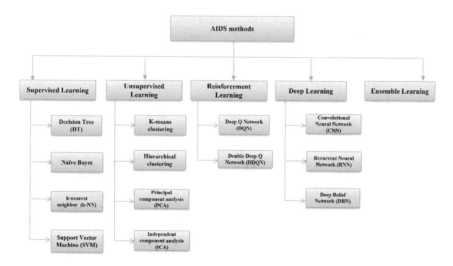

Fig. 12.5 Taxonomy of anomaly-based intrusion detection system

Machine learning algorithms

Fig. 12.6 Conceptual diagram of AIDS method

training and testing. In training phase, training data is used, and testing phase testing data is used. A training dataset consist two parts: input which consists of some feature and output with some predefined labels. In IDS, output is mainly two types namely normal and attack. Training data is a part of the entire dataset which is used to train the algorithm. Testing data is another part of entire data (excluding training data) which is used after completion of the training phase to check accuracy of the algorithm. Briefly, we can say that once training is completed an unknown data (testing data) is provided to the newly trained model. The model will give the predicted output without seeing the original output. After getting the predicted output, it is compared with the actual output, and accuracy of the model is calculated. Figure 12.6 depicts the general approach of AIDS technique using machine learning approach. Different classification algorithm can be used to train the model. Some classification algorithms are decision tree, Naïve bias, support vector machine, and k-nearest neighbor.

Decision Tree

Decision tree is one of the most popular supervised learning techniques which is used in classification problem as well as regression problem. A decision tree looks like a tree structure with three main parts. First part is internal node which consists of test attribute, second part is decision branch which denotes the outcome from the test node, and third part is leaf node which denote the class label. The tree is learned by splitting the source node to the sub node. This process is repeated for the sub node also, this recurring splitting is called recursive portioning. This partition is stopped when all subset value is same as target value. To predict the class of a dataset the algorithm starts from root node. The algorithm compares the value of root node with the real data of original dataset. Same process is continued for the next node and this process is continued until it reaches to the last or leaf node. Nancy et al. [23] used fuzzy decision tree classifier for intrusion detection in wireless sensor network. They have applied the algorithm on KDD dataset for performance evaluation and proved that the proposed method reduces false alarm rate and energy consumption. Khraisat et al. [24] proposed a hybrid IDS system by one class support vector machine and C5 decision tree classifier. They combined both SIDS and AIDS together in a one system. They have applied the hybrid algorithm on two datasets ADFA and NSL-KDD. Finally, they proved that proposed algorithm performed better as compared to AIDS and SIDS in term of false alarm rate accuracy.

Naïve Bayes

This is one of the simplest classification algorithm out of all algorithms. Naïve Bayes algorithm is based on Bayes theorem. The algorithm gives the result by applying conditional probability formulae. The algorithm is used mainly in high-dimensional dataset for text classification. The first term of algorithm Naïve because it assumes that each feature is independent of each other. Each feature is separately participated to identify a class. The second term is Bayes as it comes from Bayes theorem. The algorithm is the most efficient because of its efficient calculation and conditional independence assumption feature. Although accuracy will fall if the individual independence assumption is false or incorrect. Talita et al. [25] used naïve Bayes and particle swarm optimization together for features selection. They used KDD-CUP dataset to evaluate the model and the algorithm achieved 99.12% accuracy. Bhosale et al. [26] proposed a Modified Naïve Bayes for IDS using hybrid feature selection algorithm and basic naïve bayes classifier. They compared the result with SVM, KNN, ANN and claimed that the modified method performed better than other classifiers. Yang et al. [27] proposed artificial bee colony-based naïve Bayes method for intrusion detection system. They also compared the simulation result with other state or the art and concluded that the result outperformed others.

k-Nearest Neighbor Classification (k-NN)

k-NN algorithm is the most basic supervised learning algorithm technique which is used in classification as well regression problems. The applications of this algorithm are in different areas such as data mining, pattern recognition, and intrusion detection system. K-NN is a non-parametric learning technique which means the algorithm does not make assumption on the data. Figure 12.7 describes the k-nearest neighbor approach where X indicates unlabeled data that need to be labeled as intrusion or normal. Here we consider $k = 5$, and depending upon the majority of the vote here X is labeled as normal as there are three classes that are normal and two classes are intrusion.

To implement the algorithm, training and testing both datasets are required at the first stage of the algorithm. Then k nearest data point is chosen, k can be any integer. The distance between test data and each row between train data is calculated. Distance may be Euclidean, Hamming, or Manhattan but generally, Euclidean distance is used in most of the applications. Then calculated distance is sorted and chooses top k rows from the sorted list. Finally, most frequent class out of k point is assigned as final class. The problem of this algorithm is to choose an effective distance out of many techniques. Alfeilat et al. [28] concluded from their research that there is not a single distance metric which is best for all application and all types of datasets. The accuracy may vary according to the application. So it is the task of the researcher to choose an appropriate distance metric according to the particular application. Liu et al. [29] used KNN and arithmetic optimization algorithm together to build a hybrid model

Fig. 12.7 k-nearest
neighbor classification

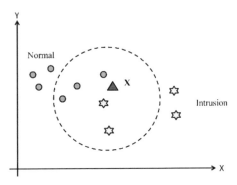

for wireless sensor network. They also used the leavy flight algorithm to optimize the result. They claimed that the proposed model gave significant result in WSN-DS dataset. Wazirali [30] tuned the parameter of KNN and used five-fold cross validation on NSL-KDD dataset, and they found that the algorithm performed better than basic KNN. Senthilnayaki et al. [31] proposed a modified KNN using fuzzy rough set for feature selection. They used KDD dataset to evaluate the model, and they claimed that the algorithm selects significant features from a large number of features.

Support Vector Machine (SVM)

Support vector machine is a machine learning algorithm which can be used in classification and as well as regression problem, but it gives better result in classification than regression. Mainly, SVM is two types: linear SVM and nonlinear SVM. Liner support vector machine is used in linearly separable data which means the dataset is classified into two classes by a single straight line. Non-linear SVM is used in non-linearly classified data, which means only a single straight line cannot divide the classes. The algorithm creates a boundary or line in N-dimensional space that easily distinguish different classes and the boundary line is known as hyperplane. Hyperplane dimension is determined by the total number of features. It means if the total number of features is two then hyperplane is a one-dimensional plane. If there are two features then hyperplane is two-dimensional plane and so on. It will be difficult to imagine the hyperplane when the number of features is more than three. SVM uses a kernel function that converts a two-dimensional space to a multi-dimensional space which means it converts non-separable to a separable problem. The kernel actually uses a highly complex transformation function which finds a procedure to separate the data from the actual feature and label. SVM uses different types of kernel separation functions such as liner, Gaussian, hyperbolic, and polynomial. Huang et al. [32] used SVM with different meta-heuristics algorithms and they found that the hybrid system achieved significant accuracy. Kabir et al. [33] invented a novel technique, i.e., Least Square Support Vector Machine in short LS-SVM for intrusion detection system. They tested the technique against some benchmark dataset KDD 99 and claimed that the algorithm gave a very good performance in terms of accuracy. Safaldin et al. [34] proposed an intrusion detection system for wireless sensor network based on a variant of gray wolf optimization and SVM. They claimed that the algorithm increased accuracy and reduced execution, false alarm rate.

Unsupervised Learning in AIDS

Unsupervised machine learning techniques use datasets without any labels. In the supervised model, labeled dataset is used to train the dataset. But in unsupervised technique finding a hidden pattern from the data itself by comparing the features of datasets is done. This means the data is grouped by their learning technique. In an intrusion detection system, if we assume a dataset with normal and with intrusion. At first, unsupervised technique try to find similarities between the dataset. When all records are in a clustered form then the small size of cluster is labeled as intrusion, and comparatively big size cluster is labeled as normal because the occurrence of normal data should be more than intrusion. Normal and intrusion data is not similar so they fall into the different cluster.

K-means Clustering

This clustering technique aims to rearrange n data points to k clusters. According to the nearest mean each data object is grouped into a cluster. It is an iterative technique so the same thing is repeated until the no of iterations is completed and at the final iteration, we get out the final clusters. The total number of clusters or k points depends on user. Figure 12.8 explains the working principle K-means clustering algorithm.

Sukumar et al. [35] proposed an IDS system using a variant of the genetic algorithm. The result shows that the proposed algorithm gives better result than normal K-means clustering. Al-Yaseen et al. [36] proposed a hybrid model based on multi-level SVM and modified K-means clustering for intrusion detection system to improve the efficiency of unknown and known attack. They compared their proposed method with other machine learning methods and found that the proposed method's accuracy was 95.75%. Tahir et al. [37] proposed a hybrid technique by combining three algorithms namely k-means clustering, naïve Bayes, and discrete technique for intrusion detection. They used ISCX 2012 to check the efficiency of the algorithm, and the algorithm gave better accuracy and detection rate.

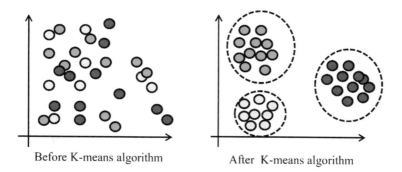

Before K-means algorithm   After K-means algorithm

Fig. 12.8 K-means clustering

Hierarchical Clustering

The algorithm builds a tree-like hierarchy of cluster hierarchical. Cluster is of two types namely agglomerative and divisive. Horng et el. [38] proposed a novel IDS based on hierarchical clustering and SVM. They used KDD Cup 1999 to evaluate the proposed model. The model performed better compared to other existing algorithms, especially for probe and DoS attack. Sangve et al. [39] applied hierarchical clustering and k-means clustering technique on NSL-KDD dataset and they proved that hierarchical clustering performs better than k-means clustering technique.

Agglomerative is a bottom-up hierarchical clustering. In every iteration, a data point is merged to its nearest data point until one cluster is formed. Song et al. [40] used agglomerative hierarchical clustering for features grouping in an intrusion detection system. They tested the method on KDD CUP 99 dataset, and the result showed that the method gives better accuracy with the properly selected feature. Mazarbhuiya et al. [41] proposed an agglomerative hierarchical clustering technique to measure similarity for numeric categorical attributes.

Divisive hierarchical clustering is opposite to agglomerative clustering. All data point is taken in the first iteration then in every iteration data point as separated from another cluster. This process will continue until n data points are formed.

Principal Component Analysis (PCA)

PCA is an unsupervised learning technique which is used for dimension reduction. The algorithm converts correlated feature into a set of linearly uncorrelated features, and this feature is called the principal component. The algorithm is very popular in data analysis and image compression. The most common problem in intrusion detection system is the huge number of irrelevant data. Hadri et al. [42] used PCA and fuzzy PCA for the extraction of relevant information from the original traffic data. They used KDDcup99 for their experiment, and they showed that fuzzy PCA achieves better accuracy than PCA to detect U2R and DoS attacks. Waskle et al. [43] proposed an IDS using random forest and PCA. They used PCA to extract useful information from the dataset and random forest to classify the attacks. They showed that the proposed model achieved 96.78% accuracy and outperformed decision tree, support vector machine, and naïve Bayes. Bhattacharya et al. [44] proposed a PCA firefly algorithm for intrusion detection. They have used XGBoost algorithm for classification. They claimed that the model performed better than other machine learning algorithms.

Independent Component Analysis (ICA)

ICA is a type of machine learning technique that separates independent signal from mixed signal. In other word, it can be said that independent component analysis extracts a single sound although multiple noises are there. Srinoy et al. [45] proposed a hybrid PCA meta-heuristics fuzzy system for intrusion detection system.

Reinforcement Method

Reinforcement method is another part of machine learning. Unlike supervised learning, there is no labeled data in the training dataset. In this process, agent can learn by the environment through trial-and-error method. The goal of the agent is to improve the performance by collecting positive rewards. Deep reinforcement learning is the combination of reinforcement learning and deep learning. It has input layer, output layer, and some hidden layer.

Deep Q Network (DQN)

DQN is a reinforcement learning technique. The aim of the model is to obtain the Q function. The Q function gives the maximum reward for a specific state. Depending upon Q function, policy function is calculated, which is necessary for taking any action.

Double Deep Q Network (DDQN)

DDQN is nearly same as DQN only the difference is in Q function. IN DQL, only one Q function is used, but in DDQN, two Q function is used. One queue evaluates the current Q function and other Q evaluates the target Q function. Martin. et al. [46] applied four deep reinforcement learning technique on NSL-KDD, AWID dataset and found that DDQN perform better compared to other techniques. They used AWID and NSL-KDD datasets to evaluate the performance of the DQN based model, DDQN, and another two algorithms. It has been found that DDQN perform best out of them.

Deep Learning (DL)

Deep learning is a subset of machine learning where a machine can learn by experience by replacing human intelligence. Deep learning is based on artificial neural network (ANN), here the term neural is inspired by biological neuron. Like biological neuron, artificial neuron is also connected to each other in different layer. ANN consists of input, node, weight, and output. The basic structure of an ANN mainly divided into three layer namely, input layer, hidden layer, and output layer. The hidden layer is in between the input and output layer where all complex mathematical function is calculated. We have discussed few deep learning algorithm in next section.

Convolutional Neural Network (CNN)

Convolutional neural network (CNN) or ConvNets is a type of artificial neural network (ANN) which is used for image classification. Yann LeCun, a post-doctoral computer science researcher first invented CNN in 1980. It is a supervised deep learning algorithm. CNN has four layers namely convolution layer, ReLU layer, pooling layer, and fully connected layer. In convolution layer, the image generates a feature map after passing through some filters called kernels. The second layer is called the pooling layer. The main function of this layer is to reduce the feature map size or volume to make the system computationally fast. This method is known as subsampling or down sampling. Pooling is many types like max pooling, average pooling, global pooling, etc., max pooling chooses the most significant value from the feature map and it is the most popular method as it gives the best result. Average pooling calculates the average of each feature map. Rectified linear unit or ReLU is the next step of pooling. This layer performs element-wise operation and assigns all negative pixels to zero. Like a normal neural network, this layer takes inputs from the previous layer and produces output as one-dimensional array as final classes. Some popular CNN architectures are ResNet, AlexNet, GoogleNet, and ZFNet. Vinayakumar et al. [47] applied CNN in intrusion detection system on popular dataset KDDCup 99 and they showed that CNN perform better than other deep learning algorithm. Kim et al. [48] proposed a CNN-based network intrusion detection system on KDD dataset which consists of four kinds of attacks namely, Dos, remote to local (R2L), user to root (U2R), and probing. They claim that their proposed model is special for detection of DoS attacks. They compared the results with recurrent neural network and concluded that CNN perform better than RNN. Riyaz et al. [49] presented a deep learning algorithm using CNN in wireless network for efficient intrusion detection. They used conditional random and correlation coefficient-based classifier with existing CNN to achieve best accuracy.

Recurrent Neural Network (RNN)

The abbreviation of recurrent neural network is RNN. It is a type of artificial neural network where current input depends on previous output. In normal neural network, input is not depended on output. Application of RNN is pattern recognition of text, speech, and handwriting where a stream of data is use. Previous output is stored in a state vector which is further used is to compute the current output. RNN uses previous input and current input to calculate new output. Disadvantage of RNN is data training is very difficult in RNN due to gradient vanishing and exploring problem. Solution of this problem is long short-term memory (LSTM). LSTM stores information for a long time. It remembers previous result and which is useful for time series prediction. Recurrent neural network which consists of three layers namely input layer, hidden layer, and output layer. Yin et al. [50] proposed a deep learning model based on RNN for intrusion detection system for binary and multi class classification. They compared the result with other classification algorithms such as ANN, SVM, J48, random forest, and they concluded that their proposed model gives better accuracy. Tang et al. [51] proposed a Gated Recurrent Unit Recurrent Neural Network (GRU-RNN)-based IDS for SDN system here SDN stands for software-defined network which is a cost-effective dynamic network. They used NSL-KDD dataset to test their proposed model and they achieved 89% accuracy. Kim et al. [52] combined LSTM with RNN for intrusion detection system. They also used NSD-KDD dataset and got significant result. Khan [53] proposed a novel algorithm namely HCRNNIDS which is combination of CNN and RNN. Here, CNN executes convolution of local feature and RNN extracts local feature from the dataset. They performed tenfold cross validation on the CSE-CIC-DS2018 and obtained 97.75% accuracy.

Deep Belief Network (DBN)

DBN is a type of neural network with multiple layers. Connection between top two layers are directed connection and with other layer are undirected connections. DBN is consists stack layers from Restricted Boltzmann Machine (RBM). It is used to solve the supervised learning for regression and classification also unsupervised learning to reduce feature dimensionality. Two steps are required to train a deep belief network is layer-by-layer training and other one is fine tuning. Unsupervised learning method is used in the layer-by-layer training, and in second step, back propagation is used for fine tune of the parameters. Tian et al. [54] proposed a modified version of DBN for intrusion detection system. They used Kullback–Leibler for divergence and Gaussian distribution to the existing DB network. They used NSL-KDD and UNSW-NB15 dataset to evaluate their model and achieved better accuracy and lower FPR. Alom et al. [55] applied DBN in NSL_KDD dataset and they achieved 97.5% accuracy only in fifty iterations. Zhang et al. [56] proposed a new algorithm based on a variant of genetic algorithm and DBN. They used NSL_KDD dataset to simulate their model. Experimental results show that it improved the recognition rate of intrusion and reduced the complexity of neural network. Wei et al. [57] proposed a hybrid DBN which includes inertia weight, learning factor based on PSO algorithm; they also used self-adjusting genetic operator to improve the search ability of PSO. They found that the hybrid algorithm reduces the time complexity of the original DBN algorithm and achieved better accuracy.

Ensemble Classifier

An ensemble method is a combination of multiple classifiers. The model gives better result than a single model. Gaikwad et al. [58] proposed bagging ensemble method using RPPTree. They used NSL_KDD dataset to evaluate their model. They conclude that the hybrid model selects relevant feature from the dataset and improves classification accuracy. Jabbar et al. [59] proposed a new ensemble method RFAODE for intrusion detection system. The model is a combination of random forest (RF) and average one-dependence estimator (AODE). They use Kyoto dataset for evaluation of the model and concluded that their proposed method gives best accuracy. Fitni et al. [60] proposed an ensemble classifier after combining gradient boosting, decision tree, and logistic regression. They used CSE-CIC-IDS2018 dataset to evaluate their proposed model, and they showed that their proposed model achieved 98.8% accuracy.

AIDS Deployment Techniques in IoT

Based on deployment strategy of IDS in IoT, it is classified into three types such as centralized, distributed, and hierarchical IDS.

Centralized Intrusion Detection System

In centralized IDS system, the IDS is placed on central IoT devices. Here, IoT device collects all information and sent to the boundary switch. IDS checks the packets between IoT device and network. By checking only network packets that pass-through boundary switch is not enough to identify anomalies that may cause harm to the IoT device. Centralized IDS monitors network traffic that is extracted from different network source such as NetFlow and packet capture.

Distributed Intrusion Detection System

Distributed IDS is formed by many IDS in a big system of IoT network, where a system check the other system performance. Here, all IDS can communicate with each other, and a central sever helps for packet analysis and event response. Many IDS are deployed in a distributed manner. Distributed IDS has many advantages over centralized IDS. The main advantage of the distributed system is as the whole IoT system acts as a single system, so here identifying of an intrusion is easier which increases the attack detection. Distributed IDS has the capability of early detection of IoT botnet. After detection of such type of attack, it cleans the system that has infected by the IoT botnet and stops to spread further.

Hierarchical Intrusion Detection System

In hierarchical IDS, the entire network is divided into different clusters. Those sensor nodes are close to each other fit to the same cluster. Each cluster has a leader namely cluster head. The task of the cluster head is network analysis node by node.

AIDS Validation Techniques

Validation in IDS is a process by which an IoT IDS model is checked and determined if the system is accurate or not. Different theoretical, hypothetical, and empirical techniques are used by the researchers to validate their method. Performance of an IDS system is measured by some matrices which are discussed below.

True Positive Rate (TPR)

TPR is the ratio of the number of correctly predicted attacks to total numbers of attacks. If all classes are correctly detected by an IDS then the TPR is 1. Mathematically, TPR is expressed by

$$TPR = \frac{TP}{TP + FN}$$

Here, true positive (TP) is the positive outcome that the model predicted correctly including actual and predicted. False negative (FN) are those outcomes that the model predicted incorrectly.

False Positive Rate (FPR)

FPR is the ratio of the false positive instances to the total number of actual negative instances. High FPR indicates low performance of the system.

$$FPR = \frac{FP}{FP + TN}$$

False positive (FP) means normal instances but classified incorrectly as attack and true negative (TN) means actually normal instance and also predicted as a normal.

False Negative Rate (FNR)

FNR is the ratio of the false negative instances to the total number of actual positive instances. The system failed to identify attack instances and classified them as normal.

$$FNR = \frac{FN}{FN + TP}$$

True Negative Rate (TNR)

TNR is the ratio between true negative and total number of actual negative instances.

$$TNR = \frac{TN}{TN + FP}$$

Accuracy

It is the ratio of total number of correctly classified instances (normal instances as normal and attack as attack) to the total number of instances.

$$Accuracy = \frac{TP + TN}{TP + TN + FP + FN}$$

Confusion Matrix

Confusion matrix is also used to evaluate the performance of an IDS. Figure 12.9 shows a confusion matrix of a two-class classifier where each row of confusion matrix represents actual class of an instance and each column denotes predicted class of an instance.

Fig. 12.9 Confusion matrix

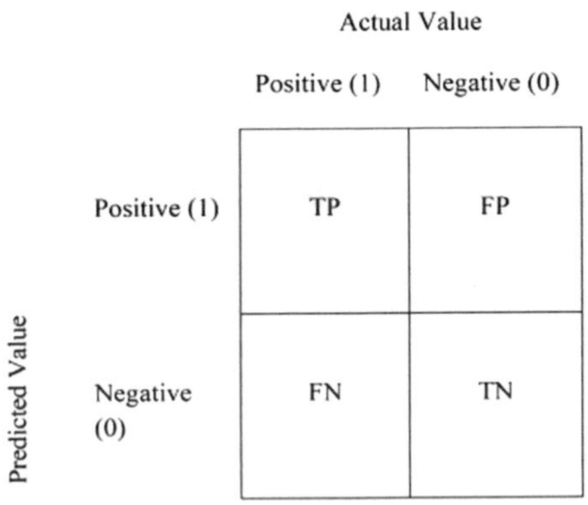

Conclusion

There is a wide variety of serious security concerns and malicious activities that might compromise healthcare systems based on IoT. In this work, we presented a comprehensive analysis of the advantages and deficiencies of existing approaches to IoT intrusion detection system approaches, deployment approaches, and validation methods. However, the IoT's design may make it difficult for such methods to identify all IoT intrusions. To address these IoT security challenges, we reviewed the findings of previous studies and dove into the state-of-the-art methods for enhancing IoT IDS performance. Finally, we hope that by summing up the current state of this massive and rapidly evolving field of study, this review will prove useful to security researchers who are keen on creating new intrusion detection systems (IDS) to deal with issues of Internet of Things (IoT) security in the framework of IoT communication.

References

1. Saif, S., Gupta, R., Biswas, S.: A complete secure cloud-based WBAN framework for health data transmission by implementing authenticity, confidentiality and integrity. Int. J. Adv. Intell. Paradigms **20**(1–2), 171–189 (2021)
2. Saif, S., Gupta, R., Biswas, S.: Implementation of cloud-assisted secure data transmission in WBAN for healthcare monitoring. In: Advanced Computational and Communication Paradigms, pp. 665–674. Springer, Singapore (2018)
3. Ahmed, M.I., Kannan, G.: Secure end to end communications and data analytics in IoT integrated application using IBM Watson IoT platform. Wireless Pers. Commun. **120**(1), 153–168 (2021)
4. Saif, S., Saha, R., Biswas, S.: On Development of MySignals based prototype for application in health vitals monitoring. Wireless Pers. Commun. **122**(2), 1599–1616 (2022)
5. Ammar, M., Russello, G., Crispo, B.: Internet of Things: a survey on the security of IoT frameworks. J. Inf. Secur. Appl. **38**, 8–27 (2018)
6. Becker, R.: Cyber Attack on German Steel Mill Leads to Massive Real World Damage. PBS Magazine (2015). Accessed 29 Nov 2022. [Online]. Available: http://www.pbs.org/wgbh/nova/next/tech/cyber-attack-german-steel-mill-leads-massive-real-world-damage/
7. Robert, L.M., Michael, A.J., Tim, C.: Analysis of the cyber attack on the Ukrainian power grid. Electr. Inf. Sharing Anal. Center, Washington, DC, USA (2016). [Online]. Available: https://ics.sans.org/media/E-ISACSANSUkraineDUC5.pdf
8. Wagstaff, J.: All at sea: global shipping fleet exposed to hacking threat (2014). Accessed 29 Nov 2022. [Online]. Available: http://reut.rs/1rnmjdI
9. Lima., J.: IoT security breach forces kitchen devices to reject junk food (2015). Accessed: 29 Nov 2022. [Online]. Available: https://www.cbronline.com/news/iot-security-breach-forces-kitchen-devices-to-reject-junk-food-4544884/
10. Starr, M.: Fridge Caught Sending Spam Emails in Botnet Attack. CNET Magazine, San Francisco, CA, USA (2015.) Accessed: 29 Nov 2022. [Online]. Available: https://cnet.co/2oPzNJC
11. Heightened DDoS threat posed by MIRAI and other bot-nets. Cybersecurity Infrastruct. Security Agency, Rep. TA16-288A (2016). Accessed: 29 Nov 2022. [Online]. Available: https://www.us-cert.gov/ncas/alerts/TA16-288A

12. Saif, S., Das, P., Biswas, S., Khari, M., Shanmuganathan, V.: HIIDS: hybrid intelligent intrusion detection system empowered with machine learning and metaheuristic algorithms for application in IoT based healthcare. Microprocess. Microsyst. 104622 (2022)
13. Saif, S., Karmakar, K., Biswas, S., Neogy, S.: MLIDS: machine learning enabled intrusion detection system for health monitoring framework using BA-WSN. Int. J. Wireless Inf. Netw. 1–12 (2022)
14. Velliangiri, S., Manoharn, R., Ramachandran, S., Krishnasamy, V., Rajasekar, V.R., Karthikeyan, P., et al.: An efficient lightweight privacy preserving mechanism for industry 4.0 based on elliptic curve cryptography. IEEE Trans. Industr. Inf. (2021)
15. Sangeetha Francelin, V.F., Daniel, J., Velliangiri, S.: Intelligent agent and optimization-based deep residual network to secure communication in UAV network. Int. J. Intell. Syst. (2022)
16. Khraisat, A., Gondal, I., Vamplew, P., Kamruzzaman, J., Alazab, A.: A novel ensemble of hybrid intrusion detection system for detecting internet of things attacks. Electronics 8(11), 1210 (2019)
17. Mohd, N., Singh, A., Bhadauria, H.S.: A novel SVM based IDS for distributed denial of sleep strike in wireless sensor networks. Wireless Pers. Commun. 111(3), 1999–2022 (2020)
18. Neshenko, N., Bou-Harb, E., Crichigno, J., Kaddoum, G., Ghani, N.: Demystifying IoT security: an exhaustive survey on IoT vulnerabilities and a first empirical look on Internet-scale IoT exploitations. IEEE Commun. Surv. Tutorials 21(3), 2702–2733 (2019)
19. Masdari, M., Khezri, H.: A survey and taxonomy of the fuzzy signature-based intrusion detection systems. Appl. Soft Comput. 92, 106301 (2020)
20. Kumar, V., Sangwan, O.P.: Signature based intrusion detection system using SNORT. Int. J. Comput. Appl. Inf. Technol. 1(3), 35–41 (2012)
21. Dixit, U., Gupta, S., Pal, O.: Speedy signature based intrusion detection system using finite state machine and hashing techniques. Int. J. Comput. Sci. Issues (IJCSI) 9(5), 387 (2012)
22. Khraisat, A., Alazab, A.: A critical review of intrusion detection systems in the internet of things: techniques, deployment strategy, validation strategy, attacks, public datasets and challenges. Cybersecurity 4(1), 1–27 (2021)
23. Nancy, P., Muthurajkumar, S., Ganapathy, S., Santhosh Kumar, S.V.N., Selvi, M., Arputharaj, K.: Intrusion detection using dynamic feature selection and fuzzy temporal decision tree classification for wireless sensor networks. IET Commun. 14(5), 888–895 (2020)
24. Khraisat, A., Gondal, I., Vamplew, P., Kamruzzaman, J., Alazab, A.: Hybrid intrusion detection system based on the stacking ensemble of c5 decision tree classifier and one class support vector machine. Electronics 9(1), 173 (2020)
25. Talita, A.S., Nataza, O.S., Rustam, Z.: Naïve bayes classifier and particle swarm optimization feature selection method for classifying intrusion detection system dataset. J. Phys.: Conf. Ser. 1752(1), 012021 (2021)
26. Bhosale, K.S., Nenova, M., Iliev, G.: Modified naive bayes intrusion detection system (MNBIDS). In: International Conference on Computational Techniques, Electronics and Mechanical Systems (CTEMS), pp. 291–296. IEEE (2018)
27. Yang, J., Ye, Z., Yan, L., Gu, W., Wang, R.: Modified naive bayes algorithm for network intrusion detection based on artificial bee colony algorithm. In: IEEE 4th International Symposium on Wireless Systems within the International Conferences on Intelligent Data Acquisition and Advanced Computing Systems (IDAACS-SWS), pp. 35–40. IEEE (2018)
28. Abu Alfeilat, H.A., Hassanat, A.B., Lasassmeh, O., Tarawneh, A.S., Alhasanat, M.B., Eyal Salman, H.S., Prasath, V.S.: Effects of distance measure choice on k-nearest neighbor classifier performance: a review. Big data 7(4), 221–248 (2019)
29. Liu, G., Zhao, H., Fan, F., Liu, G., Xu, Q., Nazir, S.: An enhanced intrusion detection model based on improved kNN in WSNs. Sensors 22(4), 1407 (2022)
30. Wazirali, R.: An improved intrusion detection system based on KNN hyperparameter tuning and cross-validation. Arab. J. Sci. Eng. 45(12), 10859–10873 (2020)
31. Senthilnayaki, B., Venkatalakshmi, K., Kannan, A.: Intrusion detection system using fuzzy rough set feature selection and modified KNN classifier. Int. Arab J. Inf. Technol. 16(4), 746–753 (2019)

32. Huang, W., Liu, H., Zhang, Y., Mi, R., Tong, C., Xiao, W., Shuai, B.: Railway dangerous goods transportation system risk identification: comparisons among SVM, PSO-SVM, GA-SVM and GS-SVM. Appl. Soft Comput. **109**, 107541 (2021)
33. Kabir, E., Hu, J., Wang, H., Zhuo, G.: A novel statistical technique for intrusion detection systems. Futur. Gener. Comput. Syst. **79**, 303–318 (2018)
34. Safaldin, M., Otair, M., Abualigah, L.: Improved binary gray wolf optimizer and SVM for intrusion detection system in wireless sensor networks. J. Ambient. Intell. Humaniz. Comput. **12**(2), 1559–1576 (2021)
35. Sukumar, J.A., Pranav, I., Neetish, M.M., Narayanan, J.: Network intrusion detection using improved genetic k-means algorithm. In: International Conference on Advances in Computing, Communications and Informatics (ICACCI), pp. 2441–2446. IEEE (2018)
36. Al-Yaseen, W.L., Othman, Z.A., Nazri, M.Z.A.: Multi-level hybrid support vector machine and extreme learning machine based on modified K-means for intrusion detection system. Expert Syst. Appl. **67**, 296–303 (2017)
37. Tahir, H.M., Said, A.M., Osman, N.H., Zakaria, N.H., Sabri, P.N.A.M., Katuk, N.: Oving K-means clustering using discretization technique in network intrusion detection system. In: 3rd International Conference on Computer and Information Sciences (ICCOINS), pp. 248–252. IEEE (2016)
38. Horng, S.J., Su, M.Y., Chen, Y.H., Kao, T.W., Chen, R.J., Lai, J.L., Perkasa, C.D.: A novel intrusion detection system based on hierarchical clustering and support vector machines. Expert Syst. Appl. **38**(1), 306–313 (2011)
39. Sangve, S.M., Thool, R.C.: ANIDS: anomaly network intrusion detection system using hierarchical clustering technique. In: Proceedings of the International Conference on Data Engineering and Communication Technology, pp. 121–129. Springer, Singapore (2017)
40. Song, J., Zhu, Z., Price, C.: Feature grouping for intrusion detection system based on hierarchical clustering. In: International Conference on Availability, Reliability, and Security, pp. 270–280. Springer, Cham (2014)
41. Mazarbhuiya, F.A., AlZahrani, M.Y., Georgieva, L.: Anomaly detection using agglomerative hierarchical clustering algorithm. In: International Conference on Information Science and Applications, pp. 475–484. Springer, Singapore (2018)
42. Hadri, A., Chougdali, K., Touahni, R.: Intrusion detection system using PCA and fuzzy PCA techniques. In: International Conference on Advanced Communication Systems and Information Security (ACOSIS), pp. 1–7. IEEE (2016)
43. Waskle, S., Parashar, L., Singh, U.: Intrusion detection system using PCA with random forest approach. In: International Conference on Electronics and Sustainable Communication Systems (ICESC), pp. 803–808. IEEE (2020)
44. Bhattacharya, S., Maddikunta, P.K.R., Kaluri, R., Singh, S., Gadekallu, T.R., Alazab, M., Tariq, U.: A novel PCA-firefly based XGBoost classification model for intrusion detection in networks using GPU. Electronics **9**(2), 219 (2020)
45. Srinoy, S., Kurutach, W., Chimphlee, W., Chimphlee, S., Sounsri, S.: Computer Intrusion Detection with Clustering and Anomaly Detection, Using ICA and Rough Fuzzy
46. Lopez-Martin, M., Carro, B., Sanchez-Esguevillas, A.: Application of deep reinforcement learning to intrusion detection for supervised problems. Expert Syst. Appl. **141**, 112963 (2020)
47. Vinayakumar, R., Soman, K.P., Poornachandran, P.: Applying convolutional neural network for network intrusion detection. In: International Conference on Advances in Computing, Communications and Informatics (ICACCI), pp. 1222–1228. IEEE (2017)
48. Kim, J., Kim, J., Kim, H., Shim, M., Choi, E.: CNN-based network intrusion detection against denial-of-service attacks. Electronics **9**(6), 916 (2020)
49. Riyaz, B., Ganapathy, S.: A deep learning approach for effective intrusion detection in wireless networks using CNN. Soft. Comput. **24**(22), 17265–17278 (2020)
50. Yin, C., Zhu, Y., Fei, J., He, X.: A deep learning approach for intrusion detection using recurrent neural networks. Ieee Access **5**, 21954–21961 (2017)
51. Tang, T.A., Mhamdi, L., McLernon, D., Zaidi, S.A.R., Ghogho, M.: Deep recurrent neural network for intrusion detection in sdn-based networks. In: 4th IEEE Conference on Network Softwarization and Workshops (NetSoft), pp. 202–206. IEEE (2018)

52. Kim, J., Kim, J., Thu, H.L.T., Kim, H.: Long short term memory recurrent neural network classifier for intrusion detection. In: International Conference on Platform Technology and Service (PlatCon), pp. 1–5. IEEE (2016)
53. Khan, M.A.: HCRNNIDS: hybrid convolutional recurrent neural network-based network intrusion detection system. Processes **9**(5), 834 (2021)
54. Tian, Q., Han, D., Li, K.C., Liu, X., Duan, L., Castiglione, A.: An intrusion detection approach based on improved deep belief network. Appl. Intell. **50**(10), 3162–3178 (2020)
55. Alom, M.Z., Bontupalli, V., Taha, T.M.: Intrusion detection using deep belief networks. In: National Aerospace and Electronics Conference (NAECON), pp. 339–344. IEEE (2015)
56. Zhang, Y., Li, P., Wang, X.: Intrusion detection for IoT based on improved genetic algorithm and deep belief network. IEEE Access **7**, 31711–31722 (2019)
57. Wei, P., Li, Y., Zhang, Z., Hu, T., Li, Z., Liu, D.: An optimization method for intrusion detection classification model based on deep belief network. IEEE Access **7**, 87593–87605 (2019)
58. Gaikwad, D.P., Thool, R.C.: Intrusion detection system using bagging ensemble method of machine learning. In: International Conference on Computing Communication Control and Automation, pp. 291–295. IEEE (2015)
59. Jabbar, M.A., Aluvalu, R.: RFAODE: a novel ensemble intrusion detection system. Procedia Comput. Sci. **115**, 226–234 (2017)
60. Fitni, Q.R.S., Ramli, K.: Implementation of ensemble learning and feature selection for performance improvements in anomaly-based intrusion detection systems. In: IEEE International Conference on Industry 4.0, Artificial Intelligence, and Communications Technology (IAICT), pp. 118–124. IEEE (2020)

Chapter 13
War Strategy Algorithm-Based GAN Model for Detecting the Malware Attacks in Modern Digital Age

S. Rudresha, Alim Raza, Vivek Anand, Himanshu Payal, Kundan Yadav, and Balasubramanian Prabhu Kavin

Introduction

One of the most significant risks to information security is the proliferation and sophistication of malicious software [1, 2]. While the cybersecurity sector works tirelessly to keep tabs on malware and even profit from it in a number of different ways, cybercriminals show no signs of slowing down or even reducing the frequency of their assaults. The polymorphism [3], metamorphism [4], code obfuscations [5], etc., backdoors, miners, spyware, and info stealers are forms of malware employed by attackers targeting organisations. Typical examples of information stealers that use malspam (also known as spam) to infect computers include Emotet [6] and TrickBot [7]. Malicious spam often includes malware in the form of malicious attachments or links. Many varieties of malware may now steal confidential data from businesses. WannaCry isn't the only recent virus that has made headlines; others include Kovter, ZeuS, Dridex, IcedID, Gh0st, Mirai, and many more. Attackers and security researchers are still locked in a technological arms race to develop the best methods for detecting malware.

For hiding their tracks from signature-based security systems, researchers have shown that hackers rely heavily on polymorphism and metamorphism. Software tools are used to extract the scripts and analyse the API calls manually, which is how this issue is currently being tackled. Due to the time and effort required, [8] introduced a

S. Rudresha
Department of Mechanical Engineering, J.S.S Academy of Technical Education Noida, C-20/1, Sector 62, Noida U.P.-201301, India

A. Raza · V. Anand · H. Payal · K. Yadav · B. P. Kavin (✉)
Data Science and Business Systems, SRM Institute of Science and Technology, Kattankulathur, India
e-mail: ceaserkavin@gmail.com

© The Author(s), under exclusive license to Springer Nature Singapore Pte Ltd. 2023
V. Sarveshwaran et al. (eds.), *Artificial Intelligence and Cyber Security in Industry 4.0*, Advanced Technologies and Societal Change,
https://doi.org/10.1007/978-981-99-2115-7_13

system that can automatically extract API requests and analyse their harmful properties in four simple steps. First, the virus must be unzipped. Secondly, the binary executable is deconstructed. The third step requires pulling API calls. Analysing the API's call graph and other statistical features is the fourth step [9]. Machine learning algorithms (MLAs) such support vector machines (SVMs) using n-gram features taken from huge samples of both benign and malicious executables and tenfold cross validations were used to improve this technique. A later timeframe saw the execution of a comparison analysis of many conventional detection, and the proposal of a framework for the detection of zero-day malware. Similarity-based mining and algorithms [10] are used to process harmful code variations based on the sequence of API calls and the frequency with which they occur. Extensive experimental research was performed on a massive dataset, and a unified framework was developed for extracting characteristics from malware binaries. As the number of malware assaults from unknown sources has grown, researchers have been working to enhance MLAs for malware detection [11]. This is the impetus for the present study.

Features are the backbone of (MLAs); hence, feature engineering, feature selection, and feature representation are essential. A model is trained using a feature set associated with a certain class in order to generate a boundary between benign and malicious software [12]. Using this division, malicious software may be located and placed in the correct family. The fundamental problem with conventional machine learning-based malware detection systems is their reliance all of which need extensive domain level knowledge [13]. Worse, an attacker who is aware of the properties of the infection may easily evade detection.

Machine learning has a subset called deep learning, which aims to improve knowledge representations by learning the input on numerous layers. Convolutional neural networks are where much of the deep learning-based improvements in computer vision may be found (CNN). Models trained with deep learning use millions of parameters as they learn and adjust a wide variety of complicated characteristics across several convolutional layers [14]. As a consequence, the model becomes overfit after just a few iterations, and it cannot generalise properly, leading to subpar results. CNNs that have been trained on a large, well-described dataset may then apply that expertise to the task of detecting and classifying malware pictures. Learning one model may improve performance on a different learning activity; this is the primary notion behind transfer learning. In order to construct a CNN, the input must go through a series of layers, each one more complex than the one before. There is a chance that the input data will be lost before it spreads the output network [15]. While ResNet and other CNNs solve this issue, they do so by creating shorter connections between the layers.

In this study, we describe a new tactic to malware classification using a deep learning model dubbed the PATE outline and then use it to GANs to produce synthetic data. In this research, we suggest a brand-new measure for assessing the simulated data. For the GAN, WSA chooses the optimal solution. The associated research is discussed in Chap. 2. In Sects. "Proposed System" and "Results and Discussion", respectively, we see a short description of the suggested model and a validation study of the proposed model. Section "Conclusion" closes the scientific contribution with a discussion of the findings.

Related Works

A method based on static investigation and current developments in the classification field has been proposed by Obaidat et al. [16]. In specifically, Jadeite takes a Java bytecode file and parses it for its Inter-procedural Control Flow Graph (ICFG), which it then prunes and transforms into an adjacency matrix. Once this matrix is complete, Jadeite uses an object identification technique in a deep (CNN) classifier to assess whether or not the picture is harmful. In addition, Jadeite extracts a supplementary collection of characteristics from the Java malware programme to enhance the precision of malware categorization. Consolidating these characteristics with the retrieved pictures, they serve as inputs to the CNN classifier. The experimental findings show that Jadeite can identify both well-known and recently discovered real-world harmful Java applications with a high degree of accuracy (98.4%) compared to existing Java malware detection methodologies.

The MAPAS malware detection method proposed by Kim et al. [17] is very accurate and flexible in its use of system resources. Convolutional neural networks are used by MAPAS to analyse the behaviour of malicious apps based (CNN). In contrast to other approaches, MAPAS does not rely on a CNN-generated classifier model but rather uses CNN to find shared characteristics in malware's API call graphs. For effective malware detection, MAPAS makes use of a lightweight classifier that determines the degree of similarity between the API call graphs of apps to be categorised and the API call graphs utilised for harmful activities. We deploy a prototype of MAPAS and carefully assess its efficacy and efficiency. And, we compare MAPAS to MaMaDroid, a cutting-edge method for detecting Android malware. Here, we see that compared to MaMaDroid, MAPAS is 145.8% quicker at app classification and consumes roughly ten times less RAM. Furthermore, MAPAS has a greater success rate in identifying unknown malware (91.27%) than MaMaDroid (84.99%). What's more, MAPAS has a high rate of success in detecting malware of all kinds.

A deep learning digital data and picture features on multiple dimensions at the same time has been suggested by Lian et al. [18]. This model incorporates the simultaneous learning of three sub-models as well as the ensemble learning of a fourth. It is possible to use the four sub-models in edge computing settings and process them in parallel on separate devices. The model is able to learn and predict across several modalities in an adaptive fashion. This model has a maximum detection rate of 97.01% and a false alarm rate of just 0.63%. The experimental findings validate the benefits and efficacy of the suggested approach.

To efficiently identify malware variations, Kim and Cho [19] offer a hybrid deep generative model that uses both global and local data. Local features are extracted from the binary code sequences when the virus is being transformed into an image to efficiently represent global characteristics using a pre-defined latent space. The malware detector receives a combined feature set consisting of two features, each of which is derived from the underlying data and has its own unique properties. The suggested model, which makes use of both attributes, achieves state-of-the-art performance, with an accuracy of 97.47%. By analysing the created malware's CAM findings, we can determine what areas of the malware's code have an impact on the detection results, confirming the use of virtual malware production in enhancing detection performance.

A unique robust dynamic analytic approach for malware identification using fine-grained behavioural features, i.e. control flow traces, was presented by Qiang et al. [20]. Control flow traces are translated into byte sequences, and then a malware classifier is constructed. Among known malware, the proposed classifier has a maximum detection rate of 95.7%, while among unknown malware, it has a detection rate of 94.60%. In the meanwhile, early experimental findings show that packing has a special interaction with existing behaviour-based malware classifiers, resulting in even poorer presentation than static classifiers in certain cases. However, the projected classifier beats the API call-based classifier when it comes to robustness, as it maintains performance despite interference from an uneven packing distribution has improved resistance to hostile samples and a minimum detection rate of 83%.

To identify malicious Java code, Feng et al. [21] provide a new methodology called BejaGNN, which combines static analysis, the word embedding procedure, and a graph neural network. To be more precise, BejaGNN uses static analysis methods to identify ICFGs in Java source code, which it then uses to prune unnecessary or irrelevant code. To then learn semantic representations for Java bytecode instructions, work embedding methods are used. In the end, BejaGNN constructs a classifier based on a graph neural network to identify harmful Java code. BejaGNN outperforms state-of-the-art Java malware detection methods by a wide margin, as shown by experimental findings on a publicly available Java bytecode benchmark, proving the efficacy of graph neural networks in this setting.

Static analysis of disassembled programme code by García-Soto et al. [22] includes aspects suggestive of malware, such as API calls and permissions that are automatically learnt by the network. Because of this, classifying malware is no longer dependent on using characteristics that had to be manually developed. Our results on the Drebin and AMD benchmark datasets highlight the usefulness of this multi-view design for integrating data from many sources. Finally, deep learning architectures provide state-of-the-art outcomes in automated malware detection while decreasing reliance on feature engineering and domain experience. Multi-view architectures outperform single-view ones because they expose the learner to multiple data sources at once, allowing for the acquisition of a richer feature set. An essential parameter for possible deployment in the real world, the model achieves state-of-the-art detection performance in a difficult zero-day scenario by minimizing false positives by an average of 77%.

Data saved [23] in the cloud servers used by the IoT are extremely susceptible to a wide variety of assaults. Analysis to date suggests that roughly 23% of IoT devices are vulnerable to attack. Cloud data is extremely susceptible to attacks, which might reduce economic development by as much as 15%. This research provides a framework integrating the Jaya algorithm and genetic algorithm to achieve optimal detection of intrusion in the IoT network, keeping in mind the aforementioned security of the IoT devices. Because it is a parameter-less algorithm, the JA doesn't need fine-tuned settings. The GA, on the other hand, is a meta-heuristic strategy for solving complex functions that yields acceptable results. Improving key metrics including accuracy, recall, and F-score, the suggested method has undergone comprehensive examination, and the results are promising.

The need [24] of developing intrusion detection systems has grown in step with the rapid expansion of data being transmitted from IoT devices to end-user devices. The role of intrusion detection systems cannot be overstated in the context of the modern smart home, smart city, agricultural, and business enterprises. The intruders use the IoT sensor gadget as a conduit for their attacks, which begin with the intruders' crate attacks. Many deep learning models have been developed and implemented in the IoT context to identify malicious actions. This research study provides an overview of the deep supervised learning model, deep unsupervised learning model, and dataset employed by the intrusions detection system in an Internet of Things (IoT) setting. At last, the intrusion detection system research gap in the IoT setting is introduced.

In order to select [25] the best features, a layered paddy crop optimization (LPCO) technique is proposed here. In addition, the proliferation of smart gadgets results in a flood of data that, using a capsule network (CN) classification scheme, can be categorised as either benign or malicious. The suggested method is tested on the NSL KDD, UNSW NB, CICIDS, CSE-CIC-IDS, and UNSW Bot-IoT benchmark datasets for network traffic. Experiments show that the presented method outperforms the state-of-the-art base classifiers and feature selection techniques. In terms of performance, the proposed strategy outperforms the current best practises.

Proposed System

There are three stages to the process: (1) global feature modelling (feature extraction at the picture level), (2) local feature modelling (feature extraction at the code level), and (3) malware detection. By reusing a model that has already successfully completed comparable training tasks in a different domain, this technique improves the efficiency of training. One way to look at it is as a transmission of skills to another person. Malware codes in binary format are converted into malware pictures for inclusion in the proposed model as part of the global feature modelling procedure. In order to effectively deal with malware obfuscation, it is necessary to generate a malware image, which involves defining the latent space within which the properties of the virus are contained. The feature extractor is improved by the virtual virus that is created from a predetermined latent space. Meanwhile, the malware codes are converted into a series of binary codes before being used in local feature modelling. At last, the combined characteristics from the extracts are sent into the detector.

Fig. 13.1 Examples of generated malware images

Dataset Description

The research utilised the dataset from the Challenge [19] to validate the performance of both developing and identifying malware. Each of Ramnit, Kelihos version 3, Simda, and Kelihos version 1 may be used to launch a distributed denial of service (DDoS) attack, steal information, and gain access to the infected machine and network. The letters V, S, T, and G are a Trojan horse because they mask the genuine objectives of the user. Although they have the appearance of legitimate programmes, running them will result in harm to the computer as they will execute malicious code. Displaying advertising inside the user's computer, Lollipop (L) is adware that generates revenue for its creator. Obfuscator. Some approaches, known as ACY (O.ACY) (Fig. 13.1).

We just utilised the binary data provided, not the assembly code. We standardised the hexadecimal values to the range of 0–1 by converting them to decimal. Similar to how Nataraj accomplished it, the malicious code is then rendered as a picture, which is referred to as a malware image. Since the original malware graphics were too big to utilise in our perfect, we shrank them to 128 by 128, with a representative example shown in Fig. 13.2. Table 13.1 details the prevalence of several forms of malware. There were 9780 train datasets and 1088 test datasets.

We altered the amount of hexadecimal values in a single row in order to create a picture from the code. The formula for determining the size of the converted row and column is provided by.

$$C = 2^{\log \sqrt{16k/\log 2 + 1}}; R = 16k/C \qquad (13.1)$$

Adam optimiser is used for training all the models, and the weights are initialised using a certain technique.

Detection of Malware Using PATE-GAN

The suggested approach is based on the GAN and PATE models. To make our discriminator differentially private, we swap out the GAN discriminator for a PATE mechanism; however, we still need the student version to facilitate back-propagation to the generator. Due to a lack of publicly accessible data, we make adjustments to the

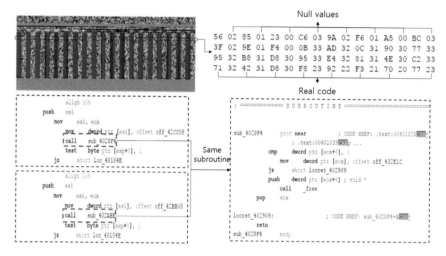

Fig. 13.2 Malware obfuscation used as an example. Modifications include **a** moving subroutines around and **b** inserting null values

Table 13.1 Overview of the malware database used in this research

Malware type	# test	# training	Description
Tracur	63	688	TrojanDownloader
Kelihos ver. 1	40	358	Backdoor
Ramnit	153	1387	Worm
Lollipop	229	2249	Adware
Vundo	40	435	Trojan
Simda	7	35	Backdoor
Gatak	115	898	Backdoor

The data is divided 9:1 across training and testing sets

student's implementation, emphasising that the training paradigm given in [23] is not suitable for this context. The dataset is split into k pieces before training begins, $D_1, \ldots, D_k$ with $|D_i| = \frac{|D|}{k}$ for $\forall i$.

Generator

The generator, G, is as in the standard GAN outline. Officially it is a function $G(\cdot; \theta_G) : [0; 1]^d \rightarrow U$, parametrised by θ_G that takes random noise, $z \sim \text{Unif}([0, 1]^d)$, as input and outputs a vector in $U = X \times Y$. The loss with discriminator. Given n i.e. trials of $\text{Unif}([0, 1]^d)$, $z_1, \ldots, z_n$, the empirical loss of G at θ for fixed S is distinct by

$$\mathcal{L}_G(\theta_G; S) = \sum_{j=1}^{n} \log(1 - S(G(z_j; \theta_G))) \tag{13.2}$$

To indicate the G-over-U distribution, we shall use the symbol P_G.

Discriminator

In the canonical GAN architecture, one discriminator (D) is trained directly adversarially with the generator (G), with each iteration seeing either G or D attempt to reduce their loss relative to the other. However, in our suggested paradigm, the PATE mechanism takes the role of D. This implies that we will be introducing one student discriminator, S, and discriminators, $T_1, \ldots, T_k$. The adversarial training is now asymmetrical, with the instructors being educated to reduce their loss relative to G, while G is trained to reduce its loss relative to S, and S is trained to reduce its loss relative to the teachers.

Teacher-Discriminators

Formally, the teacher-discriminators are functions $T_1(\cdot; \theta_T^1); \ldots; T_k(\cdot; \theta_T^k) : U \to [0, 1]$ each parametrised by θ_T^i. As input, either an actual sample or a sample generated at random is supplied to the instructors. Teachers are taught how to properly categorise these students.

Given n i.i.d. samples of Unif($[0, 1]^d$), $z_1; \ldots; z_n$, we define the empirical loss i with weights θ_T^i, where it is optimally selected by WSA (described in Sect. "Mathematical Model of the War Approach") for fixed G by

$$\mathcal{L}_T^i(\theta_T^i) = -\left[\log T_i\left(u, \theta_T^i\right) + \sum_{j=1}^{n} \log(1 - T_i(G(z_j); \theta_T^i)) \right] \tag{13.3}$$

In a typical GAN framework, the discriminator is trained alongside the teachers, but in this case, each teacher only sees the actual data that corresponds to its own partition.

Student-Discriminators

Our key contribution is to use the student-discriminator (hence referred to as the "student") in this context. The conventional student model only provides a differential privacy guarantee with regard to the student. To teach the student without access to public data, we must suggest a new technique in our context, which places all emphasis on creating synthetic data.

We first note that the standard PATE_mechanism would be used to train the student in the paradigm described in [23]. This means that the student would be trained using similar to that used to generator. We think about the consequences of solely using examples produced by the instructor for training.

When student just on created samples, we are really training it on the distribution we need it to achieve well on, so we first note that the discriminator is only graded on samples from the generator itself during training and not on the actual data. However, we highlight that if the student only ever sees unrealistic instances from the generator (i.e. constructed samples that phoney), then the generator will never have any data to utilise to progress its generated samples. Consequently, it is crucial to instruct the pupil using fictitious yet plausible instances. This leads us to realise that if $Supp(PU) \subset Supp(PG)$ realistic.

In order to ensure $Supp(PU) \subset Supp(PG)$, To begin, we use Xavier initialization to randomly set the parameters of the generator once the input has been normalised into [0, 1]. As a result, that $Supp(P) \subset [0; 1]^d \subset G([0; 1]^d) = G(Supp(Z)) = Supp(G(Z))$ when $Z \sim Unif([0; 1]^d)$.

We create our training $Unif([0, 1]^d)$, $z_1, \ldots, z_n$, generating n trials using the generator, $\hat{u}_1, \ldots, \hat{u}_n$ with $\hat{u}_j = G(z_j)$, and using PATEλ, setting $r_j = PATE_\lambda(\hat{u}_j)$: We train the student, $S(\cdot; \theta_S) : U \rightarrow [0; 1]$, to exploit the standard teacher-labelled data, i.e.

$$\mathcal{L}_S(\theta_S) = \sum_{j=1}^n r_j \log S(\hat{u}_j; \theta_S) + (1 - r_j) \log(1 - S(\hat{u}_j; \theta_S)) \tag{13.4}$$

Although the above technique does not seem to rely on the number of instructors a priori, it is worth noting that for a given, a larger number of teachers results in a lesser amount of noise being introduced to the teacher-labelled dataset. This provides a trade-off since having too few teachers may cause the output to be meaningless due to noise, while having too many instructors can cause the output to be meaningless due to insufficient data being used to educate each teacher. The solution to this issue lies in striking a balance. We train each instructor in our studies with d samples, where d is input space, consisting of both actual and artificial data. However, when several teachers are aggregated (even if in a noisy fashion), even though a single teacher doesn't have much of an impact, the resultant classifier is fairly useful. Since the mechanism assumes that the presence of a single sample can completely flip a teacher's vote, we still need to add the same noise regardless of the number of teachers used, so our differential privacy guarantees are strengthened.

In order to train $G, T_1, \ldots, T_k$, and S, we conduct n_T updates on all instructors at the beginning of each iteration of G, and then n_S updates on the student at the end of each iteration. For as long as our privacy requirement, must be met, we repeat the generator iteration process.

To determine the algorithm's privacy, we use a moments accountant approach described in [23] to establish a data-dependent privacy guarantee during execution. The Appendix contains the full description of what it is and the main outcomes we rely on. We designate PATE-moment GAN's accountant as (13.2). The moments accountant allows for a more reasonable privacy cost to be assigned to accessing the noisy aggregate of the teachers when the teachers have a more solid agreement. This is due to the fact that when instructors are in agreement, the impact of any one teacher (and hence any one sample) is much less compared to when the votes (n_0 and n_1) are equally divided.

Mathematical Model of the War Approach

Every soldier has a fair shot at becoming King or Commander each time around, based on their combative strength. On the battlefield, the King and the Commander are both leaders. Other warriors will follow the King and the Commander's lead on the battlefield. The King or the Commander may find himself in a bind if the enemy deploys a strong enough soldier (Local Optima) to ambush the Leaders. Combatants will not only follow the King's or Commander's orders but also rely on coordinated movement strategies to prevent this.

Attack Strategy

Both offensive and defensive plans have been modelled. Each soldier in the first example adjusts his location in accordance with the changes made by the King and the Commander. The monarch takes a strategic high ground from which to unleash a devastating assault on the enemy. Thus, the fittest or most aggressive soldier is considered the king. At the outset of the conflict, all troops will be treated equally regardless of rank or size. If the plan is carried out effectively, the soldier will be promoted. The ranks and weights of all troops, however, will be adjusted as the conflict continues depending on how well the plan is working. The King, the Army's commander, and the troops are all in very close proximity to one another as the war's end draws closer.

$$X_i(t + 1) = X_i(t) + 2 \times p \times (C - K) + \text{rand} \times (W_i \times K - X_i(t)) \quad (13.5)$$

where $X_i(t + 1)$ is the new position, X_i is the previous C position, and the king, W_i is the weight.

The coloured circles position. If $W_i > 1$, then position of $W_i \times k - X_i(t)$ is in addition to the monarch, the modern soldier's rank now exceeds that of commander. If W_i is less than one, then the present location of $W_i \times k - X_i(t)$ is between the monarch and the soldier. When compared to before, the soldier's current location is more accurate. In the latter phase of a conflict, when W_i approaches 0, the soldier's updated location draws near that of the commander.

Rank and Weight Updation

Each search agent's location update is contingent on the dynamic between the King's location, the Commander's location, and the soldier's rank. Each soldier's standing in the ranks is determined by his prior performance in battle, as described by Eq. (13.5), which in turn affects the relative importance of each soldier's W_i. Each soldier's rank indicates the degree of proximity the soldier has to the objective. Note that in other competing algorithms like GWO, WOA, GSA, and PSO, the weighting factors would change linearly; however, in the present suggested WSO method, the weight (W_i) varies exponentially as a factor of.

If the soldier's fitness (attack force) at the novel position (F_n) is lower than in the old position (F_p), the soldier will return to the old position.

$$X_i(t + 1) = (X_i(t + 1) \times (F_n < F_p) + X_i(t)) \times (F_n < F_p) \qquad (13.6)$$

If the soldier is successful in making the position update, his or her rank R_i will rise.

$$R_i = (R_i + 1) \times (F_n \geqslant F_p) + (R_i) \times (F_n < F_p) \qquad (13.7)$$

Based on the rank, the novel weight is intended as:

$$W_i = W_i \times \left(1 - \frac{R_i}{\text{Max\_iter}}\right)^a \qquad (13.8)$$

Defence Strategy

Positions of the King, of the armed forces, and a randomly selected soldier form the basis of the second strategy update. While changes to rankings and weights are handled in the same way.

$$X_i(t + 1) = X_i(t) + 2 \times p \times (K - X_{\text{rand}}(t)) + \text{rand} \times W_i \times (c - X_i(t)) \quad (13.9)$$

Due to the inclusion of the random soldier's position, this battle tactic investigates a larger search area than its predecessor. When W_i is large, the troops update their position by taking big steps. W_i troops move incrementally while location is updated for tiny values.

Replacement/Relocation of Weak Soldiers

Finding the least fit troops in each cycle. Numerous methods of replacement have been tried and evaluated. The weak soldier may be swapped out with a random one, as shown in for a simpler solution (13.10).

$$X_w(t + 1) = \text{Lb} + \text{rand} \times (\text{Ub} - \text{Lb}) \tag{13.10}$$

The second tactic, outlined in, is moving the weak soldier to a position closer to the median of the whole army on the battle field (13.11). As a result, the algorithm's convergence behaviour is enhanced by this method.

$$X_w(t + 1) = -(1 - \text{randn}) \times (X_w(t) - \text{median}(X)) + K \tag{13.11}$$

Salient Features of the Proposed Algorithm

(i)   The suggested algorithm strikes a nice compromise between discovery and profit-making, as will be shown.
(ii) There is a different value assigned to each solution (soldier) depending on his position in the hierarchy.
(iii) If a soldier is able to increase his fitness throughout the updating process, his weight will be adjusted accordingly. Consequently, the particle's location relative to the King's and commander's position is the only determinant of the weight update.
(iv) Nonlinear changes in the weights are to be expected. There is a broad range of weight values in the first few iterations, and a smaller range in the final few. This expedites the process of reaching the global optimal value.
(v)   The procedure for updating a position consists of two phases. This enhances the power to search for the global optimal solution.
(vi) The suggested approach is easy to understand and implement, reducing the amount of work needed on the computer.

Exploration and Exploitation

The two basic requirements for every metaheuristic optimization method are exploration (for global optima) and exploitation (for convergence). This algorithm will be more stable and productive if these two phenomena can be balanced well. The exploitation represented by an attack plan, and the exploration represented by a defensive strategy. Other key elements that affect the proposed algorithm's exploration and exploitation include:

(i)   Let's start with the "rand" variable, which may have any value between 0 and 1. This "random" variable determines whether the soldier's next action will be focused on exploration or exploitation. For a second, the parameter r allows the user to tailor the goal function to the value selected. Experiments conducted on a variety of test functions suggest that for unimodal functions, a low value of r in the range of (0–0.5) is optimal, whereas for multimodal functions, values in the range of (0.5–1) are preferable. Last but not least, sending the search agent in the direction of Xrand makes the procedure more exploratory, leading it to the global optimum by way of the more prominent regions of the search space.

(ii)   The W_i factor, in the end, steers the search agent in the most optimal direction. W_i motivates the search agents to travel the world over in an exploratory capacity, and later, as the search process near its conclusion, the search agents will become more aggressive in their pursuit of resources.

Over time, a soldier's equipment will change to best suit his or her needs. The more physically fit a soldier is, the less weight he or she will have to carry, and vice versa for less fit soldiers. Each soldier charges into combat with a series of massive, individually weighted strides. Warriors take baby steps towards triumph as the conclusion of the struggle comes near, and the significance of each step varies with the state of the conflict. The strategy's inherent randomness means that the soldiers won't be able to reliably follow the king's instructions and will instead go in whichever direction they choose to decide upon. In this way, the algorithm becomes more inquisitive. After a battle, military forces decide where to go next (prominent search space). The King and the Army Commander are situated near the goal, while the army is arrayed around them. Based on Eqs. (13.5) and (13.9), we may infer that the combined effort is making slow but steady progress towards the target. For this reason, we might postulate that the algorithm is also capable of using the information it collects.

Results and Discussion

There were 70% of the data used for training, and 30% was used for validating. The samples were split between the train and test files such that 30% of the total were used for testing. The tests were run on a Linux machine equipped with an Intel® Xeon(R) CPU E3-1226 v3 running at 3.30 GHz 4, 32 GB of RAM, and an NVIDIA GM107GL Quadro K2200/PCIe/SSE2 graphics card. Hyperparameters such 100 epochs, 0.0001 learning rate, and 32-person batches were used in the performance tests. Using the Python programming language and the Keras v0.1.1 deep learning toolkit, the suggested deep neural network model was built. The tests were conducted using 32-by-32-pixel and 64-by-64-pixel binary images as input. For resized 64 × 64 photos, it has been shown that the information is preserved, and that the predicted accuracy is higher.

Classification predictions may be evaluated using four distinct measures.

This binary picture has been correctly identified as malware, an example of a true positive (TP), which is the prediction that a remark belongs to a class and it really does fit to that class.

A binary picture that is classed as not malware (negative) and is truly not malware is an example of a true negative (TN), which to a class and, in this case, the class actually does not correspond to the observation (negative).

The term "false positive" refers to the situation in which an incorrect classification is made for an observation; for example, a binary picture is labelled "virus," although it is really safe (negative).

When a binary image is incorrectly labelled as not malware (negative) when it is, in fact, malware, this is called a false negative (FN).

Four primary metrics are used to evaluate the quality of a classification system: accuracy (Acc), precision (Pr), recall (Re), and $F1$ score. Accuracy is defined as the ratio of successful forecasts to total predictions. The accuracy of a classifier may be measured by comparing the actual number of positive results with the expected number of positive outcomes. In other words, recall is the percentage of true positives that were really detected as such. The $F1$-score is the mathematical mean of the recall and accuracy scores. Both the classifier's accuracy (how many examples it properly labels) and its stability depend on this metric. In Eq., we get the mathematical equation for four different metrics (13.12–13.15).

$$Accuracy(ACC) = \frac{TN + TP}{TP + TN + FN + FP} \times 100 \qquad (13.12)$$

$$F\text{- measure}(F\text{-}M) = \frac{2TP}{(2TP + FP + FN)} \times 100 \qquad (13.13)$$

$$Precision(PR) = \frac{TP}{(FP + TP)} \times 100 \qquad (13.14)$$

$$Recall(RC) = \frac{TP}{(FN + TP)} \times 100 \qquad (13.15)$$

Table 13.2 displays the results of a comparison between several deep learning (DL) approaches to malware detection. The effectiveness of the projected malware detection approach is evaluated by comparing its performance analysis to that of malware detectors based on pretrained DL replicas like CNN and its variations.

In the analysis of accuracy, the LeNet has 90.82%, AlexNet has 94.63%, VGGNet has 93.12%, MobileNet has 95.39%, and proposed model has 96.87%. The reason for better performance is that the hyper-parameter tuning process of GAN model is carried out by using WSA. When all the techniques are tested with precision, the AlexNet and VGGNet has nearly 95% to 96%, LeNet has 92.43%, MobileNet has

Table 13.2 Comparative analysis of proposed model with existing techniques

Algorithm	Accuracy (%)	Precision (%)	Recall (%)	F-score (%)
LeNet	90.82	92.43	89.79	93.57
AlexNet	94.63	96.16	92.66	97.27
VGGNet	93.12	95.31	92.55	96.90
MobileNet	95.39	97.23	94.11	98.48
Proposed	96.87	99.65	95.39	99.52

97.23% and proposed model has 99.65%. The proposed model has 95.39% of recall and 99.52% of *F-M*, where the existing techniques such as AlexNet have 92.66% of RC and 97.27% of *F-M*, VGGNet has 92.55% of RC and 96.90% of *F-M* and MobileNet has 94.11% of RC and 98.48% of *F-M*. Figures 13.3, 13.4, 13.5 and 13.6 present the graphical analysis of proposed model with existing techniques.

The computation cost and memory cost analysis with accuracy for different techniques is provided in Table 13.3.

In the analysis of memory, the proposed model has 237.89 MB, the MobileNet has 461.10 MB, VGGNet has 370.88 MB, AlexNet has 270.87 MB, and LeNet has 289.11 MB. While testing the models, the computation cost for malware detection for each model are considered. For instance, the LeNet has 0.18G, AlexNet has 0.19G, MobileNet has 1.03G, VGGNet has 1.14G, and proposed model has only 0.17G.

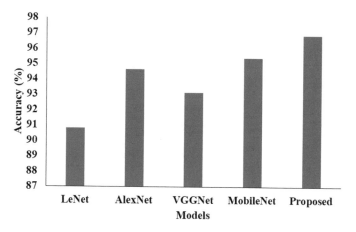

Fig. 13.3 Accuracy comparison

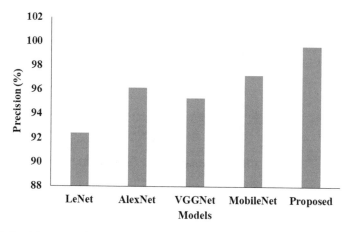

Fig. 13.4 Precision comparison

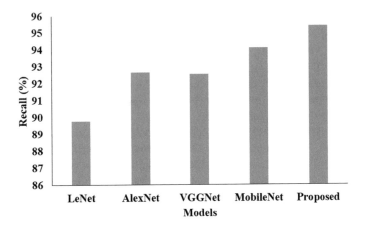

Fig. 13.5 Recall comparison

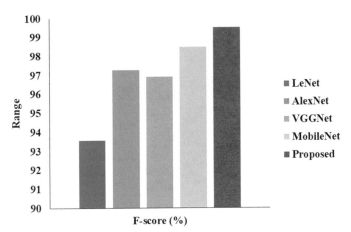

Fig. 13.6 *F*-score comparison

Table 13.3 Computational and memory cost for techniques

Model	Accuracy (%)	Size (MB)	MACs(G)
Proposed	**96.87**	237.89	0.17
MobileNet	95.57	461.10	1.03
VGGNet	92.96	370.88	1.14
AlexNet	94.27	270.87	0.19
LeNet	91.92	289.11	0.18

Bolded accuracy value is the outcome of this proposed work

Table 13.4 Training and testing time for apiece perfect

Model	Training time (s)	Testing time (Image/s)
LeNet	2004	38.7
AlexNet	2103	17.6
VGGNet	2706	15.9
MobileNet	1911	41.1
Proposed model	1795	10.3

Table 13.4 shows the comparative analysis of various models for training and testing the images.

When the training the whole malware images of dataset, the proposed model took 1795s, MobileNet took 1911s, LeNet took 2004s and AlexNet took 2103s. When compared with all techniques, the VGGNet has high training time, i.e. 2706s. The same model has only 15.9 s for testing the images, and this is due to only 30% of data are used for testing. The proposed model has 10.3 s for testing, AlexNet has 17.6 s and LeNet has 38.7 s, where the MobileNet has high testing time (41.1 s).

Conclusion

Through the use of a GAN and an optimization model, the research suggests a method for detecting and categorising malware that is both effective and practical. Differentially private synthetic data is created using a unique approach. We conducted a number of tests to show that our technology can generate high-quality synthetic data while still providing rigorous differential privacy assurances. To successfully implement PATE in a GAN context, we relied on the foundational work done by the original GAN framework. For optimum weight selection, WSA is used, and performance is measured using the Kaggle dataset for malware detection. The suggested model successfully recognised the vast majority of the malware samples, demonstrating its robustness in the face of countermeasures designed to prevent malware infections. The packaged executables are not run or unpacked in the recommended approach. The outcomes of the experiments show that the suggested method can easily and effectively categorise malware samples into their respective families. The suggested detection approach eliminates the need for human feature engineering while maintaining performance levels competitive with those of existing systems based on machine learning.

References

1. Tahir, R.: A study on malware and malware detection techniques. Int. J. Educ. Manage. Eng. **8**(2), 20 (2018)
2. Ashawa, M.A., Morris, S.: Analysis of android malware detection techniques: a systematic review (2019)
3. Alqahtani, E.J., Zagrouba, R., Almuhaideb, A.: A survey on android malware detection techniques using machine learning algorithms. In: Sixth International Conference on Software Defined Systems (SDS), pp. 110–117. IEEE (2019)
4. Kouliaridis, V., Barmpatsalou, K., Kambourakis, G., Chen, S.: A survey on mobile malware detection techniques. IEICE Trans. Inf. Syst. **103**(2), 204–211 (2020)
5. Kasthuri, S., Nisha Jebaseeli, A.: Social network analysis in data processing. Adalya J. (UGC CARE—B J. —Web Sci.) **IX**(2), 260–263. Impact Factor 5.3. ISSN: 1301–2746
6. Amro, B.: Malware detection techniques for mobile devices (2018). arXiv:1801.02837
7. El Merabet, H., Hajraoui, A.: A survey of malware detection techniques based on machine learning. Int. J. Adv. Comput. Sci. Appl. **10**(1) (2019)
8. Chennam, K.K., Muddana, L., Aluvalu, R.K.: Performance analysis of various encryption algorithms for usage in multistage encryption for securing data in cloud. In: 2nd IEEE International Conference on Recent Trends in Electronics, Information and Communication Technology (RTEICT), pp. 2030–2033. IEEE (2017)
9. McDole, A., Abdelsalam, M., Gupta, M., Mittal, S.: Analyzing CNN based behavioural malware detection techniques on cloud IaaS. In: International Conference on Cloud Computing, pp. 64–79. Springer, Cham (2020)
10. Kumar, R., Zhang, X., Wang, W., Khan, R.U., Kumar, J., Sharif, A.: A multimodal malware detection technique for Android IoT devices using various features. IEEE access **7**, 64411–64430 (2019)
11. Sreekumari, P.: Malware detection techniques based on deep learning. In: IEEE 6th International Conference on Big Data Security on Cloud (BigDataSecurity), IEEE International Conference on High Performance and Smart Computing (HPSC) and IEEE International Conference on Intelligent Data and Security (IDS), pp. 65–70. IEEE (2020)
12. Sallow, A.B., Sadeeq, M., Zebari, R.R., Abdulrazzaq, M.B., Mahmood, M.R., Shukur, H.M., Haji, L.M.: An investigation for mobile malware behavioral and detection techniques based on android platform. IOSR J. Comput. Eng. (IOSR-JCE) **22**(4), 14–20 (2020)
13. Savenko, O., Nicheporuk, A., Hurman, I., Lysenko, S.: Dynamic signature-based malware detection technique based on API call tracing. In ICTERI Workshops, pp. 633–643 (2019)
14. Hwang, J., Kim, J., Lee, S., Kim, K.: Two-stage ransomware detection using dynamic analysis and machine learning techniques. Wireless Pers. Commun. **112**(4), 2597–2609 (2020)
15. Sabhadiya, S., Barad, J., Gheewala, J.: Android malware detection using deep learning. In: 3rd International Conference on Trends in Electronics and Informatics (ICOEI), pp. 1254–1260. IEEE (2019)
16. Obaidat, I., Sridhar, M., Pham, K.M., Phung, P.H.: Jadeite: a novel image-behavior-based approach for java malware detection using deep learning. Comput. Secur. **113**, 102547 (2022)
17. Kim, J., Ban, Y., Ko, E., Cho, H., Yi, J.H.: MAPAS: a practical deep learning-based android malware detection system. Int. J. Inf. Secur. 1–14 (2022)
18. Lian, W., Nie, G., Kang, Y., Jia, B., Zhang, Y.: Cryptomining malware detection based on edge computing-oriented multi-modal features deep learning. China Commun. **19**(2), 174–185 (2022)
19. Kim, J.Y., Cho, S.B.: Obfuscated malware detection using deep generative model based on global/local features. Comput. Secur. **112**, 102501 (2022)
20. Qiang, W., Yang, L., Jin, H.: Efficient and robust malware detection based on control flow traces using deep neural networks. Comput. Secur. 102871 (2022)
21. Feng, P., Yang, L., Lu, D., Xi, N., Ma, J.: BejaGNN: behavior-based Java malware detection via graph neural network (2022)

22. García-Soto, E., Martín, A., Huertas-Tato, J., Camacho, D.: Android malware detection through a pre-trained model for code understanding. In: International Conference on Ubiquitous Computing and Ambient Intelligence, pp. 1055–1060. Springer, Cham (2023)
23. Velliangiri, S., Joseph, I.T., Pandiaraj, S., Jancy, P.L., Madhubabu, C.: An enhanced security framework for IoT environment using Jaya optimisation-based genetic algorithm. Int. J. Internet Technol. Secur. Trans. **13**(1), 11–25 (2023)
24. Balaji, R., Deepajothi, S., Prabaharan, G., Daniya, T., Karthikeyan, P., Velliangiri, S.: Survey on intrusions detection system using deep learning in IoT environment. In: International Conference on Sustainable Computing and Data Communication Systems (ICSCDS), pp. 195–199. IEEE (2022)
25. Narayanavadivoo Gopinathan, B.A., Sarveshwaran, V., Ravi, V., Chaganti, R.: LPCOCN: a layered paddy crop optimization-based capsule network approach for anomaly detection at IoT edge. Information **13**(12), 587 (2022). Accessed 21 Nov 2016

Chapter 14
ML Algorithms for Providing Financial Security in Banking Sectors with the Prediction of Loan Risks

T. R. Mahesh [D], **V. Vinoth Kumar** [D], **H. K. Shashikala** [D], and **S. Roopashree** [D]

Introduction

Financial markets are maturing as a result of rapid economic development, and the phenomena of individuals acting as financiers have become frequent. If credit risk management and control are done correctly, lending to individuals (personal loans) can be a profitable business. The loan risk management of individuals consists of three components: risk assessment (i.e., determining whether or not to issue the loan based on the applicant's ability to repay and credit payback), repayment tracking, and breach of contract, with risk assessment being the most important [1]. Currently, relevant information is based on the application, as well as the relevant rules that must be followed while deciding whether or not to provide loans. Banking organizations, from small to large, rely on the activity of lending out loans to generate profits and keep their operations running smoothly during times of financial constraint [2].

There are numerous risks associated with current bank loans. Understanding what risk means is very much necessary for risk analysis in case of bank loans. Risk is

---

T. R. Mahesh (✉) · H. K. Shashikala
Department of Computer Science and Engineering, Faculty of Engineering and Technology, JAIN (Deemed-to-be University), Bangalore, India
e-mail: t.mahesh@jainuniversity.ac.in

H. K. Shashikala
e-mail: hk.shashikala@jainuniversity.ac.in

V. Vinoth Kumar
School of Information Technology and Engineering, VIT University, Vellore, Tamil Nadu, India
e-mail: pvkumar243@gmail.com

S. Roopashree
Department of Computer Science and Engineering, RV Institute of Technology and Management, Bengaluru, India
e-mail: roopashaily@gmail.com

defined as the likelihood of specific outcomes or the uncertainty of those outcomes particularly a current negative danger to achieving a current monetary activity [3]. Prediction and description are the two most significant aims for data mining. Prediction entails using some of the data set's variables to forecast unknown values for other variables. The goal of description is to uncover patterns in data that can be interpreted by humans. The process of finding a hidden pattern from a vast amount of data and using it to make write decisions is known as data mining [4]. The derived knowledge should be novel, not obvious, as well as applicable in the field in which it was acquired. It's also the process of extracting valuable data from unstructured data.

Organization of the paper: The related work section contains an overall summary of different techniques of credit risk analysis employed previously. Section "Methodology" provides the dataset used for training and testing the model, as well as the recommended approach in this paper, which includes data analysis, dataset cleaning, and a walkthrough of the model's three classifiers. The model evaluation and performance comparison are presented in Sect. "Conclusion". Finally, the last section of the report finishes with closing remarks and suggestions for further research.

Related Work

In the financial and banking industries, data mining has been used to undertake a variety of studies. This section briefly discusses some of the approaches used in loan risk management and their identification.

The authors [5] compared the performance of regression analysis and neural networks in predicting which farmers will default on their home administration loans and which farmers will repay the loans according to the appointment. This study used unstable data to show that neural networks with improved logistic regression can categorize farmers into two groups: those who repay their loans on time and those who do not.

Several approaches have been presented to predict credit default. The method's application is determined by the complexity of banks and financial institutions, as well as the amount and kind of loan [6]. Discrimination analysis has been a popular method. Although this method employs a scoring function to aid decision-making, some researchers have questioned the validity of discriminate analysis due to its restricted assumptions of normalcy and independence among variables. To tackle the drawbacks of other inefficient credit default models, artificial neural network models were developed. Banks and financial institutions can get more realistic credit risk projections by employing AI-based and deep learning scoring models that use clients' credit histories and the power of big data. Credit can then be approved by the appropriate persons, and better price alternatives can be provided to those who deserve it [7, 8].

The authors [9] provided an effective instrument for reducing the number of unauthorized borrowers with a beneficial impact on the financial institution. Furthermore, the study provided insight into agricultural loan data to assist decision-makers and improve their ability to manage the lending operations of lending farms, reducing the time and expense of checking loan ambiguity and assisting loan officers in a critical role with consumers.

The authors [10] developed a ML model that aids in the assessment of credit risk that provides the identification of loan defaulters. They also conducted a comparison of other ML methods and concluded that the adjusted support vector machine as well as RFE with cross-validation combination produced the most accurate results of all the models. The authors' proposed five-step approach is also introduced in the study. Other tree-based or regression models can be outperformed by support vector machines. Furthermore, their work has demonstrated that removal with cross-validation really outperforms the remaining models in the dispute over which dimensionality reduction strategy to utilize.

The authors [11] showed a machine and deep learning-based approach to credit risk analysis. The authors worked on developing binary classifiers that may forecast loan default likelihood using these models. According to the authors, the selection of algorithms, relevant traits, parameters, and determine the final conclusion. The paper explains the algorithms employed, the norms being used to categorize the outputs, how to deal with imbalanced datasets, findings, associated parameters as well as criteria for each model [10]. The paper also explains, how ML and DL algorithms are being used to sanction loan for an individual of an approval based on whether or not they meet the majority of the criteria. To forecast the correct customers, the authors built a logistic regression model that employed a Kaggle dataset. This method saves time as well as decreases human error, but such models must be created carefully, or banks will suffer significant losses if incorrect predictions are made.

The authors [12] discussed a comparative examination of numerous algorithms, including their benefits and drawbacks. Support vector machines (SVMs) are a well-known approach for solving binary classification issues. The margin of the hyperplane for solving linear samples is known as the decision boundary. The authors of this [13] mentioned how the banking industry's main function is to lend money to those who are in need of it. The authors of this research examined various credit risk analysis approaches. In this study, the authors discussed with respect to credit risk management strategies. The authors also used various types of classifiers to assess and compare their accuracies. This paper paves the way for more credit risk management research in the financial area utilizing machine learning classifiers.

Methodology

Lending Club has worked with WebBank, a traditional bank, since 2008, to receive promissory notes, issue loans, and transfer notes to Lending Club. The bank pattern, as seen in Fig. 14.1, was essentially the same as the promissory note design. Lending Club, on the other hand, has not taken on the position of the lender of record, avoiding the interest rate ceiling. The upper interest rate limit varies by state in the United

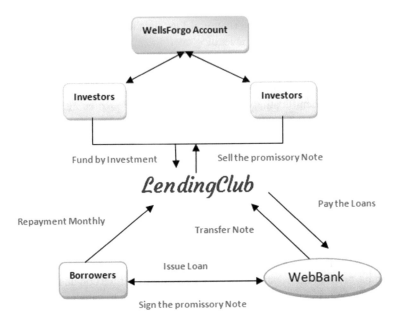

Fig. 14.1 Process of bank pattern

States, and the goal of this limit is to safeguard borrowers. Lending Club is thought to be able to lower its business costs by implementing this technique.

Lending Club was registered with the Securities and Exchange Commission (SEC) until October 2008. The company switched to a security-based business strategy. Borrowers and investors have a different connection under this new model since borrowers acquire the securities issued by Lending Club directly, while investors receive rewards based on the investment performance. To summarize, the relationship between borrowers and investors is no longer one of debtor and creditor. The proposed methodology is shown in Fig. 14.2.

Data Pre-processing

Many classification algorithms strive to collect only pure examples for learning and make the boundary between each class as clear as possible to improve prediction. Most classifiers find it significantly more difficult to learn how to classify synthetic cases close to the boundary than those far from it. Based on these results, the authors present an enhanced pre-processing method (A-SMOTE) for imbalanced training sets [14, 15]. SMOTE is used to produce the balanced data [16]. The k-nearest neighbor selected at random, and the random data would then be combined to create synthetic data. The SMOTE method is explained in the below section.

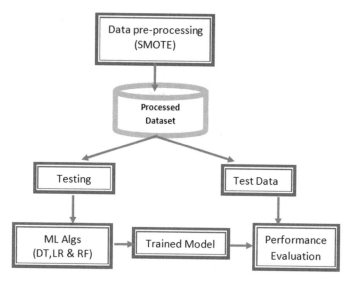

Fig. 14.2 Proposed process

Step A: Using Eq. (14.1) synthetic instance is created

$$N = 2 * (r - z) + z, \tag{14.1}$$

where r denoted class samples (which are majority), z denotes minority class samples and N is newly created synthetic instance.

Step B: The below mentioned steps are carried out to remove the outlier that is, noise.

If, $\hat{S} = \{\hat{S}_1, \hat{S}_2, \hat{S}_3, \ldots \hat{S}_n\}$ is a new instance received by Step A, then we will calculate the distance among $\hat{S}_i$ with original minority S_m, $\mathrm{Min_{Rap}}(\hat{S}_i, \hat{S}_m)$ defined using Eq. (14.2).

$$\mathrm{Min_{Rap}}(\hat{S}_i, \hat{S}_m) = \sum_{k=1}^{z} \sum_{j=1}^{M} \sqrt{(\hat{S}_i^{(j)} - S_{mk}^{(j)})^2}, \tag{14.2}$$

where:

$\mathrm{Min_{Rap}}(\hat{S}_i, \hat{S}_m)$ are samples rapprochement and as per Eq. (14.2), L is calculated using Eq. (14.3).

$$L = \sum_{i=1}^{n} (\mathrm{Min_{Rap}}(\hat{S}_i, S_m)), \tag{14.3}$$

Step C: Calculate distance between $\hat{S}_i$ and every original majority S_a, $\mathrm{Maj_{Rap}}(\hat{S}_i, S_a)$, described using Eq. (14.4).

$$\mathrm{Maj_{Rap}}(\hat{S}_i, S_a) = \sum_{i=1}^{r} \sum_{j=1}^{M} \sqrt{(\hat{S}_i^{(j)} - S_{al}^{(j)})^2}, \quad (14.4)$$

$\mathrm{Maj_{Rap}}(\hat{S}_i, S_a)$ are samples rapprochement and as per Eq. (14.4), H is computed using Eq. (14.5).

$$H = \sum_{i=1}^{n} (\mathrm{Maj_{Rap}}(\hat{S}_i, S_a)) \quad (14.5)$$

Entropy. Entropy is nothing but the measure of disorder. The following Eqs. (14.6) and (14.7) was used by Claude E. Shannon to put this link between probability and heterogeneity or impurity into mathematical form:

$$H(X) = -\sum (p_i * \log_2 p_i) \quad (14.6)$$

$$\mathrm{Entropy}(p) = -\sum_{i=1}^{N} p_i \log_2 p_i \quad (14.7)$$

The likelihood of a category's uncertainty or impurity is given as the log to base 2 (p_i). The number of potential categories is indicated by the index since our issue is a binary categorization, $i = 2$ in this instance. The graph 14.3 shows how his equation is represented by a symmetric curve. The probability of the occurrence is plotted on the x-axis, and the heterogeneity or impurity H is shown on the y-axis (X), and it is shown in Fig. 14.3.

Fig. 14.3 For a Bournoulli trial ($X = \{0,1\}$) the graph of entropy versus $\Pr(X = 1)$. The Highest $H(X) = 1 = \log(2)$

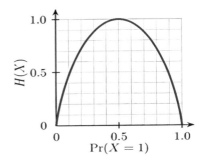

Table 14.1 The chosen attributes

Attribute/Feature	Description	Type
loan_amnt	The total amount of the loan that the borrower has applied for	Numeric
instalment	If the loan is originated, the monthly payment due by the borrower	Numeric
grade	LC assigned loan grade	Numeric
open_acc	In a borrower's credit file, the number of open credit lines	Numeric
total_pymnt	The borrower's total repayment	Numeric
total_rec_int	Borrower's interest rate on loans	Numeric
home_ownership_ANYor_MORTGAGE	The borrower's house ownership status at the time of registration	Nominal
verification_status_Not Verified	Indicates if LC confirmed the co-borrowers' combined income, whether it was not checked, or whether the income source was validated	Nominal
application_type_Individual	If the loan is an individual application or a combined application with two co-borrowers, this field indicates which	Nominal
purpose_major_purchase	The borrower's choice of category for the loan request	Nominal
purpose_renewable_energy		Nominal
purpose_small_business		Nominal
purpose_vacation		Nominal
term_36 months	The total number of loan payments. The months are measured in months and can range from 36 to 60. Binary discretization is used as a pretreatment	Nominal

Data Modeling

The dataset utilized in this article comes from Lending Club for the first quarter of 2019. It has almost 115,000 unique user loan data with 102 attributes. The missing values are then filled in via mode interpolation, and the duplicate or nonsensical characteristics are removed, leaving 15 attributes, which are listed in Table 14.1.

Logistic Regression

LR models were provided by the statistics department. This method has been tweaked for binary classification. The LR takes the value and reduces it to a number between 0 and 1. This strategy can be used to handle problems when we need to forecast for a variety of reasons. The LR function's definition is as depicted below:

$$h\theta(X) = \frac{1}{1 + e^{-(\beta_0 + \beta_1 X)}} \tag{14.8}$$

Decision Tree (DT)

By learning straightforward decision rules derived from previous data, a decision tree is used to build a training model that may be used to predict the class or value of the target variable (training data). In decision trees, we begin at the tree's root when anticipating a record's class label. We contrast the root attribute's values with that of the attribute on the record. We follow the branch that corresponds to that value and move on to the next node based on the comparison [17, 18]. Decision trees classify examples by arranging them in a downward spiral from the root to a leaf or terminal node, with the leaf or terminal node indicating the example's classification.

Every node in the tree serves as a test case for a certain attribute, and every edge descending from the node represents various potential solutions to the test case. Every subtree anchored at the new node goes through this recursive procedure once again [19].

Random Forest (RF)

The core idea behind ensemble methods based on randomization is to "incorporate random perturbations into the learning procedure to build multiple alternative models from a single learning set L and then to aggregate the predictions of those models to make the ensemble prediction" [20]. Otherwise said, "Growing an ensemble of trees and letting them choose the most popular class has led to notable gains in classification accuracy. These ensembles are frequently grown by creating random vectors that control how each tree in the ensemble grows". When creating a random tree, there are three major options available. These are: (1) the technique for dividing the leaves; (2) the kind of predictor to apply to each leaf; and (3) the technique for adding unpredictability to the trees. Using a bootstrapped or sub-sampled data set to generate each tree is a typical method for adding unpredictability to a tree. By training the forest's trees in this manner on slightly different sets of data, variations between the trees are introduced. The optimal split at a given node can alternatively be chosen randomly, though investigations show that bagging typically produces superior results when noise is a factor.

Gini Index is computed using (14.9).

$$G(X_i) = \sum_{j=1}^{J} \Pr(X_i = L_j)(1 - \Pr(X_i = L_j))$$

$$= 1 - \sum_{j=1}^{J} \Pr(X_i = L_j)^2 \tag{14.9}$$

Methodology

Various performance measures, including precision, recall, accuracy as well as $f1$-measure are used. Precision denotes what amount of the total positive anticipated is genuinely positive. The ratio of true positives (TPs) to the total number of TPs and true negatives is called as precision (TNs). Precision is a good statistic to utilize when the costs of False Positives (FNs) are high. Using the equation, the precision is calculated using Eq. (14.10).

$$\text{Precision} = \frac{\text{TPs}}{\text{TPs} + \text{FPs}} \tag{14.10}$$

The recall is computed as shown below. The real positive rate is the same as the recall. The metric recall counts how many valid positive predictions were made out of all the possible ones. The equation is used to calculate the recall is shown in (14.11).

$$\text{Recall} = \frac{\text{TPs}}{\text{TPs} + \text{FNs}} \tag{14.11}$$

We need a statistic that takes precision and recall into account. The harmonic mean of precision and recall is used to get the $F1$ score. When the dataset is balanced, Precision and Recall function quite well. The $F1$-score is a precision and recall measure that is computed using Eq. (14.12).

$$F1 \text{ Score} = 2 * \frac{\text{precision*recall}}{\text{precision} + \text{recall}} \tag{14.12}$$

The performance of various ML algorithms is shown in Table 14.2 and Fig. 14.4.

A ROC, i.e., receiver operating characteristics a two-dimensional graphical display in statistics that shows how well a binary classifier system performs. The performance of the classifier can be intuitively represented by the ROC curve. The FPR and TPR are computed as shown in Eqs. (14.13) and (14.14).

Table 14.2 Performance of various ML techniques

ML techniques	Recall	Precision	$F1$-score
LR	91.25	91.11	91.17
DT	90.53	89.16	89.83
RF	94.6	93.54	94.04

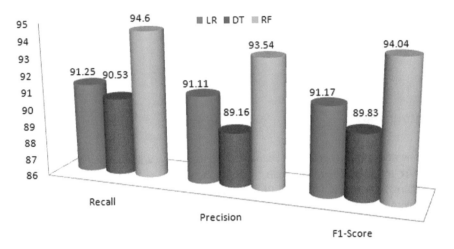

Fig. 14.4 Performance evaluation of ML algorithms

$$FPR = \frac{FP}{FP + TN} \qquad (14.13)$$

$$TPR = \frac{TP}{TP + FN} \qquad (14.14)$$

Assume that the ROC is built by connecting points with $\{(x_1, y_1), (x_2, y_2), (x_3, y_3), \ldots (x_m, y_m)\}$ coordinates in a sequential manner. The AUC is computed using Eq. (14.15).

$$AUC = \frac{1}{2} \sum_{i=1}^{m-1} (x_{i+1} - x_i).(y_i + y_{i+1}) \qquad (14.15)$$

AUC values range from 0.5 to 1.0, with a larger AUC value indicating better performance. The AUC values of different ML algorithms are shown in Table 14.3 and Fig. 14.5.

Accuracy is calculated using the formula (14.16).

$$Accuracy = \frac{TNs + TPs}{TNs + TPs + FPs + FNs} \qquad (14.16)$$

Table 14.3 Accuracy of various ML techniques

ML techniques	Accuracy (%)	AUC
LR	93.21	0.843
DT	94.14	0.935
RF	98.24	0.984

Fig. 14.5 AUC of ML algorithms

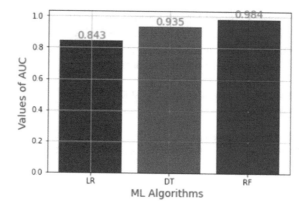

Fig. 14.6 Accuracy of different ML algorithms

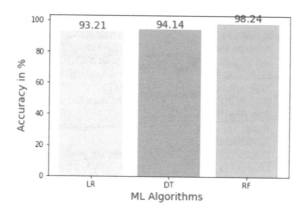

The performance of all the mentioned algorithms is also calculated. It was found that accuracy of LR, DT, and RF are 93.21, 94.14, and 98.24%. The RF method is the best with an accuracy of 98.24% compared to other algorithms. The accuracy performance of the all these ML algorithms is depicted in Fig. 14.6.

Conclusion

Machine learning can assist banks in anticipating the future of a loan and its condition, allowing them to act early in the loan's life cycle. Banks can use machine learning to reduce the number of problematic loans and avoid significant losses. Using the methods described in this work, a bank can quickly find the essential information from a huge amount of data sets. This results in a very good loan prediction system and also, it reduces the amount of bad loans. The random forest along with LR and DT approaches are used in this work to create a model for forecasting loan default in the dataset of lending club, and all the results are compared. The experiment reveals that the RF classifier outperforms the other two algorithms in predicting loan default with an accuracy of 98.24% and has high generalization ability. In a future study, an attempt can be made to conduct tests on datasets of larger size or adjust the model to obtain state-of-the-art performance.

Data Availability The dataset used for the findings will be shared by the corresponding author upon request.

Conflicts of Interest
The authors declare that there is no conflict of interest regarding the publication of this paper.

References

1. Madaan, M., Kumar, A., Keshri, C., Jain, R., Nagrath, P.: Loan default prediction using decision trees and random forest: a comparative study. In: IOP Conference Series: Materials Science and Engineering, vol. 1022, no. 1, p. 012042. IOP Publishing (2021)
2. Li, J., Liu, H., Yang, Z., Han, L.: A credit risk model with small sample data based on G-XGBoost. Appl. Artif. Intell. 1–17 (2021)
3. Strahan, P.E.: Borrower risk and the price and nonprice terms of bank loans. FRB New York Staff Rep (90) (1999)
4. Tomar, D., Agarwal, S.: A survey on data mining approaches for healthcare. Int. J. Bio-Sci. Bio-Technol. **5**(5), 241–266 (2013)
5. Hamid, A.J., Ahmed, T.M.: Developing prediction model of loan risk in banks using data mining. Mach. Learn. Appl.: Int. J. **3**(1), 1–9 (2016)
6. Chaitanya Reddy, P., Chandra, R.M.S., Vadiraj, P., Ayyappa Reddy, M., Mahesh, T.R., Sindhu Madhuri, G.: Detection of plant leaf-based diseases using machine learning approach. In: IEEE International Conference on Computation System and Information Technology for Sustainable Solutions (CSITSS), pp. 1–4 (2021). https://doi.org/10.1109/CSITSS54238.2021.9683020
7. Trustorff, J.H., Konrad, P.M., Leker, J.: Credit risk prediction using support vector machines. Rev. Quant. Financ. Acc. **36**(4), 565–581 (2011)
8. Roopashree, S., Anitha, J., Mahesh, T.R., Vinoth Kumar, V., Viriyasitavat, W., Kaur, A.: An IoT based authentication system for therapeutic herbs measured by local descriptors using machine learning approach. Measurement **200**, 111484 (2022). ISSN 0263-2241.https://doi.org/10.1016/j.measurement.2022.111484
9. Shoumo, S.Z.H., Dhruba, M.I.M., Hossain, S., Ghani, N.H., Arif, H., Islam, S.: Application of machine learning in credit risk assessment: a prelude to smart banking. In: TENCON 2019–2019 IEEE Region 10 Conference (TENCON), pp. 2023–2028. IEEE (2019)
10. Mahesh, T.R., Vinoth Kumar, V., Vivek, V., et al.: Early predictive model for breast cancer classification using blended ensemble learning. Int J SystAssurEngManag (2022). https://doi.org/10.1007/s13198-022-01696-0
11. Sheikh, M.A., Goel, A.K., Kumar, T.: An approach for prediction of loan approval using machine learning algorithm. In: International Conference on Electronics and Sustainable Communication Systems (ICESC), pp. 490–494. IEEE (2020)
12. Velliangiri, S., Manoharn, R., Ramachandran, S., Krishnasamy, V., Rajasekar, V.R., Karthikeyan, P., et al.: An efficient lightweight privacy preserving mechanism for industry 4.0 based on elliptic curve cryptography. IEEE Trans. Ind. Inf. (2021)
13. Hussein, A.S., Li, T., Yohannese, C.W., Bashir, K.: A-SMOTE: a new preprocessing approach for highly imbalanced datasets by improving SMOTE. Int. J. Comput. Intell. Syst. **12**(2), 1412 (2019)

14. Sarveshvar, M.R., Gogoi, A., Chaubey, A.K., Rohit, S., Mahesh, T.R.: Performance of different machine learning techniques for the prediction of heart diseases. In: International Conference on Forensics, Analytics, Big Data, Security (FABS), vol. 1, pp. 1–4. IEEE (2021)
15. Jha, K.K., Jha, A.K., Rathore, K., Mahesh, T.R.: Forecasting of heart diseases in early stages using machine learning approaches. In: International Conference on Forensics, Analytics, Big Data, Security (FABS), vol. 1, pp. 1–5. IEEE (2021)
16. Zhu, L., Qiu, D., Ergu, D., Ying, C., Liu, K.: A study on predicting loan default based on the random forest algorithm. Procedia Comput. Sci. **162**, 503–513 (2019)
17. Mahesh, T.R., Vivek, V., Kumar, V.V., Natarajan, R., Sathya, S., Kanimozhi, S.: A comparative performance analysis of machine learning approaches for the early prediction of diabetes disease. In: International Conference on Advances in Computing, Communication and Applied Informatics (ACCAI), pp. 1–6 (2022). https://doi.org/10.1109/ACCAI53970.2022.9752543
18. Djeundje, V.B., Crook, J.: Identifying hidden patterns in credit risk survival data using generalised additive models. Eur. J. Oper. Res. **277**(1), 366–376 (2019)
19. Teles, G., Rodrigues, J.J.P.C., Rabê, R.A., Kozlov, S.A.: Artificial neural network and Bayesian network models for credit risk prediction. J. Artif. Intell. Syst. **2**(1), 118–132 (2020)
20. Sangeetha, V.F., Daniel, J., Velliangiri, S.: Intelligent agent and optimization-based deep residual network to secure communication in UAV network. Int. J. Intell. Syst. (2022)

Chapter 15
Machine Learning-Based DDoS Attack Detection Using Support Vector Machine

V. Kathiresan, Vamsidhar Yendapalli, J. Bhuvana, and Esther Daniel

Introduction

In this information age plenty of data generated every second, storing, managing and retrieving the data is a big challengeable task in this era. About 463 Exabyte's of data is going to be generated per day in the year 2025 as per the prediction. In addition to this securing the data from the intrusion and information leakage is the highly essential and most challengeable task.

The cybercrime and data theft increased more than 600% during the pandemic [1, 2]. Ransomware attack increased to the large extend. The companies are investing a lot in cyber security and intrusion deduction. Malware attack, phishing attack, password attack, man in the middle attack and ransomware attack are quiet common attacks.

Especially the distributed denial of service attack is the one which makes the service inactive. It makes the particular server or resource inactive or inoperative. DDoS attack effects the organisation to the large extend [3, 4]. The cost of DDoS attack is maximum when compared to the other attacks. This book chapter focuses the DDoS attack since it is taking a significant role in the cyber-attacks.

Machine learning systems are the one which is learning from the data and doing classification or prediction according to the need [5, 6]. This book chapter focuses on

V. Kathiresan (✉) · V. Yendapalli
Department of Computer Science and Engineering, School of Technology (GST), GITAM University (Deemed to be University), Bengaluru, Karnataka, India
e-mail: xyzkathir@gmail.com

J. Bhuvana
Department of CS and IT, Jain (Deemed to be University), Bangalore, India
e-mail: J.bhuvana@jainuniversity.ac.in

E. Daniel
Karunya Institute of Technology and Sciences, Coimbatore, India

© The Author(s), under exclusive license to Springer Nature Singapore Pte Ltd. 2023   329
V. Sarveshwaran et al. (eds.), *Artificial Intelligence and Cyber Security in Industry 4.0*, Advanced Technologies and Societal Change,
https://doi.org/10.1007/978-981-99-2115-7_15

utilising the machine learning techniques in the cyber security. Intrusion detection is the one, by which the traffic coming to the internal network from outside the world can be classified as genuine or intrusion, based on the characteristics of the traffic includes destination port, flow duration, total forward packet, total backward packets, flow packets, etc.

Support vector machine is the machine learning technique. It is the machine learning-based classification model which classifies the given input based on the hyper plan. The data points that lie one side of the hyperplane belong to one class. The data points that lie other side of the hyperplane belong to another class [7, 8].

In this work machine learning's support vector machine is used to detect the distributed denial of service attack. The machine learning model is created and trained with the network traffic-based intrusion deduction data which is taken from Canadian Institute for Cyber Security. The dataset contains 79 attributes including destination port, flow duration, total forward packet, total backward packets, etc.

The model is getting trained with the labelled data which is having DDoS attack—yes as one label and DDoS attack—no as another label. Once the model got trained with the training data whenever the new traffic is coming it will detect whether it is an authorised traffic or intrusion. The dataset is divided into training set and testing set. Training set has been used for training purpose. The testing set is used to assess the quality of the model. The metrics precision, recall, $F1$-score and support are used in order to assess the quality of the SVM-based machine learning model.

Previous Work

Mihoub et al. in the year 2022 [9] proposed a method to detect the distributed denial of service attack using the machine learning classification model random forest. Random forest is a machine learning classification method which is commonly used for classification. Regression operation also can be performed by using random forest, but random forest performs well with classification. This proposed method is functioning based on looking back enabled method.

Liu et al. in the year 2020 proposed a method to detect the intrusion in the wireless sensor network [10]. Modified and improved KNN is used to detect the distributed denial of service in the wireless sensor networks. KNN is the well-known classification algorithm in machine learning.

Mahajan et al. in the year 2022 proposed a method to detect distributed denial of service attack using deep learning [11]. The 5G technology is enabling high data rate. In high data rate communication the chances for distributed denial of service attack also more. Here deep learning is utilised to solve the problem of distributed denial of service attack detection. Once the attack is detected then it is handled by mitigation policies. Model's performance is measured by the metrics.

Tonkal et al. in the year 2021 proposed a method to detect the distributed denial of service attack in software-defined network (SDN) [12]. The dataset which is used in this approach contains 23 features. The approach is using KNN, decision tree and artificial neural network approach too. The dataset contains the traffic details of Transfer Control Protocol (TCP), User Datagram Protocol (UDP) and Internet Control Message Protocol (ICMP). It contains both normal traffic and attack traffic. After using the algorithms KNN, decision tree and artificial neural network, it has been proved that decision tree is good in performance basis.

Kumar et al. proposed a method in the year 2011 to detect distributed denial of service attack [13]. Neural classifier is the machine learning-based classification method which is used for detection. KDD Cup, DARPA 1999, DHRPA 2000 or the datasets which are used for the training of the model. In this method the falls positive and the false negative are taken into the consideration. Method has been proved that it is having less false positive and false negative.

Zekri et al. [14] proposed a method using machine learning-based C4.5 algorithm to detect the distributed denial of service (DDoS) attack. C4.5 is the decision tree-based machine learning classification algorithm. The DDoS attack detection is done for cloud computing environment in order to secure the cloud environment from intrusion. Cloud resources can be utilised effectively when security is enabled properly.

He et al. proposed denial of service (DoS) (2017) [15] attack detection method in network using naïve Bayes method. Naïve Bayes method is a machine learning-based classification algorithm. Traditionally DoS attack detection happened based on threshold value. In order to improve the efficiency in detection machine learning-based naïve Bayes classification algorithm proposed. It enhances the security in cloud-based environment.

De Miranda et al. [16] proposed fuzzy logic and machine learning-based algorithm for distributed denial of service (DDoS) attack. In this proposed method reduction of quality (RoQ) attack is targeted. The K-nearest neighbour (KNN) is the classification algorithm based on machine learning which supports DDoS attack detection in the proposed method. Algorithm's performance is proved based on $F1$-score metrics.

Aamir and Zaidi proposed a clustering-based semi-supervised ML method for DDoS attack detection (2021) [17]. This proposed method uses clustering techniques rather than classification method. Since it is a clustering method unlabelled data and partially labelled data can be used. Agglomerative clustering is the clustering method which is used to do the clustering with respect to DDoS detection. Accuracy is measured to assess the quality of the model.

Aysa et al. [18] proposed a method to detect DDoS attack detection for IoT or wireless sensor network (WSN). The method is based on machine learning. Machine learning-based decision tree approach is used to detect the DDoS attack. Decision tree is a classification method which provides solution for DDoS detection. Model's performance is measured by the accuracy metrics of the classification model. It prevents the abnormal traffic in the wireless sensor networks.

Distributed Denial of Service Attack

In the year 2021 the distributed denial of service attack grown 31%. DDoS attack affects the organisation to the large extend. The cost or loss involved in the DDoS attack is high when compared to the other attacks. The service or server which undergone the DDoS attack will become inoperative. It will slowly move operative state to inoperative state. The attack affects all the resources including software and hardware resources. Edge network devices are the target for the DDoS attackers. Monitoring the network traffic and detecting the attack is the best way of detecting the intrusion. Even after that attack DDoS attack makes the system more vulnerable.

Application Layer Attack

The main objective of application layer attack is making the application inoperative. After identifying the vulnerabilities in the application. The attack happened against the application and make application inoperative. It makes the application unable to provide service to its uses. It is done by sending millions of requests with exception. Keep on sending the handshake message even after dialogue over. This kind of attempts makes the server irresponsive to the original user query.

Protocol Attack

Protocol-based distributed denial of service attack is different from application layer attack. Protocol attack is hard to find. There are lot of complications involved in identifying the protocol-based DDoS attack [19, 20]. The vulnerability in the network protocol is identified and utilised in this attack. It is hard to identify based on the complexity in the protocol. The close monitoring of the network traffic and analysis of streams in depth can increase the probability of identifying the protocol attack. Border Gateway Protocol is the one which undergone the protocol attack. Protocol attack is not the one which is frequently happened, but the impact of the attack is high.

Syn Flood

Syn flood attack takes significant role in the DDoS attack. The attacker sends repeated Syn message or packet to the server and makes the server irresponsive. Making the server busy in replying is the objective of the attack. Once the server is receiving frequent more number of Syn messages or Syn packets in order to process all the requests the entire server resources became busy. Making server resources busy and making it irresponsive from the user is the objective of the attack [21].

In the normal three-way handshake of the TCP. The client is sending synchronisation message to the server. The server is replying back with the Syn and acknowledgement message, then again client is sending acknowledgement message. After the three-way handshake the packet will start getting transfer. During Syn flood attack the attacker will send the spoofed Syn packet continuously to the server. The server will became busy by replying the acknowledgement for the packets received.

Volumetric Attacks

Volumetric attack targets the internal network. The attacker targets the internal network and creates the malicious traffic inside the internal network. Due to the artificially created traffic the service which is being delivered to the user or client will get interrupted. The main objective of volumetric attack is consuming the bandwidth of the internal network and makes the server or distributed system inoperative. The volumetric attack finds the vulnerability in DNS in order to occupy the traffic by sending the malicious code.

SVM

Support vector machine is the machine learning-based classification model. SVM mainly focuses binary classification. It supports multiclass classification also. SVM classifies the classes based on support vectors and hyperplane. Hyperplane is the one which classifies the given data into multiclasses. It is a supervised learning algorithm. SVM supports both classification and regression. It is more famous for classification [22].

Hyperplane divides the data environment into two classes. The objective of hyperplane is to have the maximum margin. Positive side of the hyperplane contains 1 class, and negative side of the hyperplane contains another class. Support vector machine is getting trained based on hinge loss function.

$$\text{hingeloss} = \arg\min \sum_{i=0}^{n-1} \max(0, 1 - y_i(w^T x_i + w_0))$$

Hyperplane

Hyperplane divides the data object into classes. Hyperplane is having margin. SVM's objective is to increase the size of margin or having the maximum margin size. When the data environment is one-dimensional dot is the hyperplane which divides the data into two. When the distribution is two-dimensional a line can divide the entire distribution into two classes. In three-dimensional environment hyperplane is a plane. In multi-dimensional, it is called as hyperplane. The data points which are closed to the hyperplane is called support vectors. Support vectors decides the size of the margin. The objective of SVM is having bigger margin.

Support Vectors

Support vectors are the data points which are very close to the hyperplane. Margin of the hyperplane is decided by the support vectors in the environment. Every data point in the environment is represented by vector. If the data environment is three-dimensional environment then each vector contains three elements. The values in the vector are directly proportional to the dimension of the data environment.

Linear SVM

Linear SVM is used to do the classification on linearly separable data. The data is categorised into two types, the first one is linearly separable, and the second one is linearly inseparable. Based on the nature of the data distribution it is classified into linearly separable and linearly inseparable. Linear SVM is applied on the data which is linearly separable. In this the hyperplane divides the data linearly into two classes with respect to binary classification. If dimension is two then hyperplane is the line. If the number of dimension is three the hyperplane is plane. In the case of multi-dimensional environment the hyperplane is segregating the classes [23].

Nonlinear SVM

All the time the data is not convenient to separate it linearly. If the data is not ready to linearly separable, it should be converted into linearly separable data. In order to apply the SVM the linearly inseparable data should be converted into linearly separable data [24].

Increasing the dimension is the one way to make linearly inseparable data into linearly separable data. Kernel trick in SVM is used to make linearly inseparable data into linearly separable data. x is considered as the original independent variable. $\emptyset(x)$ is the independent variable after applying the kernel tricks [25].

Deduction of DDoS Through SVM: (SVMBD)

In the proposed method SVMBD is detecting the DDoS attack based on the SVM algorithm. The dataset which is used for the training and model creation has been downloaded from Canadian Institute for Cyber Security. The Canadian Institute for Cyber Security (CIC) is providing the data in order to enhance the research in cyber security. The data, which is used, contains more than 10 k tuples including both the classes. The dataset contains two classes that are normal traffic and intrusion traffic. The dataset contains 78 attributes; it is a high-dimensional dataset. The data undergone the pre-processing techniques removes the missing values and changes the values into finite values.

The attributes of the dataset include destination port, flow duration, total forward packets, total length of forward packets, total length of the backward packet, etc. The model is getting trained with the dataset and tested with the unknown data. The quality and the performance of the model is tested based on the well-known metrics accuracy, precision, recall and $F1$-score. Figure 15.1 represents the flow diagram of the proposed SVMBD.

SVM-Based Model Creation

After pre-processing the dataset with 10,741 tuples and 78 attributes, it is splitted into X and Y. X is stated as independent variable, and Y is stated as dependent variable. Here the number of attributes in the independent variable is 77. Y is stated as dependent variable. Here the attribute 'Label' is considered as dependent variable. Based on the list of X, the Y is going to be identified.

Training the Model

Once the entire dataset is divided into independent variable and dependent variable and Y. Both X and Y are splitted into training data and testing data. The entire X is splitted into X-train and X-test. The same way entire Y is splitted into Y-train, Y-test. The purpose of dividing the entire data into training set and testing set is to evaluate the model.

The model should be evaluated based on the unforeseen data. The 15% of testing data taken into the consideration for evaluating the model. This 15% of data is unknown to the model because the data is not present in the training of the model.

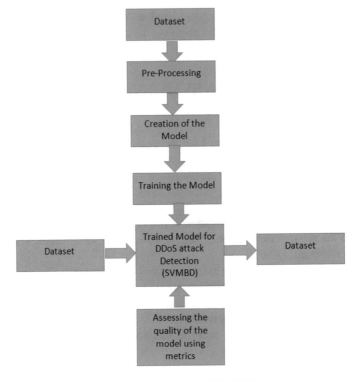

Fig. 15.1 SVM-based DDoS attack detection scheme (SVMBD)

The model is getting trained with X-train and Y-train datasets. X-train and Y-train datasets is 85% of the original dataset.

Testing the Model

About 85% of the original data is taken into the consideration for training purpose. Remaining 15% of the data is allocated for testing. It is used to assess the quality of the model. Testing data is used to validate the model [26].

If the model is tested by the training data itself then it is not an effective way of testing. In order to enable the quality testing unforeseen data should be used. The model's performance is measured in an effective way by validating the model using unforeseen test data.

Table 15.1 Performance evaluation before scaling

Accuracy	Precision	Recall	$F1$-score
0.85	0.86	0.86	0.86

Assessing the Quality of the Model Using Metrics

The model's performance is measured by comparing the Y-test data and Y-pred data. The metrics accuracy, precision, recall and $f1$-score are used to measure the quality of the model.

Accuracy is the main metric to access the quality of the classification model.

$$Accuracy = \frac{True\ positive + True\ Negative}{Total\ number\ of\ Predictions}$$

Accuracy of our proposed SVMBD algorithm is 85% (before scaling). It is stated in Table 15.1. Precision is another one metric which is used to access the quality of the model.

$$Precision = \frac{TP}{TP + FP}$$

TP   true positive
FP   false positive.

Precision of proposed SVMBD algorithm is 86% (before scaling).
Recall is another metric which is conveying the proportion of correctly classified positives.

$$Recall = \frac{TP}{TP + FN}$$

Recall of our proposed SVMBD algorithm is 86% (before scaling).
$F1$-score is also quite famous metrics which is assessing the quality of the machine learning model.

$$F1\text{- score} = 2*\left(\frac{Precision*Recall}{Precision + Recall}\right)$$

$F1$-score of the proposed model SVMBD this 86% (before scaling).
Scaling is the mechanism which is related to pre-processing, which enhances the quality of data. Scaling makes the data more convenient to training of the machine learning model. Scaling operation performed on the data which is used to detect DDoS. After the scaling, performance of the model is assessed again using the metrics accuracy, precision, recall and $F1$-score. Improved performance achieved with 97% accuracy. It is stated in Table 15.2. Table 15.3 and Fig. 15.5 stated the comparison

Table 15.2 Performance evaluation after scaling

Accuracy	Precision	Recall	F1-score
0.97	0.95	0.98	0.96

Table 15.3 Performance before scaling versus after scaling

	Before scaling	After scaling
Accuracy	0.85	0.97
Precision	0.86	0.95
Recall	0.86	0.98
F1-score	0.86	0.96

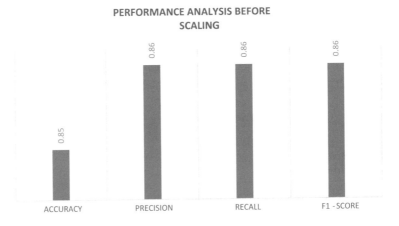

Fig. 15.2 Performance analysis before scaling

before scaling and after scaling. Figure 15.4 represents the detailed performance analysis after scaling. Figure 15.2 analysis chart represents the performance analysis associated with various metrics accuracy, precision, recall and score before scaling. Figure 15.3 analysis chart represents the performance analysis associated with various metrics accuracy, precision, recall and score after scaling.

Conclusion

Establishing and enhancing the cyber security in all the field is highly essential in today's digital era. This book chapter focused the distributed denial of service attack detection. The impact of distributed denial of service attack is more, the attack is creating more data loss. It creates more reputation problem for the organisation since the service given by the organisation stops due to the attack. The machine learning technique has been utilised here in order to detect the DDoS attack.

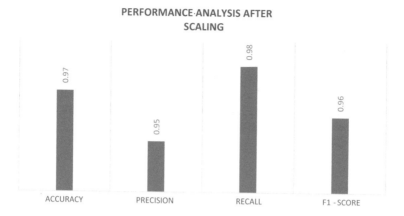

Fig. 15.3 Performance analysis after scaling

```
After Scaling:
                precision    recall  f1-score    support

            0      1.00       0.96      0.98        1229
            1      0.90       0.99      0.94         383

     accuracy                          0.97        1612
    macro avg      0.95       0.98      0.96        1612
 weighted avg      0.97       0.97      0.97        1612
```

Fig. 15.4 Detailed performance analysis after scaling

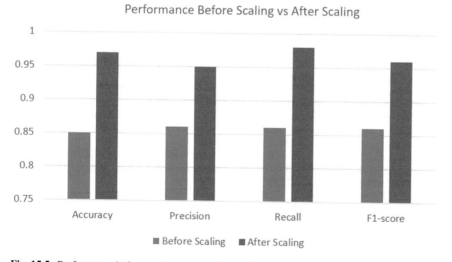

Fig. 15.5 Performance before scaling versus after scaling

Support vector machine-based machine learning model has been created and trained with the dataset downloaded from Canadian Institute for Cyber Security. The machine got trained with more the 10 K tuples which are having both the labels intrusion and genuine traffic. Once the model got trained the performance of the model is measured using the well-known classification metrics. Performance is measured before scaling the data as well as after the scaling. The model's performance has been proved with 97% of accuracy. The model is now ready to receive the traffic and classify whether it is an intuition or normal traffic.

References

1. Buil-Gil, D., Miró-Llinares, F., Moneva, A., Kemp, S., Díaz-Castaño, N.: Cybercrime and shifts in opportunities during COVID-19: a preliminary analysis in the UK. Eur. Soc. **23**(sup1), S47–S59 (2021)
2. Monteith, S., Bauer, M., Alda, M., Geddes, J., Whybrow, P.C., Glenn, T.: Increasing cybercrime since the pandemic: concerns for psychiatry. Curr. Psychiatry Rep. **23**, 1–9 (2021)
3. Deshmukh, R.V., Devadkar, K.K.: Understanding DDoS attack and its effect in cloud environment. Procedia Comput. Sci. **49**, 202–210 (2015)
4. Sadre, R., Sperotto, A., Pras, A.: The effects of DDoS attacks on flow monitoring applications. In: IEEE Network Operations and Management Symposium, pp. 269–277. IEEE (2012)
5. Khanzode, K.C.A., Sarode, R.D.: Advantages and disadvantages of artificial intelligence and machine learning: a literature review. Int. J. Libr. Inf. Sci. (IJLIS) **9**(1), 3 (2020)
6. Attaran, M., Deb, P.: Machine learning: the new 'big thing' for competitive advantage. Int. J. Knowl. Eng. Data Min. **5**(4), 277–305 (2018)
7. Yuan, R., Li, Z., Guan, X., Xu, L.: An SVM-based machine learning method for accurate internet traffic classification. Inf. Syst. Front. **12**, 149–156 (2010)
8. Shetty, S., Rao, Y.S.: SVM based machine learning approach to identify Parkinson's disease using gait analysis. In: International Conference on Inventive Computation Technologies (ICICT), vol. 2, pp. 1–5. IEEE (2016)
9. Mihoub, A., Fredj, O.B., Cheikhrouhou, O., Derhab, A., Krichen, M.: Denial of service attack detection and mitigation for internet of things using looking-back-enabled machine learning techniques. Comput. Electr. Eng. **98**, 107716 (2022)
10. Liu, G., Zhao, H., Fan, F., Liu, G., Xu, Q., Nazir, S.: An enhanced intrusion detection model based on improved kNN in WSNs. Sensors **22**(4), 1407 (2022)
11. Mahajan, N., Chauhan, A., Kumar, H., Kaushal, S., Sangaiah, A.K.: A deep learning approach to detection and mitigation of distributed denial of service attacks in high availability intelligent transport systems. Mobile Netw. Appl. 1–21 (2022)
12. Tonkal, Ö., Polat, H., Başaran, E., Cömert, Z., Kocaoğlu, R.: Machine learning approach equipped with neighbourhood component analysis for DDoS attack detection in software-defined networking. Electronics **10**(11), 1227 (2021)
13. Kumar, P.A.R., Selvakumar, S.: Distributed denial of service attack detection using an ensemble of neural classifier. Comput. Commun. **34**(11), 1328–1341 (2011)
14. Zekri, M., El Kafhali, S., Aboutabit, N., Saadi, Y.: DDoS attack detection using machine learning techniques in cloud computing environments. In: 3rd International Conference of Cloud Computing Technologies and Applications (CloudTech), pp. 1–7. IEEE (2017)
15. He, Z., Zhang, T., Lee, R.B.: Machine learning based DDoS attack detection from source side in cloud. In: IEEE 4th International Conference on Cyber Security and Cloud Computing (CSCloud), pp. 114–120. IEEE (2017)

16. de Miranda Rios, V., Inácio, P.R., Magoni, D., Freire, M.M.: Detection of reduction-of-quality DDoS attacks using Fuzzy Logic and machine learning algorithms. Comput. Netw. **186**, 107792 (2021)
17. Aamir, M., Zaidi, S.M.A.: Clustering based semi-supervised machine learning for DDoS attack classification. J. King Saud Univ.-Comput. Inf. Sci. **33**(4), 436–446 (2021)
18. Aysa, M.H., Ibrahim, A.A., Mohammed, A.H.: IoT DDoS attack detection using machine learning. In: 4th International Symposium on Multidisciplinary Studies and Innovative Technologies (ISMSIT), pp. 1–7. IEEE (2020)
19. Yuan, J., Mills, K.: Monitoring the macroscopic effect of DDoS flooding attacks. IEEE Trans. Dependable Secure Comput. **2**(4), 324–335 (2005)
20. Srivastava, A., Gupta, B.B., Tyagi, A., Sharma, A., Mishra, A.: A recent survey on DDoS attacks and defense mechanisms. In: Advances in Parallel Distributed Computing: First International Conference on Parallel, Distributed Computing Technologies and Applications, PDCTA 2011, Tirunelveli, India, September 23–25, 2011. Proceedings, pp. 570–580. Springer Berlin Heidelberg (2011)
21. Bogdanoski, M., Suminoski, T., Risteski, A.: Analysis of the SYN flood DoS attack. Int. J. Comput. Netw. Inf. Secur. (IJCNIS) **5**(8), 1–11 (2013)
22. Noble, W.S.: What is a support vector machine? Nat. Biotechnol. **24**(12), 1565–1567 (2006)
23. Joachims, T.: Training linear SVMs in linear time. In: Proceedings of the 12th ACM SIGKDD International Conference on Knowledge Discovery and Data Mining, pp. 217–226 (2006)
24. Suykens, J.A.: Nonlinear modelling and support vector machines. In IMTC 2001 Proceedings of the 18th IEEE Instrumentation and Measurement Technology Conference. Rediscovering Measurement in the Age of Informatics (Cat. No. 01CH 37188), vol. 1, pp. 287–294. IEEE (2001)
25. Hofmann, M.: Support vector machines-kernels and the kernel trick. Notes **26**(3), 1–16 (2006)
26. Erickson, B.J., Kitamura, F.: Magician's corner: 9. Performance metrics for machine learning models. Radiol.: Artif. Intell. **3**(3) (2021)

Chapter 16
Artificial Intelligence-Based Cyber Security Applications

Sri Rupin Potula, Ramani Selvanambi, Marimuthu Karuppiah⊙, and Danilo Pelusi

Introduction

Past two decades formed the roots for innovation, research, and development in technology. Population all around the globe got connected by smart devices. People at various locations are given privilege to work this indeed corresponds to the installation of various software, upgrading our machines to latest hardware. This all deals with transfer of information from one to another via network. Hackers take advantage of the various system conditions such as hardware, software, and network traffic in order to gather sensitive information and cause damage to the system software. As a result of this, protection is to be given for system which prevents from identity theft of an individual or at any corporate sector such as companies [1]. Precious time has arrived to protect the computer applications from various network intrusions and is given utmost priority this is where computer security or cyber security accounts.

In order to secure systems from threats and vulnerabilities and to effectively deliver the right services to users, numerous measures, methods, and techniques are

S. R. Potula · R. Selvanambi
School of Computer Science and Engineering, Vellore Institute of Technology, Vellore, Tamil Nadu 632014, India
e-mail: srirupin.23@gmail.com

R. Selvanambi
e-mail: ramani.s@vit.ac.in

M. Karuppiah (✉)
School of Computer Science and Engineering and Information Science, Presidency University, Bengaluru, Karnataka 560064, India
e-mail: marimuthume@gmail.com

D. Pelusi
Faculty of Communication Sciences, University of Teramo, Teramo, Italy
e-mail: dpelusi@unite.it

© The Author(s), under exclusive license to Springer Nature Singapore Pte Ltd. 2023
V. Sarveshwaran et al. (eds.), *Artificial Intelligence and Cyber Security in Industry 4.0*, Advanced Technologies and Societal Change,
https://doi.org/10.1007/978-981-99-2115-7_16

used to preserve information systems. As a result, both external and internal system threats are to be addressed. The main goal of cyber security is to ensure safety in all the devices by protecting them from various malicious activities and trying to analyze the incident that took place and solve the issue with proper remedy. These risks will have a significant influence on the routine operation of the systems.

With growth in technology, traditional ways of manual testing or indications of threats are critical as cyber criminals always find out the ways to exploit the resources and always find out new ways to gather the information they need. So, we must possess latest technology which is smarter than human intelligence to detect the known attacks as well as to identify the unknown attacks. One such technology is Artificial Intelligence.

Artificial Intelligence provides the experts the ways to identify, classify, and analyze threats caused by various attacks. This helps them to form patterns and dependencies of various attributes for any given attack because of which early-stage remedies or identification of attacks are done which indeed helps to see various tools used by cyber criminals. The importance of Artificial Intelligence in growing at a faster rate in cyber security applications. The main use of AI and its components benefits the organization to enhance Network Security, Detecting Advanced Malware, and Increasing Data Privacy [2].

Scope of Study

This chapter "Artificial Intelligence-based cyber security applications" focuses on latest advances in Artificial Intelligence that ensure cyber security. The abundance availability of data created a space for growth of machine learning and Neural Networks to find patterns and come up with remedial solutions. It is almost important to focus on various techniques, methods and applciations of Artificial Intelligence. In general, concentrating on various methods such as regression, neural networks, classification techniques in machine learning for detecting false systems is also important. Artificial Intelligence cyber security strategies by having glances of software and tools used by Enterprises for cyber security and also analyzing the performance metrics of models are used thereby suggesting solutions to enhance performance attributes.

Understanding Artificial Intelligence

In general, AI systems replicates human intelligence, as these machines are trained to behave and act like humans. The term AI can also correlate with the devices that really have abilities of human mind to solve real-world problems [3]. AI has advanced quickly in a variety of fields, including medical, communication technology, online applications, security, knowledge acquisition, and reasoning, thanks to its capacity

Fig. 16.1 Components of artificial intelligence

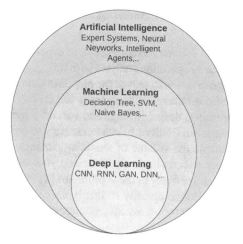

to identify past mistakes and prevent making them happen again. Because of its wide diversity, Artificial Intelligence can also be divided as below and can be understood from Fig. 16.1.

Basic Components of AI

- Machine Learning: Machine learning is the process of gathering data from various resources, applying algorithms on the data and applying statistical techniques to predict the output, and even updating the output so that the performance can be improved.
- Deep Learning: Deep Learning using computer systems that analyze the data which is similar to how humans think and perform. This Artificial Intelligence technology is achieved via neural networks. Deep learning being a part of machine learning tries to mimic the human brain by using interconnected neurons or nodes in a layered framework. This might be used by computers to create an adaptive system that enables them to continuously learn from errors and improve. Deep Learning algorithms strive to more precisely handle challenging problems.
- Neural Networks: In Deep learning, neural networks is a model with more neural layers and many hidden layers. Neural networks are based upon the neurons that humans have and its action when encountered by an issue. They try to behave similar to human brain function and even handles large volumes of data. So, typically a neural network has many interconnected neurons in order to act and think like a human brain.

AI for Cyber Security

As the volume and complexity of cyber-attacks rise, Artificial Intelligence is enabling security operation analysts with limited funding to anticipate threats. Artificial Intelligence helps to reduce the response time dramatically and offer rapid insights about various attacks on daily basis and its cutting-edge technologies including machine learning, deep learning, and natural language processing collects threat information from millions of academic papers, online resources, and news. Growing issues include the volume of attacking channels and the volume of the data available. In order to fight back, an AI-based cyber security management system should be able to resolve many of these problems [4]. A self-learning system can be properly developed technologically so that it can collect data continually and autonomously from all of your company information systems.

Traditional System and AI Systems in Cyber Security

Firewalls, antivirus software, and other classic cyber security solutions are currently ineffective when compared with the wide variety of technology available today to perform a cyber-attack. The fundamental problem with traditional systems is that they are frequently managed by a limited group of qualified security specialists, where enormous amount of data cannot be managed effectively. A significant role for intelligent cyber security services and management, however, can be played by Artificial Intelligence depending because of its ability to handle problems. In order to build cyber security computers more automated and intelligent than the local conventional security systems, we concentrate on AI-driven cyber security [5–7].

AI's Prevalence and Use in Cyber Security

In cyber security, uses of AI are growing steadily. Security professionals struggle with cyber-attack mitigation and defense for a number of reasons. There are too many tasks for them to handle. The volume of analysis and evaluation jobs they must complete is unfathomably large. Never enough time is available (time deficit). The amount of time needed to respond to cyber-attacks must be kept to a minimum. Too much data must be handled (data overload). Having tons of data at your disposal is excellent, but it still needs to be organized and analyzed. Without access to AI cyber security technologies, effective time management could not be achieved. There is a persistent lack of skilled workers in the sector. The required talents are not always easily accessible, as we have observed inside the sector for a while now. As a result, AI is used in all security products. Some businesses think AI can automate the reaction to security breaches since it enables quicker detection, analysis, and response, speeding up the process while also improving accuracy [8].

History of AI in Cyber Security Applications

Classic security systems generally make use of signatures or attack indicators to identify previously known assaults, yet 90% of the attacks are undiscovered and they use different signatures, as a result vulnerabilities, threats, and network breaches are multiplying day by day. Additionally, the majority of traditional models consistently attempt to comprehend the network environment using established policies [9]. The Artificial Intelligence techniques serve to increase threat detection, threat response time, mobile endpoint analysis, security checks, security and crime prevention, and vulnerability assessment.

Gathering of Cyber Security Data for Training Models

Any AI solution's fundamental building block begins with the dataset that will be used for the investigation. Numerous datasets are available across many academic disciplines. To recognize various cyber-attacks across various disciplines, a variety of datasets are readily available. 7 classes have been created from the public datasets based on their content [10, 11].

- DARPA Dataset: The DARPA dataset was created in 1998 at the MT Lincoln Laboratory as a network-based dataset. The training set consists of seven weeks' worth of network-based assaults. Two-week network-based attacks are also included in the test data.
- KDD 99 Dataset: The DARPA 98 dataset initiative served as the foundation for the KDD Cup 99 dataset. Attacks such as "DoS," "Remote to Local (R2L)," "User to Root (U2R)," and "probing" are detected. This dataset has around 5 million lines and is made up of 7 weeks' worth of network activity. This being one of the most often used datasets for evaluating intrusion detection methods.
- NSL-KDD Dataset: The NSL-KDD dataset can be treated as a latest dataset that cleared all the issues addressed with KDD 99. Unlike KDD 99 dataset, this dataset does not have any repetitive records. There are sufficient amount of records in it. Since duplicate records are eliminated, NSL-KDD dataset consists of 150,000 records where KDD 99 to have 5 million entries in it. For detection of intrusion detection techniques, the dataset can be better divided to training and testing datasets. The same classes and attributes are used by NSL-KDD and KDD CUP 99. This dataset works for DoS, R2L, U2R, and probing attacks [10].
- ISOT Dataset: The previous harmful and non-malicious datasets are combined to create the ISOT Botnet dataset. The ISOT Ransomware Detection dataset consists

with over 420 Gigabytes of malware and benign program execution traces. Two traffic captures included this dataset which are "malicious DNS data for nine different botnets and benign DNS for 19 different well-known software programs" [12].

- Malicious URLs Dataset: As the name itself translates, this dataset is made up of 2.4 million URLs and 3.2 million characteristics. There are two types of datasets available which are MATLABS and SVM-light. The file url.mat in Matlab format provides the features which are referred to as Feature-Types. The Feature-Types is a text file that includes the features which are real values in SVM-light format.
- ISOT Fake News Dataset: This dataset is made up of thousands of true and inaccurate stories that were found on various credible news websites and websites that Politifact.com has identified as unreliable. The dataset includes both fake news and legitimate content. Real-world sources were used to compile this dataset, and reliable articles were found through crawling Reuters.com.
- Dynamic Malware Analysis Kernel and User-Level Calls: This dataset is made by gathering of information from Cuckoo and a kernel driver. After executing about thousand malicious and a thousand accurate samples, the results are combined to make this dataset. API-calls from both accurate and inaccurate data are saved in the subfolders of the Kernel Driver folder.

Artificial Intelligence for Cyber Security

Latest innovations in AI made cyber security dependent on them, because of their robust nature to handle large volumes of data and identification of various attacks which are ranges from malware to phishing. Because of their nature to learn and gather information from the previous occurred attacks, they tend to avoid happening in the mere future. Apart from preventing suspicious attacks to happen, alerts regarding new attacks are given.

Expert Systems for Cyber Security

An expert system uses Artificial Intelligence to try and influence the behavior and decision-making nature of a person or group of persons (experts). Expert Systems make use of the relevant data from its Knowledge Base, and it further analyzes the gathered data in the most relevant form to answer users query. The knowledge base of an expert system is expanded by experts in a particular field. It is used in a variety of industries, including accountancy, programming, gaming, and medical diagnostics. An expert system is foundationally an AI software that makes use of knowledge from knowledge base in order to answer user queries. Since the knowledge base is made up of human knowledge, to build a knowledge base it needs a human expert to manage. At this point, it can be deduced that expert systems are computer programs

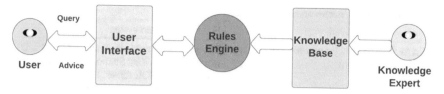

Fig. 16.2 Architecture of expert systems

created to address complicated problems in a certain field at a level of knowledge and expertise well above that of individuals [13].

Modeling Expert Systems for Threat Prevention

Any expert system is made up of five main parts which are Knowledge Base, which contains specialized information and recommendations to help with the problem, and Inference Engine, whose main job is to gather relevant data from knowledge bases in order to discover a solution for the user as in Fig. 16.2. The knowledge acquisition and learning module's user interface enables the non-expert user to communicate with the expert systems in order to tackle a challenging challenge. This has to do with the expert systems' knowledge being expanded by the accumulation of vast volumes of data. An Expert System user, knowledge base editor, knowledge engineer, inference mechanism, and system of explanations should all be included in every expert system. All expert systems are built upon a knowledge base that is specific to a given issue area [14]. Data is continuously entered and gathered while an expert system is being built and operated. An expert system's knowledge base contains facts about a certain subject along with heuristics that reflect the mental processes of a subject-matter expert. Expert systems can represent knowledge in many different ways that includes "case-based reasoning, frame-based knowledge, object-based knowledge, and rule-based knowledge." A knowledge base system, a decision support system, an intrusion detection system and an inference engine are the components of expert systems [15].

Applications of Expert Systems

Ability of an Expert System to create knowledge bases and keep an eye on incoming traffic, expert systems are mostly used in two broad areas which are ensuring the security of sensitive data and detecting intrusions on networks. Table 16.1 describes the different expert systems designed.

Table 16.1 Expert systems and their functions

Expert system designed	Function of expert system
CSAAES: an expert system for cyber security attack awareness [16]	The authors proposed an expert system that is made of using Visual Studio framework, ASP.NET and SQL server. The authors used ASP.NET in order to make the interfaces and SQL for databases. This system can further be divided into two parts which includes the information regarding to identify the attack. Here, information related to observed behavior of the computer system are asked to identify attack. Once the system identifies the attack, the attack description along with the remedies are given to protect the systems
Graded security expert system [17]	The authors proposed a method for graded security by developing a hybrid expert system. Using dynamic programming to locate Pareto optimality curve points, the expert system enables a user to make intelligent security measure selections. The expert system provides a speedy and fair security solution for a class of well-known computer systems with a high level of comfort
User behavior pattern signature-based intrusion detection [18]	This approach suggests a rule-based study, the main goal of which is to employ signatures to distinguish between allowed and unlawful user behavior. In a rule engine, the rules are kept. User activity log contains information about user behavior. This was accomplished using a novel pattern-based intrusion detection system. If any of the pattern matches to the rules that leads to its activation and the system blocks the user and sound an alarm

Intelligent Agents

An intelligent agent is a program that can decide what to do or how to do it on basis of its surroundings, input and past experience as in Fig. 16.3. These programs can be used to gather information automatically according to a time stipulated or via response to user input. Two of the most important of these agents include perception and action. "Actuators initiate activities, whereas sensors carry out perception." Intelligent agents are composed of a hierarchy of sub-agents. These sub-agents perform low-level activities. These tools also offer proactive measures, mobility, and a language for agent communication. Intelligent agents are self-sufficient computer systems that collaborate with one another to counter cyber-attacks. Some of the studies acknowledges that the intelligent agent is primarily created to counteract DDoS by assisting cyber agents in communication and movement [19].

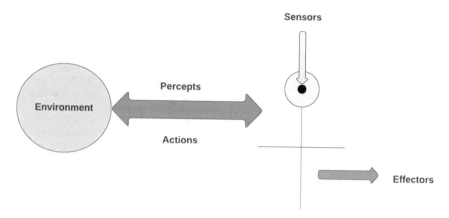

Fig. 16.3 Architecture of intelligent agents

Intelligent Agents for Defense Against Cyber-Crimes

Intelligent agents are independent, computer-generated forces that interact with one another, share knowledge, collaborate to plan, and implement appropriate responses in the event of unforeseen circumstances. Intelligent agent technology is suited for fending off cyber-attacks because of its adaptability to the environments in which they operate. Below is a discussion of a few applications of intelligent agents (Table 16.2).

Machine Learning for Cyber Security

Machine learning plays a crucial role in cyber security, and it has many applications. It is given importance to learn about machine learning and its frequently used algorithms at various platforms. Some of the applications of machine learning techniques for cyber security are there to identify new cyber threats and advancement in the antivirus software technology by integrating machine learning into it to combat against threats. Machine learning-based systems react faster when compared with many traditional systems.

Understanding Machine Learning

Machine learning enables software programs to improve their predictive abilities. Machine learning algorithms are pretrained with most accurate and available data in order to predict the output. The significance of these systems is growing daily as a result of its capacity to suggest, warn against, and stop some risky assaults from

Table 16.2 Intelligent agents and their methodology

Intelligent agent	Function of the model
Network security management with intelligent agents [20]	In this paper, a technology called multi-agent systems for network security was suggested. For intrusion detection employing intelligent agent technology, a unique model known as IA-NSM (intelligent agents for network security management) was developed. IA-NSM offers a customizable way to integrate a multi-agent system into a traditional networked environment to increase the system's defense against built-in assaults
Intelligent autonomous agents are key to cyber defence of the future army networks [21]	The author designed the architecture and addressed the key capabilities of an Intelligent agent which are strictly monitoring all the computers to detect malware. Depending upon the attack, the agent must immediately take destructive action. The agents are updated by all means such as new techniques, procedures, and software's for detecting viruses, trojans. They even added, and this system works well during wars
Multi intelligent agent based cyber attack resilient system protection and emergency control [22]	The authors of this paper suggested a "decentralized intelligent multi-agent based system for protection and emergency control technique." "An actual load rejection protection technique was implemented in the IEEE 39 bus system which has shown the cyber-attack resistance of the suggested strategy and compared with the traditional scheme"

occurring in order to improve cyber security. The choice of using machine learning in cyber security is to improve the performance of traditional systems in detection of malware attacks as these generally rely on humans. Thus, the use of machine learning makes systems more scalable and efficient. Machine learning answers major cyber threats detection with ease [23]. Machine learning algorithms are Neural Networks, Decision Tree, Support Vector Systems, and Ensemble models, and many more gave supportive results to combat cyber-crimes. To create intelligent security systems and to develop a comparable data-driven machine learning model, it is crucial to gather more information and trends in network data should be given priority.

Machine Learning Algorithms

The subcategories of machine learning algorithms are divided into supervised learning and unsupervised learning. In supervised learning, the models typically give priority to form internal patterns from the dataset to classify them into various groups and less importance is given to dependent variable. Some of the most used algorithms include K-means clustering, apriori, etc., and the most often used supervised learning strategies are classification and regression approaches. By using well-known methods, the class variables are classified or predicted.

In cyber security, classification techniques help to determine whether a specific type of attack occurs or not. The most well-known classification methods for shallow models include Decision Trees, Random Forest, Naïve Bayes, Logistic Regression, and Support Vector Machines. Unsupervised learning methods are termed as data-driven approaches that focus on identifying patterns of information in unlabeled datasets. For example, malware attacks are generally masked in various ways so that it cannot be identified but because pattern recognition machine learning can detect this. Clustering techniques comes under the same roof of unsupervised learning that helps to discover hidden patterns in the dataset and to combat various cyber assaults [24].

Machine Learning Methods

- Regression: "Regression is a method for determining how independent features or variables relate to a dependent feature or result. It is a technique for machine learning predictive modeling, where an algorithm is used to forecast continuous outcomes." Some of the machine learning techniques used in regression applications are linear regression, Polynomial regression, Ridge regression, Decision trees, SVR (Support Vector Regression), and Random forest.
- Classification: Classification techniques can be applied on structured and unstructured data. "Classification is the act of categorizing a given set of data into classes. Predicting the class of the provided data points is the first step in the procedure. The terms target, label, and classes are frequently used to describe the classes." Logistic regression (LR), K-nearest neighbors, Support vector machine, KernelSVM, Naive Bayes, Decision Tree Classification, and Random Forest Classification are classification-based machine learning techniques.
- Clustering: "The objective of clustering is to divide the population or set of data points into a number of groups so that the data points within each group are more similar to one another and different from the data points within the other groups." It is essentially grouping of objects based on how similar and unlike they are to one another. K-nearest neighbors (KNN), K-means Mixture model (LDA), DBSCan, Bayesian, Gaussian-Mixture Model, Agglomerative, and Mean-shift are a few clustering approaches.

Machine Learning Methods for Cyber Threat Prevention

- The authors of the paper in [25] used Naive Bayes Classifier and KDD 99 dataset. The features that correspond to the attacks DoS, U2R, R2L, and probing are categorized under any of the mentioned attacks and grouped together to apply Naïve Bayes classifier. They considered accuracy as an evaluation metric and obtained 99% for DoS, 90% for U2R, 90% for R2L, and 99% for probing. The use of naive Bayes helped to reduce false-positive rates.
- The authors of [26] used least square SVM model. The main use of this model is to reduce the training time on the datasets. The dataset used by the authors is KDD 99 dataset. They wish to perform understand the classification accuracy on 4 attacks. The authors of this paper have proposed to use three feature extraction techniques in order to reduce the dimensions of the dataset. By the use of this feature extraction, the dimensions are brought to 19 from 41. As a result, the data was again resampled to 7000 tuples and obtained the classification accuracy as 99% for DoS, 93% for U2R, 99% for R2L, and 99% for probing.
- The authors of the research paper in [27] replaced the SNORT decision engine with Decision Tree. SNORT is a tool that is based on signature-based approach. The authors of this paper used ID3 algorithm to make a Decision Tree classifier using by clustering the rules. The proper use of clustering the rules is to reduce the comparisons with the given input data. This approach helped to evaluate the features. They used DARPA 1999 dataset which gave a sufficient result when compared with the SNORT engine. By the use of Decision Tree, the number of rules was increased to 1581. This resulted to build a 100% faster model when used in real time.
- The authors of the paper [28] have used sequential mining technique. The dataset used by them is DARPA 1999 and 2000. They used apriori algorithms to make a pattern that tries to reduce the redundancy of alerts and maximize true positives. They even gave a visualization for users which has a support threshold of 40%. Within 20 s, the designed systems were able to detect attacks up to 93%. When it is used in real-time scenarios, good attack detection percentage was obtained with supportive threshold.
- The authors in [29] created an intrusion detection systems named ADMIT. This system varies from the traditional systems where they rely on class variables whereas this system creates its own user profiles using dynamic clustering technique. The user's data is collected and made as a sequence to build a user profile. In order to detect normal versus fraudulent behavior of users, they used K-means algorithm with subjective changes. The main objective of this paper is to reduce the false positives and attain considerable accuracy. As a result, performance of this technique is 80%.

Machine Learning Technology in Cyber Security

Machine learning technologies are changing cyber security; due to their rising demand, cyber security may benefit greatly from machine learning, which can be used to better available antivirus software, identify cyber dangers and battle online crime. So let us look at some machine learning applications in cyber security.

Machine Learning for Network Protection

Without the requirement for explicit programming, machine learning offers a security model with the capacity to autonomously research and provides results based on prior learning. The amplification of abnormalities is fundamental to many cyber security problems, including denial-of-service (DoS & DDoS) assaults, fraud, spam propagation, and spoofing components. There may be many methods for the identification of patterns in network security but one such powerful tool is machine learning which gives analysis of minute details. On the basis of the capabilities of various machine learning models, in network intrusion ML aids in gaining a thorough understanding of how a malware detection model functions. To identify distinct forms of network attacks including scanning, spoofing, and clustering for forensic analysis, classification is used and regression is used to forecast the network packet parameters and compare them with the typical ones [30].

Machine Learning for Endpoint Protection

Machine learning algorithms are used to achieve endpoint security. Machine learning algorithms are trained and tested to deploy or integrate them with the various endpoint solutions to protect various important files based upon the activities as attributes of previously occurred malicious attacks as data. Classification methods are used to divide programs into various categories which includes malware, spyware, ransomware, regression to predict the next process, and clustering techniques to secure email gateways from malware attacks [31].

Machine Learning for Application Security

Machine learning in Application Security is the defense against zero-day exploits. Protection against these attacks are given at most priority. In order identify these type of exploits, it takes months in between huge amounts of sensitive data gathered by the exploiters which is a high risk. By not just recognizing malicious behavior based on rules but also by identifying aberrant data flow and aiding in the detection of outliers, machine learning can offer a means of defense against such attacks. Application security generates a lot of data that can be effectively used to build ML models. Some of the examples include web apps, databases, ERP programs, SaaS programs, and micro services. It is nearly hard to create a universal paradigm that can successfully address all dangers in the near future. Regression methods are used to identify anomalies in HTTP requests. Classification is used to identify known attacks, and clustering is used to detect DDoS attacks and exploitation.

Machine Learning for User Behavior

User and Entity Behavioral Analytics by Gartner includes machine learning as a core competency. Corporate security teams are already using UEBA-based solutions to detect attacks and identify threats in innovative ways. UEBA uses machine learning and other analytics approaches to automatically identify threats that get past current security measures. The types of data that UEBA solutions analyze tend to separate them into several categories. Some of the enterprises concentrate on security and infrastructure logs that include data from servers, firewalls, web proxies, and authentication systems. Some enterprises concentrate on flows of packets in network traffic. Here, regression is used to identify anomalies in User actions, Classification is used to analyze the peer groups and clustering is used to identify outliers and divide user groups [32].

Machine Learning for Process Behavior

The goal of predictive process monitoring is to predict the behavior, efficiency, and results of business processes as they happen. It assists in problem-solving and resource redistribution before problems arise. Classification methods are used to identify types of attacks, and clustering are used to find outliers and also to compare business processes.

Deep Learning for Cyber Security

Deep Learning which has most advanced technology and frameworks such as PyTorch and TensorFlow plays a crucial role in cyber security. Deep Learning models which are made up of neural networks always tend to train deeper to identify complex patterns and to detect more complex attacks. Since these algorithms does not rely on known attacks and their description, instead they learn by performing in real time to identify dangerous behaviors.

Deep learning can study computer algorithms to learn and improve on its own. Deep learning uses neural networks which behave similarly like humans in terms of thought process. The complexity of neural networks was previously limited by computer power but advancements in big data analytics, which allow computers to watch, comprehend, and react to complex situations faster than people, larger, more intricate neural networks are now feasible. Current cyber security systems fall short of tackling the dynamic changes in modern attacks, particularly in terms of seeing new threats, unraveling convoluted pathways and events, and being unable to scale to the massive volume of attacks [33]. Deep learning in cyber security can address many of these problems by employing novel methods and techniques for DDoS detection, behavioral anomaly identification, malware and botnet detection, and speech recognition.

Choice of Deep Learning Over Machine Learning for Cyber Crime Detection

Generally, machine learning is long regarded as a cutting-edge method of protecting digital assets. However, it is possible to reverse-engineer ML tools to include bias or a weakness that lessens the effectiveness of its safeguards. Additionally, hackers are able to use their own ML algorithms to contaminate a cyber security solution with false datasets. Deep learning, on the other hand, does away with the need for data scientists to manually feed datasets. Deep Learning models process enormous volumes of raw data before being automatically trained into the cyber security system. Deep Learning neural networks are taught to function autonomously without human input. When it comes to successfully identifying exceedingly complicated patterns from large datasets, Deep Learning works well.

Deep Learning Methods for Cyber Security

Autoencoder: Autoencoder is used to learn data encodings. An autoencoder learns a lower-dimensional representation (encoding) for a higher-dimensional data in order to learn a higher-dimensional data in a lower-dimensional manner, frequently for dimensionality reduction. Autoencoders are employed in malware detection and classification, intrusion detection, identification of file type, network traffic analysis, and spam detection and user authentication [34].

Convolutional Neural Networks: Convolutional neural network is one of the major deep learning network architectures which directly reads from input as a result; not much importance is given for feature selection and engineering. CNNs are incredibly helpful for detecting objects, faces, and scenes in photographs as they make patterns in the images. CNNs quickly and precisely reconstruct traits. Understanding feature correlations is useful for better understanding features and rebuilding security data for better surveillance. Convolutional Neural networks are highly useful in the cases of audio, time-series, and signal data. They play a very crucial role in the fields that are related to identify objects. Convolutional neural networks are used for malware detection, drive-by download attacks, intrusion detection systems, and identification of traffic.

Recurrent Neural Networks: Recurrent neural networks (RNN) are neural networks that use sequential data. Real-time examples of RNNs Siri, Alexa, and Google translate use these. These neural networks are used in for ordinal or temporal problems in speech recognition, natural language processing, and image captioning and language translation. They use training data to fetch the details. RNNs show much deviation from other systems because of "memory" as they can alter the input and output by the use of previous information. In contrast to classic deep learning networks, which translates the fact that input and output are dependent on each other, whereas RNN's output is dependent on previous parameters. Unidirectional recurrent neural networks are unable to include future events into their predictions, despite the fact that they may be useful in predicting the outcome of a particular sequence. RNNs have applications in Intrusion Detection, Border Gateway Protocol, Malware Detection, Anomaly Detection, Custom Keystroke Verification, and DGA.

Generative Adversarial Networks: It combines two deep learning neural network technique: a Generator and a Discriminator. "The Generator Network creates fake data, and the Discriminator helps discern between true and false data. Both networks are in competition with one another as long as the Generator keeps producing artificial data that is identical to real data and the Discriminator keeps track of real and fake data." In a situation to build a picture library, the Generator network would generate simulated data to the actual photos. The result would be a deconvolution neural network. Following that, a network of image detectors would be utilized to tell real images from fake ones. The detector must improve classification accuracy since the generator would get more competent at creating deceptive images starting with a fifty percent accuracy chance. Effectiveness and speed of the network would increase as a result of this competition. Because of these characteristics, it is used in the Domain Generation Algorithm.

Deep Learning Methods in Cyber Security Applications

Intrusion Detection and Prevention System: By identifying harmful network behavior, these systems alert users and prevent unauthorized individuals from accessing the systems. Common attack formats and well-known signatures are often used to identify them. This aids in protecting against threats like data leaks. Some of the methods such as CNN and RNN are used to develop smarter Intrusion Detection and Prevention Systems [35]. These systems can analyze the traffic with greater ease, reducing the number of falsehood and helping information security teams to differentiate between malicious and true activities.

Malware Detection: Classic malware detection solution includes to build firewalls and to detect malware attacks using signature-based methods. The business systems routinely update a database of known risks to add recently found dangers to it. This approach works to detect simple threats but falls short against more complex ones. Since deep learning algorithms do not rely on recollection of well-known signatures and attack patterns, they can recognize threats that are more complicated. Instead, they become accustomed to the system and are able to spot unusual activity that can indicate the presence of malware or hostile actors.

Spam Detection: Natural language processing (NLP) being a deep learning approach is used to recognize spam and various other forms of social engineering rapidly to take appropriate action. NLP uses a variety of statistical models, organic linguistic patterns, and modalities of communication to recognize and block spam.

Analysis of Network Traffic: ANNs have demonstrated promising results in the analysis of HTTPS network traffic to look for malicious activities. They are used to handle a variety of online hazards, such as SQL injections and DOS assaults, considerably simpler.

Analyzing User Behavior: Every company should adhere to the security best practices of monitoring and evaluating user behavior. Since it goes beyond security safeguards, it is much more difficult to detect than typical malicious actions against networks using conventional techniques. For instance, insider threats occur when employees hack into the system from the inside rather than from the outside, rendering many cyber defense technologies useless in the face of such attacks [32]. UEBA is preferred against such assaults. Taking some time, it can learn the regular staff behavior patterns, see suspicious conduct that might be an insider attack like accessing the system at strange hours and sound the alarms.

AI-Powered Cyber Security Platforms for Enterprises

CrowdStrike: CrowdStrike Falcon is an AI-based detection method which is based on user and entity behavior analytics. CrowdStrike has created a cyber security tool (UEBA). With real-time protection and visibility across the company, Crowd-Strike's Falcon has reinvented security for the cloud era and prevents assaults on endpoints and workloads on or off the network. The Falcon platform from Crowd-Strike was created with the specific goal of preventing breaches through the use of a unified set of technologies that thwart all kinds of threats. In order to backstab businesses, sophisticated attackers are increasingly turning to exploits and it is difficult to detect techniques like credential theft as well as technologies that are already present in the environment or operating system of the victim, like PowerShell. In Crowd-Strike Falcon, "Next-generation antivirus (NGAV), endpoint detection and response (EDR), cyber threat intelligence, managed threat hunting capabilities and security hygiene are all combined in a small, single, lightweight sensor that is cloud-managed and delivered." Solutions for Endpoint Security, Security and IT Operations, Threat

Intelligence, Cloud Security, and Identity Protection are all offered by CrowdStrike Falcon.

Darktrace: The Enterprise Immune System (EIS) was created by Darktrace as a platform for all of its cyber security solutions. EIS employs AI techniques and leverages unsupervised machine learning to fill status rule bases. When EIS is installed on a network, the first task it must complete is creating a baseline of typical behavior. To produce this list of acceptable behavior, traffic patterns for each network, device activity on the network, and user behavior are all modeled. DarkTrace's cyber-AI loop is preventive, detective, responsive, and curative.

Cynet: In order to analyze threats and take automatic action on them, Cynet uses AI in its network threat detection systems. The philosophy of Cynet is to make using any system monitoring program as simple as protecting against advanced threats. The Cynet network protection suite was created to give enterprises without specialized cyber security staff easy access to threat defense. However, the approach is not limited to tiny businesses with insufficient workers. Large multinational corporations with tens of thousands of employees, as well as businesses with substantial costs associated with security failure, including banks, are among the clients of the service.

FireEye: Compared to the companies described above, FireEye is significantly older. It was established in 2004. It has a focus on threat research and consulting services for recovery. It took a lot of labor to complete this work, and the corporation lost money doing it. The company has expanded into the manufacturing of cyber security technologies that employ AI to monitor networks and discover anomalies through invention. With the help of this strategy and the transition from a fee-based to a subscription Software-as-a-Service model, the company has become profitable and what was starting to appear to be an overpriced novelty has become a sought-after investment. The Multi-Vector Virtual Execution engine, dynamic machine learning, instant protection, validating and prioritizing assaults, and flexible deployment choices for Enterprises or small-scale businesses are some of FireEye's main benefits. This helps Enterprises to identify unforeseen attacks, and they can be mitigated.

Check Point Software Technologies: Check Point is a developing technological business that has been able to go from startup to global status. This Israeli business has long been a pioneer in the application of AI to cyber security. The company first produced firewalls; then, in 2003 it acquired Zone Labs, the company behind the security program ZoneAlarm, to strengthen its position in that market. Instead of creating a unique AI-based threat management product, the company invested in the creation of three AI-driven platforms that support many of the core services provided by the company. These are Context-Aware Detection, Huntress, and Campaign Hunting. Check Point's software solution is used to synchronize management and security activities, safeguard user's access and the cloud. This software is used in deliver variety of services for access control (Firewall, App Control), data security (IPS, IPsec VPN), threat prevention (Antivirus, Anti-bot), and security (Firewalls, Private Cloud Virtual Firewalls).

Performance of an AI Solution

Performance metrics are essential for any Artificial Intelligence-powered cyber security system in order to accurately measure and evaluate the model we developed. We may learn from the performance measures how well the trained model predicts for the available datasets and different scenarios. The proper measure must be chosen in order to comprehend our model's behavior and make the necessary adjustments to advance model development. According to the kind of model linear regression, binary classification, etc., AI model metrics are measured and assessed. This results in a statistical table that summarizes all the metrics and serves as the foundation to evaluate the model. These are the criteria used to gauge the effectiveness of an AI model. There are different types of performance metrics [36]. They are discussed below.

Performance Analysis Metrics of Artificial Intelligence Solutions

Confusion Matrix: It is a tabular representation of the dataset's actual values and anticipated values. To avoid any "confusion," the resulting matrix is intended to improve comprehension and clearly visualize the findings of the models. There are "true positives, false positives, true negatives, and false negatives" in the confusion matrix. By the use of various algorithms or models, the confusion matrix's ultimate objective is to maximize true positives and true negatives, or correct forecasts, and minimize false positives and false negatives, or incorrect predictions.

Accuracy: Accuracy is the most popular performance metrics for evaluating algorithms. The number of accurate forecasts divided by the total number of predictions is known as accuracy. When the classes are balanced, accuracy works well. However, accuracy might not be the optimal statistic to use if there are unequal numbers of samples in each class.

Precision: Precision tells us how accurately the model identified the favorable outcomes. It measures the proportion of real positives to all positives. The amount of accurate positive forecasts divided by the total of accurate positive, and inaccurate negative predictions is known as precision. The precision value is a number between 0 and 1.

Recall: Recall is even termed as true-positive rate, calculated by dividing the total number of correct positive predictions by the total number of false-positive forecasts. The recall value is between 0 and 1. Recall is a statistic we use when our goal is to reduce false negatives.

F1-score: Both precision and recall levels are used in the $F1$-score. It is the precision and recall score's harmonic mean. The $F1$-score provides a balance between

Precision and Recall. It works best when both scores are equal. We continually aim for high recall and precision in our classifiers, but there is always a trade-off when optimizing the classifier.

Specificity: It is a measurement of the projected negative points in relation to all of the actual negative points (including false positives). Similar to recall, except for negative predictions, it really conveys the offset in the model's successful prediction of the negative values.

ROC-AUC Curve: A statistic for displaying the effectiveness of a binary classification problem is the ROC curve. A ROC curve is a plot representing the ratio of false positives to true positives. The area under ROC curve is known as the AUC score. It can be inferred that the higher the AUC score the classifier performs better. AUC score values might be between zero and one. Between true-positive rate (TPR) and false-positive rate, there is a curve called the receiver operating characteristics. By providing data on how well the model can distinguish between the classes, it is a probability curve that aids in visualizing the binary classification model. The area under this curve, or AUC, indicates how well the model can classify data. The bigger the area, the better [37]. The performance of the model is enhanced by enhancing the measurement of key parameters, such as raising AUC, minimizing log loss, raising recall, and specificity while lowering false positives and false negatives. Although the improvement in model performance appears to be very revolutionary and promising in academia, this is not the case in business.

Techniques to Enhance the Performance of Artificial Intelligence Solutions

Choice of Dataset: Choosing a quality dataset for training and testing the model is one of the fundamental building blocks of model training. In these situations, we either create new data based on the problem statements or use standardized data from resources that are already first in class. The most common technique in day-to-day situations is supplementing the data already available to meet needs. The complexity of the problem and the learning methods largely dictate the type of training datasets that help to improve the model's performance, model competence, and appraisal of the data size and application of statistical heuristics.

Feature Engineering: Data with a lot of features is often used to construct machine learning models. Another well-liked technique for improving machine learning models is engineering additional features and selecting the greatest combination of features which tries to improve the performance. To create new features that capture characteristics of the complex nonlinear function that the machine learning model is learning to emulate, feature engineering demands a high level of domain understanding. It makes sense that if the basic model contains a lot of characteristics already, this strategy would not be the best to employ. Some connected or duplicate features that do not significantly increase model performance can be removed with the use of programmatic feature selection tools [38]. The identification of robust features is aided by methods for iteratively creating and evaluating models with progressively more features or iteratively removing one feature at a time from a model trained with all the features.

Feature Selection: The process of feature selection entails choosing subset of characteristics to demonstrate how independent variables relate to the target variable. The two famous ways that are used to select features include Domain Knowledge, where we choose features that may have a greater impact on the target variable, and visualization, which makes the connections between variables easier to understand and allows you to choose your variables instantly. We also take into account the p-values, information values, and other statistical metrics while choosing the best characteristics.

Principle Component Analysis: In order to understand the underlying relationships in the dataset working upon, while still representing training data in lower-dimensional regions, it is a particular method of dimensionality reduction. The dimensions of training data can be reduced using a variety of techniques which are factor analysis, low variance, higher correlation, and backward/forward feature selection.

Use of Various Algorithms: Selecting the right machine learning algorithm is the best course of action for improving accuracy. One grows intuitive with time and experience. Different algorithms are better suited to different sorts of datasets than others. As a result, we should apply all relevant models and assess the outcomes. With practice and experience, one develops intuition to select algorithms.

Choice of the Algorithms and Its Parameters: Choosing the appropriate algorithm helps to guarantee the effectiveness of the machine learning model. The most popular algorithms are "Linear Regression," "Logistic Regression," "Decision Tree," "SVM," "Naive Bayes," "k-NN," "and K-Means," "Random Forest," "Ensemble Models," "Perceptron," and "Dimensionality Reduction Algorithms and Gradient Boosting." Parameter plays a crucial role in machine learning algorithms. The choice of parameters critical have an influence on the model that is trained. It can be inferred that machine learning model accuracy depends on the choice of parameter.

Use of Ensemble Models: This is the most typical strategy that dominates winning data science competition answers. The ensemble models combine the output of various other models that are used to give results. Some of the examples of these models include Boosting and Bagging [39]. Use of ensemble models to enhance the accuracy of your model on the dataset is always better option.

Downsides of Artificial Intelligence in Cyber Security

Hackers: Hackers are always a step ahead to gain knowledge about the latest technology and the various ways it could benefit them, being at the backend, performing analytics in order to understand the breaches in the target systems and using them for their own good will. The use of AI in threat prevention and identification is playing a very crucial role in Enterprises. But as well all know AI mimics human intelligence but the systems are not actually humans [40]. In AI-based solutions, it all depends on who trains the best algorithm for the given data. Because of this factor in enterprises by using this, one can try to analyze the sensitive information which includes the company's defense mechanisms and policies as this can help them to find vulnerability of the system and exploit them [41–50].

Data Integrity: The basic foundation of any ML and DL models is generally associated with large volumes of the data. Analysis of the data is a key factor to train and test the model on any specific algorithms to get accurate result. In Enterprises or Software Solutions, there are lots of sensitive data related to shareholders, clients, biometric details, patent technology algorithms, and many more. When this data is gathered by any agent by the use of AI techniques, the data integrity could not be achieved. So, it must be given priority to safeguard the systems knowledge or client information from vulnerabilities when AI models are trained.

Securing of Data: AI models require huge amount of data in order to train the specified model to reach better accuracy and performance. This creates a problem to effectively store and secure the data as there are many chances the data can be leaked. The leaked data can be used by many individuals for their own benefit. The data that is used by Enterprises are generally the daily transactions of many people that took place in real time. The data breaches cause various policy issues. So, latest innovations in data storage and security management which are powered by AI must be employed to secure the data.

Recognition of Activities of Users: AI-based systems try to gather the activities of the users which their login details, transaction, etc. Many third-party organization can use this information for their good. Because of latest innovations in AI, one can track and monitor the activities that are done on our systems which leads to privacy issues. Securing user activity must be main.

High Cost for Team Training: AI solution being the latest innovations, building and managing an AI team, is way expensive. Most of the AI-powered tools are still in development phases so, and days are to come to completely be dependent on those tools.

Infrastructure Problems: AI-powered solutions generally require GPUs, advanced systems with better Internet facilities. These systems can be used to their full potential when they are provided with necessary environment they require. But to maintain such environment in real time is way expensive and requires lot of funding where no corporate organization can bare; because of this reason, most of the corporates are still stuck by traditional systems, web apps to manage and ensure security to their information. So, the organizations that are trying to deploy AI must plan ways to integrate them with the existing systems so that cost can be reduced.

Integration of AI Solutions: It is said that integration of AI solutions to the existing systems is the most cost-effective way. But on the other side it is even listed that difficulties that are occurred while translating to AI systems which needs of more domain knowledge where more people lack. It can be understood that how hard it is to switch to a AI-integrated solutions without proper knowledge of it as small mistakes can lead to raise a harm to the systems. Updating to AI-integrated solutions must guarantee the safety to existing data, current system must work effectively, and the future systems must work in the way it is designed. So, to achieve AI-integrated solutions most of the steam members must have its domain knowledge.

Credit Card Fraud Detection System

Detecting credit card fraud often involves spotting erroneous purchase attempts and rejecting them rather than fulfilling the order. There are many tools and techniques that are used by organizations to identify fraud. Due to its simplicity in solving any problem, machine learning is among them the technology that is most frequently and vividly applied. The most effective techniques use machine learning algorithms since they are flexible and simple to use in any setting.

The dataset used in our study is gathered from Kaggle which is a reputed source that contains many real-time datasets. This dataset is made by credit card transactions of users in European states on September 2013. This dataset consists of a total of 285,299 tuples, and it is inferred that only 0.172% of the data consists of fraud transactions. The attributes in the dataset include features from V1 to V28, class, amount, and time. V1–V28 are the variables that have undergone principal component analysis transformation. The original attribute names are not used to ensure safety to the real features. The amount in the dataset corresponds to the amount that has used for the transaction and time corresponds to the elapsed time between the transactions. The dataset consists of class 0 and 1. 0 is used for true transaction and 1 for fraud transaction.

Objectives

The main objective of this study is to deal with unbalanced dataset, analyzing the various methods to resample the data and choose a method to resample. Furthermore, train the resampled data using various algorithms such liner, nonlinear, neural networks and ensemble learning models. A wide range of models can be used for comparative analysis, and evaluation can be done.

Exploratory Data Analysis

Exploratory data analysis is the most foundational step to analyze the data after reading it. Through this analysis, one can infer most from the dataset which includes gathering of statistical summary of the dataset, whether data is balanced or unbalanced, identify NULL values, etc.; this forms a base line to choose the algorithms which we need to train suing this dataset.

Data Imbalance

This dataset's most intriguing feature is how absolutely unbalanced the data is. The dataset contains more than 250,000 legitimate transactions, but only a small number of them are fraudulent, as seen in Fig. 16.4. Since the data is completely unbalanced, traditional models cannot be used to train as they will not yield good results. Therefore, it is crucial to talk about the effective methods for balancing the data.

Correlation of the Attributes

It is utmost important to check for the columns or the attributes that are correlated in the dataset; this is because when 2 or more attributes from the dataset are related, we can drop the others which indeed helps to reduce the dimension. So, in credit card dataset there are no highly correlated attributes. Some of the slightly correlated attributes are (V20—Amount) and (V7—Amount) from Fig. 16.5.

Resampling of the Data

Most datasets related to network intrusion detection and fraud detection in banking are not sampled well. The true cases in data used in these fields are frequently more

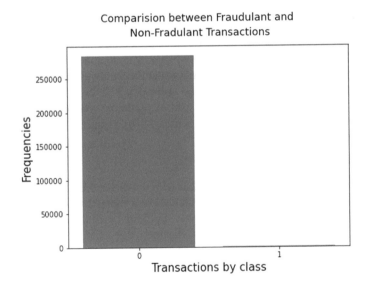

Fig. 16.4 Representation of data imbalance

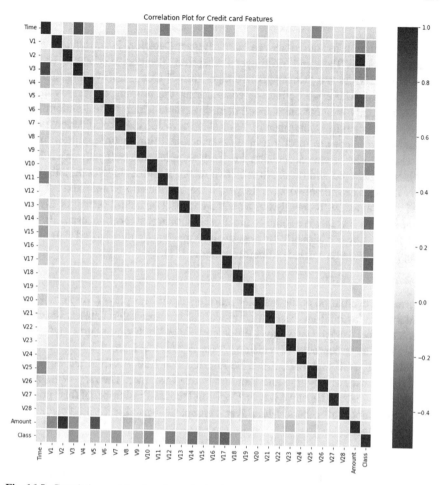

Fig. 16.5 Correlation matrix of attributes in the dataset

numerous than the fraud cases in our dataset. Unbalanced datasets do not lend themselves well to machine learning methods. Resampling the training set is crucial. Some of the most widely used techniques use "K-fold cross-validation," "ensemble different resampled datasets," "resample with different ratios," and "cluster the abundant class and creating a model." To handle this dataset, Synthetic Minority Oversampling is the method was used for resampling [51]. Through this methods, minority class's examples are oversampled to resolve the unbalanced data. The use of this model fresh data is not created; it just balances the unbalanced class. This technique is generally the most used to make new samples. SMOTE selects examples in the feature space that are close to one another, draws a line between the examples, and then creates a new sample at a location along the line [52].

Choice of Models

The model chosen for our study are Naïve Bayes, Decision Tree, Random Forest, Bagging Decision Tree, and Multilayer Perceptron. The choice of ANN, Bagging, and Decision Trees are "Cost-sensitive algorithms." Cost-sensitive algorithms are generally used where we define the costs during the training of various models. They work really well for unbalanced data. Ensemble Models like Random Forest are used. By merging a number of base classifiers for the classification system, ensemble classifiers are mostly used to enhance learning technique performance. To create the classification decision for fresh samples, the output from each base classifier is gathered and used. An effective technique for categorizing unbalanced data is artificial neural networks. However, if the imbalance was not fixed before the model was trained, it frequently has a negative impact on neural networks. Decision Tree and Naive Bayes were some of the most significant algorithms used. However, Naive Bayes can occasionally be negatively impacted by unbalanced data; hence, resampling should be carried out before training [53].

Performance Analysis

Performance evaluation of the machine learning algorithms plays a very crucial role in order to understand the best suited model for the resampled dataset. In order to understand the working of various models, we use Confusion matrix and various metrics such as "Accuracy," "Precision," "Recall," and "$F1$-score."

From Figs. 16.6 and 16.7, it can be inferred that true positives are higher and equal in the cases of Bagging and Random Forest Classifier. The ratio of high false negatives are observed in naïve Bayes. Decision Tree has the number of false positives as highest among all others. Multilayer Perceptron was able to predict the false negatives higher.

Comparison with Evaluation Metric

In comparison with the accuracy, it can be observed that Multilayer Perceptron, Random Forest, and Bagging techniques are working well as they obtained more comparative result than others. Since credit card dataset is an imbalanced dataset,

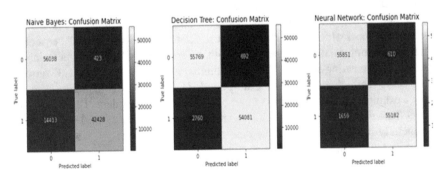

Fig. 16.6 Confusion matrix of Naïve bayes, decision tree, and neural network (Left to right)

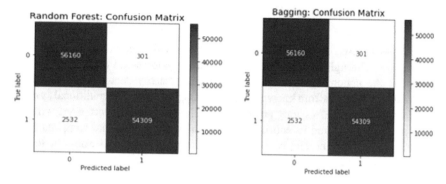

Fig. 16.7 Confusion matrices of random forest and bagging classifier (Left to right)

we generally prefer precision more over recall. Random Forest, Bagging, and Naïve Bayes have above 99%. From Recall scores, one inference we can make is that Multilayer Perceptron makes output sensitive predictions as it reduces the number of false negatives. Since all the trained machine learning models have almost similar precision and recall scores, among all of them in comparison with $F1$-score Multilayer Perceptron can be picked over Random Forest and Bagging. On a whole, the ensemble models are performing very well in all the terms (Table 16.3).

On a whole, the ensemble models are performing very well in all the terms. So, for the future use Ensemble Models can be used. Since the models such as Multilayer Perceptron, Decision Tree, and Bagging are cost-effective models, they yield better result.

Conclusion and Future Works for Credit Card Fraud Detection

It can be observed that Ensemble models works well in the case of unbalanced data. In this study, Bagging Decision Tree Classifier, Random Forest, and Multilayer Perceptron yielded good results in terms of precision. SMOTE resampling technique was implemented in order to study its behavior with different classifiers. In the future, Deep Neural Networks along with various resampling techniques can be implemented.

Table 16.3 Performance analysis metrics of all the trained models

Trained model	Accuracy	Precision	Recall	$F1$-score
Naïve Bayes	86.91	99.01	74.64	85.12
Decision tree	96.95	98.74	95.14	96.91
Random forest classifier	97.5	99.45	95.55	97.4
Bagging decision tree	96.78	99.68	93.88	96.69
Multilayer perceptron	98.0	98.91	97.08	97.99

Conclusion and Future Work

The chapter focused on cyber security applications which are based upon Artificial Intelligence. Through this chapter, it was basically understood that AI can help to determine the upcoming cyber threats as it can possibly detect a new attack by protecting the systems from known attacks. Through AI systems, traditional systems could come to an end as they are pretty much slower when compared with the latest innovation. AI and its subcomponents which are machine learning and Deep learning play a very crucial role in cyber security applications. Some of the AI techniques which are widely used are Expert Systems, and Intelligent Agents are given much focused. Some of the machine learning models such as Regression, Classification, and many more along with their areas of impact in threat detection are spotlighted. In addition, Neural Networks which are Deep learning models are given utmost significance in threat prevention. As a result, it can be concluded that machine learning and deep learning are significantly more intelligent at identifying patterns in new cyber threats. It is evident that evaluation of the trained models in machine learning and deep learning are given importance. It must also be understood that building and evaluating the systems we built plays a crucial role. Some of the latest tools for the enterprises are highlighted. One such future work can be addressing is evaluation of various Artificial intelligence platforms that are built for cyber security. To conclude, Artificial Intelligence can also be the future of cyber security.

References

1. Sun, C.C., Hahn, A., Liu, C.C.: Cyber security of a power grid: state-of-the-art. Int. J. Electr. Power Energy Syst. **99**, 45–56 (2018)
2. Tr uong, T.C., Zelinka, I., Plucar, J., Čandík, M., Šulc, V.: Artificial intelligence and cyberse-curity: past, presence, and future. In: Artificial Intelligence and Evolutionary Computations in Engineering Systems, pp. 351–363. Springer, Singapore (2020)
3. Ongsulee, P.: Artificial intelligence, machine learning and deep learning. In: 15th International Conference on ICT and Knowledge Engineering (ICT&KE), pp. 1–6. IEEE (2017)
4. Mohammed, I.A.: Artificial intelligence for cybersecurity: a systematic mapping of literature. Int. J. Innovations Eng. Res. Technol. **7**(9) (2020)
5. Anwar, S., Mohamad Zain, J., Zolkipli, M.F., Inayat, Z., Khan, S., Anthony, B., Chang, V.: From intrusion detection to an intrusion response system: fundamentals, requirements, and future directions. Algorithms **39**(2), 10 (2017)
6. Mohammadi, S., Mirvaziri, H., Ghazizadeh-Ahsaee, M., Karimipour, H.: Cyber intrusion detection by combined feature selection algorithm. J. Inf. Secur. Appl. **44**, 80–88 (2019)
7. Tapiador, J.E., Orfla, A., Ribagorda, A., Ramos, B.: Key-recovery attacks on kids, a keyed anomaly detection system. IEEE Trans. Dependable Secur. Comput. **12**(3), 312–325 (2013)
8. Abbas, N.N., Ahmed, T., Shah, S.H.U., Omar, M., Park, H.W.: Investigating the applications of artificial intelligence in cyber security. Scientometrics **121**(2), 1189–1211 (2019)
9. Zheng, Y., Li, Z., Xu, X., Zhao, Q.: Dynamic defenses in cyber security: techniques, methods and challenges. Digit. Commun. Netw. **8**(4), 422–435 (2022)
10. Kilincer, I.F., Ertam, F., Sengur, A.: Machine learning methods for cyber security intrusion detection: datasets and comparative study. Comput. Netw. **188**, 107840 (2021)

11. Sarker, I.H., Furhad, M.H., Nowrozy, R.: Ai-driven cybersecurity: an overview, security intelligence modeling and research directions. SN Comput. Sci. **2**(3), 1–18 (2021)
12. Shinan, K., Alsubhi, K., Alzahrani, A., Ashraf, M.U.: Machine learning-based botnet detection in software-defined network: a systematic review. Symmetry **13**(5), 866 (2021)
13. Buchanan, B.G., Smith, R.G.: Fundamentals of expert systems. Annu. Rev. Comput. Sci. **3**(1), 23–58 (1988)
14. Li, J.H.: Cyber security meets artificial intelligence: a survey. Frontiers Inf. Technol. Electronic Eng. **19**(12), 1462–1474 (2018)
15. Rudenko, M., Zhivago, E., Rudenko, A.: Expert System for Modeling Threats and Protecting Premises from Information Leaks (2022)
16. Rani, C., Goel, S.: CSAAES: An expert system for cyber security attack awareness. In: International Conference on Computing, Communication and Automation, pp. 242–245. IEEE (2015)
17. Kivimaa, J., Ojamaa, A., Tyugu, E.: Graded security expert system. In: International Workshop on Critical Information Infrastructures Security, pp. 279–286. Springer, Berlin, Heidelberg (2008)
18. Malek, Z.S., Trivedi, B., Shah, A.: User behavior pattern-signature based intrusion detection. In: Fourth World Conference on Smart Trends in Systems, Security and Sustainability (WorldS4), pp. 549–552. IEEE (2020)
19. Alhayani, B., Mohammed, H.J., Chaloob, I.Z., Ahmed, J.S.: Effectiveness of artificial intelligence techniques against cyber security risks apply of IT industry. Mater. Today: Proc. (2021)
20. Anwar, A., Hassan, S.I.: Applying artificial intelligence techniques to prevent cyber assaults. Int. J. Comput. Intell. Res. **13**(5), 883–889 (2017)
21. Kott, A.: Intelligent autonomous agents are key to cyber defense of the future army networks. Cyber Defense Rev. **3**(3), 57–70 (2018)
22. Wang, P., Govindarasu, M.: Multi intelligent agent based cyber attack resilient system protection and emergency control. In: IEEE Power and Energy Society Innovative Smart Grid Technologies Conference (ISGT), pp. 1–5. IEEE (2016)
23. Ford, V., Siraj, A.: Applications of machine learning in cyber security. In: Proceedings of the 27th International Conference on Computer Applications in Industry and Engineering, vol. 118. IEEE Xplore, Kota Kinabalu, Malaysia (2014)
24. Salloum, S.A., Alshurideh, M., Elnagar, A., Shaalan, K.: Machine learning and deep learning techniques for cybersecurity: a review. In: The International Conference on Artificial Intelligence and Computer Vision, pp. 50–57. Springer, Cham (2020)
25. Panda, M., Patra, M.R.: Network intrusion detection using Naive Bayes. Int. J. Comput. Sci. Netw. Secur. **7**(12), 258–263 (2007)
26. Amiri, F., Yousefi, M.R., Lucas, C., Shakery, A., Yazdani, N.: Mutual information-based feature selection for intrusion detection systems. J. Netw. Comput. Appl. **34**(4), 1184–1199 (2011)
27. Kruegel, C., Toth, T.: Using decision trees to improve signature-based intrusion detection. In: International Workshop on Recent Advances in Intrusion Detection, pp. 173–191. Springer, Berlin, Heidelberg (2003)
28. Li, Z., Zhang, A., Lei, J., Wang, L.: Real-time correlation of network security alerts. In: IEEE International Conference on e-Business Engineering (ICEBE'07), pp. 73–80. IEEE (2007)
29. Sequeira, K., Zaki, M.: Admit: anomaly-based data mining for intrusions. In: Proceedings of the Eighth ACM SIGKDD International Conference on Knowledge Discovery and Data Mining, pp. 386–395 (2002)
30. Banerjee, J., Maiti, S., Chakraborty, S., Dutta, S., Chakraborty, A., Banerjee, J.S.: Impact of machine learning in various network security applications. In: 3rd International Conference on Computing Methodologies and Communication (ICCMC), pp. 276–281. IEEE (2019)
31. Sjarif, N.N.A., Chuprat, S., Mahrin, M.N.R., Ahmad, N.A., Ariffin, A., Senan, F.M., et al.: Endpoint detection and response: why use machine learning? In: International Conference on Information and Communication Technology Convergence (ICTC), pp. 283–288. IEEE (2019)

32. Martín, A.G., Beltrán, M., Fernández-Isabel, A., de Diego, I.M.: An approach to detect user behaviour anomalies within identity federations. Comput. Secur. **108**, 102356 (2021)

33. Ferrag, M.A., Maglaras, L., Moschoyiannis, S., Janicke, H.: Deep learning for cyber security intrusion detection: approaches, datasets, and comparative study. J. Inf. Secur. Appl. **50**, 102419 (2020)

34. Roopak, M., Tian, G.Y., Chambers, J.: Deep learning models for cyber security in IoT networks. In: IEEE 9th Annual Computing and Communication Workshop and Conference (CCWC), pp. 0452–0457. IEEE (2019)

35. Choi, Y.H., Liu, P., Shang, Z., Wang, H., Wang, Z., Zhang, L., et al.: Using deep learning to solve computer security challenges: a survey. Cybersecurity **3**(1), 1–32 (2020)

36. Singh, G.A.P., Gupta, P.K.: Performance analysis of various machine learning-based approaches for detection and classification of lung cancer in humans. Neural Comput. Appl. **31**(10), 6863–6877 (2019)

37. Bradley, A.P.: The use of the area under the ROC curve in the evaluation of machine learning algorithms. Pattern Recogn. **30**(7), 1145–1159 (1997)

38. Zheng, A., Casari, A.: Feature Engineering for Machine Learning: Principles and Techniques for Data Scientists. O'Reilly Media, Inc. (2018)

39. Sagi, O., Rokach, L.: Ensemble learning: a survey. Wiley Interdisc. Rev.: Data Min. Knowl. Discov. **8**(4), e1249 (2018)

40. Schneier, B.: Invited talk: The coming AI hackers. In: International Symposium on Cyber Security Cryptography and Machine Learning, pp. 336–360. Springer, Cham (2021)

41. Karuppiah, M., Saravanan, R.: A secure remote user mutual authentication scheme using smart cards. J Inf Secur. Appl. **19**(4–5), 282–294 (2014)

42. Karuppiah, M., Saravanan, R.: A secure authentication scheme with user anonymity for roaming service in global mobility networks. Wireless Pers. Commun. **84**(3), 2055–2078 (2015)

43. Karuppiah, M., Kumari, S., Li, X., Wu, F., Das, A.K., Khan, M.K., Basu, S.: A dynamic id-based generic framework for anonymous authentication scheme for roaming service in global mobility networks. Wireless Pers. Commun. **93**(2), 383–407 (2017)

44. Kumari, S., Karuppiah, M., Li, X., Wu, F., Das, A.K., Odelu, V.: An enhanced and secure trust-extended authentication mechanism for vehicular ad-hoc networks. Secur. Commun. Netw. **9**(17), 4255–4271 (2016)

45. Karuppiah, M., Kumari, S., Das, A.K., Li, X., Wu, F., Basu, S.: A secure lightweight authentication scheme with user anonymity for roaming service in ubiquitous networks. Secur. Commun. Netw. **9**(17), 4192–4209 (2016)

46. Naeem, M., Chaudhry, S.A., Mahmood, K., Karuppiah, M., Kumari, S.: A scalable and secure RFID mutual authentication protocol using ECC for Internet of Things. Int. J. Commun. Syst. **33**(13), e3906 (2020)

47. Karuppiah, M., Das, A.K., Li, X., Kumari, S., Wu, F., Chaudhry, S.A., Niranchana, R.: Secure remote user mutual authentication scheme with key agreement for cloud environment. Mob. Netw. Appl. **24**(3), 1046–1062 (2019)

48. Maria, A., Pandi, V., Lazarus, J.D., Karuppiah, M., Christo, M.S.: BBAAS: blockchain-based anonymous authentication scheme for providing secure communication in VANETs. Secur. Commun. Netw. **2021** (2021)

49. Pradhan, A., Karuppiah, M., Niranchana, R., Jerlin, M.A., Rajkumar, S.: Design and analysis of smart card-based authentication scheme for secure transactions. Int. J. Internet Technol. Secured Trans. **8**(4), 494–515 (2018)

50. Li, X., Niu, J., Bhuiyan, M.Z.A., Wu, F., Karuppiah, M., Kumari, S.: A robust ECC-based provable secure authentication protocol with privacy preserving for industrial internet of things. IEEE Trans. Industr. Inf. **14**(8), 3599–3609 (2017)

51. Bhagat, R.C., Patil, S.S.: Enhanced SMOTE algorithm for classification of imbalanced big-data using random forest. In: IEEE International Advance Computing Conference (IACC), pp. 403–408. IEEE (2015)
52. Menardi, G., Torelli, N.: Training and assessing classification rules with imbalanced data. Data Min. Knowl. Disc. **28**(1), 92–122 (2014)
53. Tyagi, S., Mittal, S.: Sampling approaches for imbalanced data classification problem in machine learning. In: Proceedings of ICRIC 2019, pp. 209–221. Springer, Cham (2020)

Printed in the United States
by Baker & Taylor Publisher Services